ENVIRONMENTAL SOIL SCIENCE

BOOKS IN SOILS, PLANTS, AND THE ENVIRONMENT

Soil Biochemistry, Volume 1, edited by A. D. McLaren and G. H. Peterson

Soil Biochemistry, Volume 2, edited by A. D. McLaren and J. Skujiņš

Soil Biochemistry, Volume 3, edited by E. A. Paul and A. D. McLaren

Soil Biochemistry, Volume 4, edited by E. A. Paul and A. D. McLaren

Soil Biochemistry, Volume 5, edited by E. A. Paul and J. N. Ladd

Soil Biochemistry, Volume 6, edited by Jean-Marc Bollag and G. Stotzky

Soil Biochemistry, Volume 7, edited by G. Stotzky and Jean-Marc Bollag

Soil Biochemistry, Volume 8, edited by Jean-Marc Bollag and G. Stotzky

Soil Biochemistry, Volume 9, edited by G. Stotzky and Jean-Marc Bollag

Organic Chemicals in the Soil Environment, Volumes 1 and 2, edited by C. A. I. Goring and J. W. Hamaker

Humic Substances in the Environment, M. Schnitzer and S. U. Khan

Microbial Life in the Soil: An Introduction, T. Hattori

Principles of Soil Chemistry, Kim H. Tan

Soil Analysis: Instrumental Techniques and Related Procedures, edited by Keith A. Smith

Soil Reclamation Processes: Microbiological Analyses and Applications, edited by Robert L. Tate III and Donald A. Klein

Symbiotic Nitrogen Fixation Technology, edited by Gerald H. Elkan

Soil–Water Interactions: Mechanisms and Applications, Shingo Iwata and Toshio Tabuchi with Benno P. Warkentin

Soil Analysis: Modern Instrumental Techniques, Second Edition, edited by Keith A. Smith

Soil Analysis: Physical Methods, edited by Keith A. Smith and Chris E. Mullins

Growth and Mineral Nutrition of Field Crops, N. K. Fageria, V. C. Baligar, and Charles Allan Jones

Semiarid Lands and Deserts: Soil Resource and Reclamation, edited by J. Skujiņš

Plant Roots: The Hidden Half, edited by Yoav Waisel, Amram Eshel, and Uzi Kafkafi

Plant Biochemical Regulators, edited by Harold W. Gausman

ENVIRONMENTAL SOIL SCIENCE
Second Edition, Revised and Expanded

Kim H. Tan
The University of Georgia
Athens, Georgia

MARCEL DEKKER, INC.   NEW YORK • BASEL

Library of Congress Cataloging-in-Publication Data

Tan, Kim H. (Kim Howard).
   Environmental soil science / Kim H. Tan. –2nd ed., rev. and expanded.
     p. cm. – (Books in soils, plants, and the environment; vol.74)
   Includes bibliographical references (p.).
   ISBN 0-8247- 0340-5
   1. Soils—Environmental aspects. 2. Soil sciences. I. Title. II. Books in soils, plants,
and the environment; v. 74.

S591.T35 2000
631.4—dc21   99-054220

This book is printed on acid-free paper.

Headquarters
Marcel Dekker, Inc.
270 Madison Avenue, New York, NY 10016
tel: 212-696-9000; fax: 212-685-4540

Eastern Hemisphere Distribution
Marcel Dekker AG
Hutgasse 4, Postfach 812, CH-4001 Basel, Switzerland
tel: 41-61-261-8482; fax: 41-61-261-8896

World Wide Web
http://www.dekker.com

The publisher offers discounts on this book when ordered in bulk quantities. For more information, write to Special Sales/Professional Marketing at the headquarters address above.

Current printing (last digit):
10 9 8 7 6 5 4 3 2 1

PRINTED IN THE UNITED STATES OF AMERICA

PREFACE TO THE SECOND EDITION

This second edition of an essential textbook, employed by numerous universities around the world, has been completely rewritten. *Environmental Soil Science* was the textbook for many years at the University of Georgia for the senior-level undergraduate and graduate course in Environmental Soil Science. It was found very useful by a great number of students, especially those majoring in Environmental Health Science and those at the School of Forestry. The first edition of the book has been used at the Universidad Nacional del Sur, Bahia Blanca, Argentina, the University of North Sumatera, Medan, and the University of Andalas, Padang, Indonesia. I have used the book in my Environmental Soil Science course at the University of North Sumatera, where it is to be translated into the Indonesian language, and am planning to use it again at the University of Andalas.

This second edition is intended to bring the book up-to-date, by adding the most recent advances, and by incorporating suggestions and recommendations of a range of readers, including former students, teachers, professors, and scientists from universities in the United States, Indonesia, Argentina, Canada, and Europe. The organization of the book remains the same as in the first edition. However, the environmental aspect is highlighted by integrating soils with the environment. The new text relates soil with the environment by incorporating fundamental soil principles as well as state-of-the-art applications in environmental processes vital for continuation of life into a stimulating, information-packed resource. The book discusses soil properties as products of the environment, and at the same time addresses their effect on the environment. The primary use of soils for food and fiber production is examined in close relation to the consequent environmental changes.

Chapter 1 starts by presenting the major soil groups. A new soil order, the gelisol, is added, and the discussion on andisols is expanded, featuring recent advances, including coadsorption or Al-bridging processes, which I advance for accumulation of humus and P-fixation. The issue of nomenclature is addressed, and the environmental significance of distribution of soils is stressed according to latitudinal and altitudinal variations.

Chapter 2 is expanded, highlighting the key role of primary minerals not only in soils and agriculture, but also in their effect on the environment and use in industry. The discussion of the weathering of primary minerals has been rewritten, differentiating abiotic from biochemical processes and stressing the central role organisms play in decomposition and neoformation of soil minerals. The *anti-allophanic* process by soil organic matter advanced in Japan is addressed.

Chapter 3 has been overhauled, now also featuring the beneficial effect of soil organic matter and *allelophaty,* fundamental in plant toxicity and environmental and human health. New topics on humus, enzymes, plant roots, and rhizosphere have been added, underscoring the fundamental role of the macroflorae. The macrofaunae are examined in a new light, and the environmental significance of the microfaunae and microflorae is given a new dimension. The carbon and nitrogen cycles sections have been rewritten, emphasizing their significance in nature, and the text on organic compounds has been expanded showing clearly their environmental importance, using as examples the production of biofuel from organic trash containing carbohydrates and the hydrophobic properties induced by lipids. A new section featuring the key role of humic acids in production of biomedicines is presented, underscoring the role of mud baths offered at modern health spas.

The topic of soil air has been enlarged substantially in Chapter 4 to emphasize the role of soil aeration and seasonal variation in O_2 and CO_2 content in tropical and temperate region soils in response to changing respiration. Redox potentials are presented, underscoring their usefulness in soil formation and issues of proper aeration for environmental health. The biochemistry of anaerobic and aerobic soils is reassessed, and a new concept is introduced on the *critical value* of O_2 content fundamental to root growth.

Chapter 5 features new topics on movement of water in the liquid and gas phase, highlighting Darcy's law, the importance of relative humidity for microbial life, and the environmental significance of leaching and percolation as essential processes in nature. Water loss is added to stress the necessity of evapotranspiration and the importance of mulch. The section on dissolved substances and soil colloids in the soil solution is reassessed to highlight their environmental importance.

Chapter 6 is a new chapter featuring the major physical properties of soils in close relation to their environmental importance. Soil texture is presented, stressing its central role in issues on soil compaction, soil impedance and pore spaces, fundamental in root growth and penetration and percolation. Soil structure is discussed in light of biological activity, inorganic and organic cementing agents, and effectiveness of artificial soil conditioners against surface sealing. Soil density and pore spaces are examined in relation to cropping systems and the *cultivation effect* of soil macro- and microfaunae. The *Atterberg's plasticity index* forms the core of the section on soil consistence, highlighting the significance of plasticity in the environment and its fundamental role in the ceramic industry. Thermal properties in soils are examined in relation to the effect of plant canopies and the usefulness of mulch.

The electrochemical properties of soil constituents in Chapter 7 provide a clear distinction between permanent negative and permanent positive charges. The nomenclature of these charges and that of the ZPC are reassessed. A new concept of *counterion bridging* is advanced by the author in the electric double layer theory to give a better reason for flocculation of clay. Positive and negative adsorption have been added in the section on anion exchange, underscoring the significance of *exchange alkalinity* and the effect of negative adsorption on eutrophication. The environmental significance of organic substances is reexamined as illustrated by the capacities of amino acids in coadsorption or metal bridging, and those of humic acids in hydrophobic bonding with extracellular polysaccharides and in biodegradation of pesticide residues. A new section is added to stress the importance of producing *sickly* plants which are sometimes highly regarded as beautiful ornamental plants. Like the induced stunted and crooked condition of bonsai plants, plants experiencing micronutrient deficiency or toxicity often produce brilliant colors.

The primary use of soils for food and fiber production has been re-examined in Chapter 8. A new section is added addressing the beneficial and harmful effects of deforestation. Soil degradation, divided into physical, chemical, and biodegradation, is highlighted as an essential process of aging in nature. Growing old accelerated by human interference brings with it the usual harmful *sicknesses of old age*. Similarly, a new section on *desertification* is presented to underscore the natural process of sand dunes' migration that is caused by wind action but is often accelerated in arid regions by human interference. A new controversial theory is offered that considers it a myth that increased CO_2 production is causing the greenhouse effect and global warming. A section on low-input sustainable agriculture (LISA) is included, stressing differences between no-till and organic farming.

Alternative methods for food production without using soils are examined in Chapter 9, and the water and nutrient requirements in hydroponics are reassessed to highlight the significance of the transpiration ratio and threshold value, including MAC and MAL, of micronutrients. The issue of crop yields and importance of food production by hydroponics are given new dimensions in the new sections on *nutrient film technique* and *artificial soil*. The aquaculture section is rewritten, stressing the importance of the *Law of the Sea*, and the significance of salmon ranching. Channel catfish farming in the United States, where advantage is taken of cultural eutrophication, is highlighted. Growing plants not for food but as ornaments beneficial to human health is underscored by the new section on *orchid culture*.

Chapter 10, on biotechnology in food production, is enlarged to include a new section on environmentally friendly pesticides, emphasizing the need for development of inexpensive pesticides for human health, such as the successful pyrethrum coils burned as incense effective in killing mosquitoes. New concepts are introduced in the discussion of *soil biotechnology,* examining the importance for production of *biofertilizers* and *biomedicines*. The issue of genetic engineering in producing new crops for food and the public's suspicion of genetically altered crops are addressed.

Chapter 11, discussing the production of waste from agricultural and industrial operations, has been rewritten and expanded, showing now the distinction between *contamination* and *pollution* and the importance of *cometabolism*, ADI, and NOEL of pesticide residues. *Soil Remediation* and *Natural Attenuation* are new sections, underscoring physical, chemical, and bio-remediation or attenuation as fundamental but different processes in detoxification of soils and the environment. The greenhouse effect and global warming and the issue of fishkill and forest dieback are reexamined to address the controversial effects of CO_2 and acid rain.

I want to thank Dr. Harry A. Mills, Professor of Horticulture, University of Georgia, and Dr. J. B. Jones, Jr., Director, Macro-Micro International, Inc. Athens, GA, for their constructive criticism and encouragement. Special thanks are extended to Mr. Brian Kirtland, MS, PH.D. candidate Environmental Health Science, University of South Carolina, Columbia, SC; to Dr. Juan Carlos Lobartini, Departamento de Agronomia, Universidad Nacional del Sur, Bahia Blanca, Argentina, and other former graduate students for their valuable suggestions in revising the manuscript. Appreciation is extended to the various publishers, scientific societies, colleagues, and fellow scientists who gave permission to quote or reproduce figures or photographs. Special recognition also goes to the many unnamed persons who assisted or gave encouragement in the development of this second edition. Finally, I want to acknowledge the encouragement and understanding of my wife, Yelli, who always enthusiastically supported me in my work.

Kim H. Tan

PREFACE TO THE FIRST EDITION

The tremendous pressure on our soils and the environment by the industrial and agricultural expansion of the last few decades has created a new burst of questions on environmental quality. The intensive use and misuse of our soil, water, and air resources due to population expansion and the increasing level of pollution that accompanies an increasing standard of living have augmented the hazard of declining soil productivity. The use of fertilizers, insecticides, and herbicides in agriculture has expanded tremendously; and the pollution of streams and lakes and the contamination of soils and groundwater with toxic chemicals have become more widespread. These trends have resulted in an increased sense of awareness of environmental issues.

A large amount of information is, in fact, available on environmental problems, and the use of soil for disposal of organic wastes. However, most such books on soil science are based on the traditional concepts of soil conservation and erosion. Until now there has not been a book relating the principles of recent environmental issues to soil science because information on global warming, ozone depletion, and acid rain has been scant.

Environmental soil science, which can be defined as the science of soils in relation to the environment, examines how soils are the product of the environment and how their properties are closely associated with environmental factors. Those who acquaint themselves with environmental principles in soil science are better able to prevent or to recognize an environmental problem when it arises and to find ways to solve it intelligently. This book is written with the purpose of relating environmental principles and soil science in plain language, easy to comprehend by a wide range of agricultural and environmental scientists, soil remediation specialists, microbial ecologists, horticulturists, engineers, geologists, foresters, and professionals and students in related disciplines.

The organization of the book does not follow the traditional division into the physical, chemical, and biological characteristics of soil, since this book is not intended to provide another version of the many textbook existing on *basic soils*. Instead, the chapters are arranged according to the different soil constituents, e.g., solid, liquid, and gas. This format helps elucidate the natural coexistence of basic soil properties with environmental science. The soil constituents control soil characteristics and reactions, and many of these soil reactions affect the environ ment. Therefore, the text starts with an attempt to show the effect of environmental factors in the formation of different kinds of soils, and the dominant impact of climate and vegetation in determining the distribution of these soils in the world. This is followed by chapters 2 through 5, in which the soil constituents in the solid, liquid, and gas phases are discussed in terms of their interactions with the environment. The effects of environmental factors on weathering of primary minerals and formation of clay minerals are stressed in chapter 2. The organic components, biochemical reactions, and interrelationships with the environment are highlighted in chapter 3. Gaseous components, biochemical reactions in aerobic and anaerobic conditions, and their impli cations in pollution of soil and atmospheric air are featured in chapter 4. Soil water, dissolved macro- and micronutrients, and their relationship to environmental processes, e.g., eutrophication, are discussed in chapter 5. The reactions of dissolved inorganic and organic solids, and dissolved CO_2 and O_2 gas, are included as they are important in affecting environmental quality. Chapter 6 discusses the electro- chemical properties of clay and humic acids and their significance in pollution. Chapter 7 examines efforts in crop production and the consequent changes they bring to soils and the environment, whereas chapters 8 and 9 summarize alternative methods in crop production, e.g., soilless agriculture and biotechnology. The final chapter concerns soil and pollution, assessing the implications of agricultural and industrial wastes to environmental quality. Acid rain, the greenhouse effect, and the depletion of ozone are covered among others.

I want to thank Dr. Harry A. Mills, Professor of Horticulture, University of Georgia, and Dr. J. B. Jones, Jr., Director, Macro- Micro International, Inc., Athens, GA, for their review and constructive criticism. Appreciation is extended to Ms. Nickie Whitehead for reading and editing the manuscript and to Dr. Juan Carlos Lobartini and Mr. John A. Rema for their assistance with the laser printer in the development of this book. Thanks are extended to Dr. D. H. Marx, Director, Institute of Tree Root Biology, USDA—Forestry Science Laboratory, Athens, GA; Dr. M. F. Brown, University of Missouri, Columbia, MO; Tousimis Research Corp., Rockeville, MD; Potash and Phosphate Institute, Atlanta, GA, and to the various publishers, scientific societies, and fellow scientists who gave permission to reproduce photographs. Special recognition also goes to the many unnamed persons who have assisted in the development of the book. Finally, I want to acknowledge the support and understanding of my wife, Yelli, who stood by with great enthusiasm and a lot of encouragement.

Kim H. Tan

CONTENTS

Chapter 2 INORGANIC SOIL CONSTITUENTS   28

Chapter 3 ORGANIC SOIL CONSTITUENTS   80

Contents

Contents

CHAPTER 1

SOILS AND THE ENVIRONMENT

1.1 DEFINITION AND CONCEPT OF SOILS

Soil is a term understood by almost everyone, yet the meaning of this term may vary among different people and soil can be defined in many ways (Brady, 1990; Foth, 1990). The farmer, engineer, chemist, geologist, and layman bring different viewpoints or perspectives to their concepts of soil. Environmentalists may even define soil differently than soil scientists. With the introduction of the pedological concept in soil science during the beginning of the 20th century, these differences have diminished considerably. Yet, even now, a commonly accepted definition of soil is still missing, notwithstanding the fact that most people agree that soils are products of the environment. Of the several definitions of soils that can be found in the literature, the definitions of Kellogg (1941) and the USDA, Soil Survey Staff (1951; 1990) perhaps most closely reflect the relationship of soils to the environment. According to these definitions: "Soils are considered natural bodies, covering parts of the earth surface that support plant growth, and that have properties due to the integrated effect of climate and organisms acting upon the parent material, as conditioned by relief, over a period of time". Several soil scientists, notably Buol et al. (1973), may object to this definition, but the

1

definition does show the dependency of soils on several environmental factors. It may be noted that this definition does not recognize moon or lunar material as soil. Lunar material, which is not affected by organisms, is excluded from the definition, but it may qualify as a parent material or regolith (Ming and Henninger, 1989).

The above definition of soils agrees closely with an earlier formulation of soils. In his famous book, *Factors of Soil Formation*, Hans Jenny (1940) reported that soils could be characterized by the formula:

$$S = f(cl, o, p, r, t)$$

In this equation, S = soils, f = function, cl = climate, o = organisms, p = parent material, r = relief, and t = time.

Climate, organisms, parent material, relief, and time are considered the five major factors in soil formation. According to such a formulation, these environmental factors are the main variables that determine *the state of the soil*. In other words, soils are formed by the combined effect of the factors. The nature of soils can be changed only when the variables, cl, o, p, r, and t, change individually or in combination. In such a formula the factor S (soil) cannot be changed to modify such variables as climate, organisms, or parent material, which is in practice true to a certain extent. For example, a change in the nature of soils will not result in a change of the parent material. However, it may apply to the variable o, organisms. Under certain cases, a deterioration of soil conditions brings about drastic changes in vegetation and/or other organisms. It is more difficult, however, to show a clear-cut relationship with the variable cl, climate, but a change in the nature of soils may sometimes result in a change in climate. A good illustration of this type of change is the formation of desert and savannah type of climates due to deforestation and the consequent degradation of soils in tropical regions of Africa.

1.2 PEDOLOGIC CONCEPT OF SOILS

Soil science is sometimes divided into *edaphology*, the science of soils as media for plant production, and *pedology*, the science of soils as biochemically synthesized bodies in nature (Brady, 1990). The concepts of these two branches in soil science are embraced by the preceding definitions of soils. The difference between them is only in their application or use. Edaphology applies soil science mainly in crop production, as indicated earlier, whereas pedology studies the characterization, genesis, morphology and taxonomy of soils. Nevertheless, the basic concept of soils in pedology still constitutes the key for perception of soils in edaphology.

According to the pedologic concept, the soil is a three-dimensional body in nature, showing length, width, and depth. An individual soil body, briefly called soil, occurs in the landscape side by side with other soils like the pieces of a jigsaw puzzle. Each soil is considered an independent body with a unique morphology as reflected by a soil profile. The soil profile is defined by specific series of layers of soils, called soil horizons, from the surface down to the unaltered parent material. It is formed by the integrated effect of the soil formation factors.

Measured by area, the soil can be small in size or a few hectares in extent, and it is common and more convenient in research and analysis to deal with a small representative part of the soil. The smallest representative unit of a soil body is called a *pedon* (Figure 1.1). A pedon has three dimensions and is comparable in some ways to the *unit cell* of a crystal. One soil body consists of contiguous similar pedons. This group of contiguous similar pedons in one soil body is called a *polypedon*. The polypedon or soil body is bordered on all sides by other pedons with different characteristics. The size of a pedon is measured by its surface area, with the smallest measuring 1 m². A pedon is bordered on its sides by vertical sections of soils, the *soil profiles* (Figure 1.2). Each soil profile, extending from the surface down to the parent material, is composed of several soil horizons. The soil profile characterizes the pedon; hence, it identifies the soil. The horizons tell much about the soil properties. They provide information on color, texture, structure, permeability, drainage, biological activity,

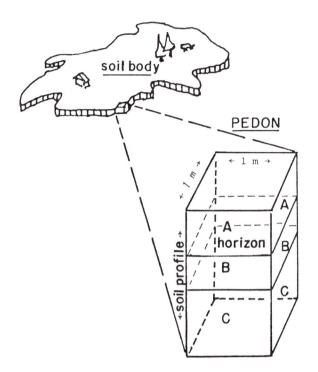

Figure 1.1 The relationship of a pedon to a soil body.

and other attributes of importance in soil characterization, formation, fertility, crop production, and engineering. Six main groups of horizons, called *master horizons*, have been identified. They are designated by the symbols O, A, E, B, C and R, respectively.

1. <u>**O**</u> horizons are organic deposits composed of dead, partially de-composed and undecomposed vegetative material. This horizon, lying on the surface above the mineral horizon, is in many cases very thin, and only in undisturbed soils covered by vegetation can it assume considerable thickness. The name for this horizon, assigned by the US Soil Taxonomy, is *histic* (Gr. *histos*, tissue) *epipedon* (Soil Survey Staff, 1990; 1992). The O horizon can be subdivided into O_i (non- to

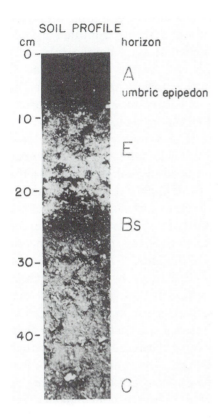

Figure 1.2 A soil profile of a sandy, mixed, frigid Entic Haplorthod (spodosol), showing the effect of the environment in its formation. The soil profile is characterized by an A horizon, rich in humus, underlaid by an E horizon, which is bleached in color because of eluviation of humus and iron. The Bs horizon is dark in color because of illuviated humus and iron. The C horizon is the parent material.

slightly decomposed), O_e (intermediately decomposed), and O_a (highly decomposed) horizons.

2. **A** horizons are the topmost mineral horizons lying below the O

horizon. They are composed of large amounts of inorganic material, e.g., sand, silt, and clay, intimately associated with humified organic matter. Because of its organic matter content, the A horizon is usually darker in color than the horizons below. In the absence of an O horizon, the A horizon is frequently the surface horizon. When the properties of the A horizon result from cultivation or related human activities, the horizon is designated by the symbol Ap (p = plow).

In the US Soil Taxonomy, a number of A horizons are considered diagnostic for classifying soils. They are called *diagnostic epipedons*, from the Greek words epi = over and pedon = soil. According to the *Keys to Soil Taxonomy* (Soil Survey Staff, 1996), the epipedon is not a synonym for the A horizon because it may include part or all of the illuvial B horizon if the darkening by organic matter extends from the surface into or through the B horizon. Seven major diagnostic epipedons are of importance in soils of the USA. The first one is the *mollic epipedon*, a thick, dark-colored A horizon, which is rich in organic matter, and has a base saturation >50% and a strong structure. *Umbric epipedon* is the second epipedon, and is similar to a mollic, except for the presence of a base saturation that is lower than 50%. The third is the *ochric epipedon*, a light colored A horizon, which is low in organic matter content and may be hard and massive when dry. The fourth epipedon, *histic epipedon*, has been defined earlier as a surface horizon (above the A horizon) rich in organic matter. This epipedon may be wet during some part of the year. The fifth diagnostic epipedon, *plaggen epipedon*, of importance in Western Europe, is a human-made surface layer, more than 50 cm thick, that has been formed by long, continued manuring. The sixth epipedon, *anthropic epipedon*, is an ill-defined epipedon, and is said to conform to all requirements of the mollic epipedon except (1) the limits on acid-soluble P_2O_5, with or without the base saturation, or (2) the length of the period during which it has available moisture. Additional data on anthropic epipedons from several parts of the world are expected to generate improvements in the definition (Soil Survey Staff, 1996). A seventh epipedon, *melanic epipedon* (Gr. *melas-anos*, black), has recently been added to meet the requirements of andisols. It is defined as a thick black horizon at or near the surface, containing high amounts of organic matter, usually associated with short-range-order minerals or aluminum-humus complexes. The intense black color is believed to be caused by the presence of *type A humic acid*, formed mainly by the decomposition of roots from graminaceous vegetation. This type of humic acid can be distinguished from that developed under a forest vegetation (*type P*) by using the *melanic index*. A melanic index of < 1.70 is indicative for the presence of type A humic acid (Tan, 1998a).

3. **E** horizons are mineral horizons located under the A horizon. They are the zones of maximum leaching or *eluviation* - zones of removal of soil constituents, e.g., clay, humus, Fe and Al compounds. E horizons are white, pale, light or bleached in color. White E horizons are called *albic* horizons in soil taxonomy. A horizons grading into E horizons are transitional horizons and carry the symbols AE. Similarly, E horizons grading into underlying B horizons are designated by the symbols EB.

4. **B** horizons are located underneath E horizons. In the absence of an E horizon, the A horizon lies directly above the B horizon. B horizons, frequently referred to as subsoil, are the zones of *illuviation* (accumulation) of soil materials removed from A and E horizons. Illuvial concentrations of silicate clays, Fe, Al, or humus alone or in combination may be present. B horizons grading into underlying C horizons are transitional horizons and carry the symbols BC.

Many of the B horizons are also diagnostic for characterizing soils. They are called *diagnostic subsurface horizons*. Some of the most important diagnostic subsurface horizons are the (1) *argillic* horizon (Bt), a B horizon enriched with silicate clays; (2) *spodic* horizon (Bh or Bhs), a horizon enriched with humus and Fe and Al oxides; (3) *cambic* horizon (Bw), a young B horizon, recently changed by physical and chemical reactions; and (4) *oxic* horizon, a highly weathered B horizon, containing Fe, Al oxides and 1:1 lattice type (kaolinitic) clays. A fifth subsurface horizon, *kandic horizon*, should be added to meet the requirements of *low activity clay soils*. It is in essence a Bt horizon with a CEC ≤16 cmol/kg clay (by $1N$ NH_4OAc pH 7) and an ECEC ≤12 cmol/kg clay (sum of bases extracted with $1N$ NH_4OAc pH 7 plus $1N$ KCL extractable Al). For a complete list of these diagnostic horizons and their characteristics reference is made to *Keys to Soil Taxonomy* (Soil Survey Staff, 1992; 1996).

5. <u>**C**</u> horizons are located under the B horizons and are considered parent materials of soils. They are mixtures of weathered rocks and minerals and are largely unaffected by soil formation processes. These materials may rest upon the rocks from which they have been formed, or they may lie upon an unrelated geologic formation.

6. <u>**R**</u> horizons are the underlying solid rock formation with little evidence of weathering.

Soils may differ from one another in the nature and arrangement of horizons. The kind and sequence of horizons determine the *soil orders* and *soil series* in Soil Taxonomy. A farm may be composed of several soil series, each of which responds differently to soil cultivation. By studying soil profile characteristics, these differences can be identified and inferences made as to the proper management practice. In general, road banks or preserved soil sections, called *soil monoliths*, can be used to examine soil profiles. One can also dig a pit, exposing a vertical section of the pedon, for this purpose.

1.3 SOIL ORDERS

As discussed in the preceding pages, soil profiles are also important in identifying *soil orders*. The order is the highest (broadest) category in the US Soil Taxonomy system (Soil Survey Staff, 1990). This classification system divides orders into *suborders*, suborders into *great groups*, great groups into *subgroups*, subgroups into *families*, families into *series*, and the latter into soil *types*. The system also recognizes *soil phases*, but these are *mapping units* and not taxonomic units. The orders, suborders, and great groups are considered the *higher categories*, whereas the families, series, and types are the *lower categories* in the US soil classification system.

Many other soil classification systems are available in the world.

Some are applicable only to soils of that country, whereas others try to attempt to cover soils worldwide. Such a system is for example the soil classification system of the Food and Agricultural Organization (*FAO*) of the United Nations. This system does not group soils into higher or lower categories, but recognizes only individual taxa. These soil taxa are the basis for the FAO soils world map, showing a distribution of about 5000 soil units. Each *soil taxon* in the FAO system can be correlated with a soil order or suborder, but sometimes it relates more closely to the great group category of the US Soil Taxonomy (Table 1.1). This will be addressed in some detail in the following subsections on the individual soil orders below. Lately, the FAO system has tried to integrate the US concept of diagnostic horizons in its system, perhaps because many of their soil scientists are US trained scientists.

Twelve groups of soil orders are now recognized in the US Soil Taxonomy (Table 1.1). Placement in a particular soil order is based on differences in soil formation processes, as reflected in the nature and sequence of soil horizons. Each soil order contains soil profiles with similar or almost similar properties. The names of these orders are derived from Latin or Greek terms and have a common suffix: *sol*, for soil. The connecting vowel "o" is supposed to be used with Greek formative elements. Of the twelve orders only spodosols and histosols, both derived from the Greek spodos and histos, respectively, carry the connecting vowel "o". The names of the remaining orders contain the connecting vowel "i". The latter is supposed to be restricted for use with Latin formative elements, such as in inceptisols, mollisols, ultisols, aridisols, and vertisols. However, entisols, alfisols, oxisols, andisols, and especially gelisols, all carrying the connecting vowel "i", are not derived at all from Latin terms. The name *gelisols* was coined even from the Greek term *gelid*, hence should be properly called *gelosols*, as the name given for these kind of soils by the FAO system. Apparently, the nomenclature in the US Soil Taxonomy has been developed more on the basis of being consistent rather than following the rules. If being consistent was the objective in the creation of the names of the soil orders, then the use of the names *spodisols* and *histisols* would serve the purpose better.

Table 1.1   Soil Orders and FAO Equivalent Names.

Orders name	Derivation/ meaning	Formative elements[1] carried to suborders	FAO equiv.
Entisols	Recent, young	ent	Lithosols
Gelisols	Gk. *gelid*=very cold, pergelic soil temp. regime	el	Gelosols
Inceptisols	L. *inceptum*=beginning	ept	Cambisols
Mollisols	L. *mollis*, soft, friable	ol	Chernozem
Spodosols	Gk. *spodos*, woodash	od	Podzols
Alfisols	Al = aluminum, fi for Fe = iron	alf	Luvisols
Ultisols	L. *ultimus*, ultimate weathering	ult	Acrisols
Oxisols	oxidation, highly oxidized	ox	Ferralsols
Aridisols	L. *aridus*	id	Xerosols
Vertisols	L. *verto*, turn, invert	ert	Vertisols
Andisols	from Andosols, Japanese an=black, do=soil	and	Andosols
Histosols	Gk. *histos*, tissue	ist	Histosols

[1]Formative elements are abbreviations from orders' name, and are used as suffixes in suborders' names, e.g., psamm*ent*, aqu*ept*, ust*ol*, hum*od*, etc.

1.3.1 Entisols

Entisols are young and shallow soils, and hence are characterized by A/C or A/R profiles. They are immature soils and have profiles in which B horizons have not yet developed. The soils do not have many horizons for various reasons, e.g., time of formation is too short, occurrence on steep slopes, actively eroding slopes, receiving frequent deposits from flooding, etc. Environmental factors not conducive for formation of soil horizons, such as the frigid climate in the permafrost areas of the arctic region, are other compelling factors. Entisols may vary from deep sand or river sediments of stratified clay beds to dry, arid lake bottoms and recent volcanic ash deposits. The soils with A/C profiles are the entisols over sand deposits. They are placed at the suborder level as psamments, which are equivalent to the FAO soil taxa *regosols*. These soils occur extensively in the coastal plains of south Georgia, Florida, and Alabama. In contrast, entisols with A/R profiles are entisols over hard rocks, called orthents at the suborder level. The FAO soil equivalent name is *lithosols*. These entisols are more common in the Rocky Mountains and other regions where rock formations can be found, such as the Blue Ridge Mountains, and the Piedmont Plateau.

The fertility of entisols varies considerably depending on the conditions from very low to very high. For instance, the alluvial floodplains of the Mississippi River are composed of fertile entisols, whereas the coastal plains of the United States have entisols with low fertility because of their high contents of quartz sand. The presence of water for irrigation and periodic flooding, contributing to a continuous buildup of nutrient supply for crop production, are some of the reasons for the formation of fertile entisols in the Mississippi River deltas and floodplains. Similarly, the entisols alongside the Ganges and Indus rivers in India, Bangladesh and Pakistan, and the Mehkong, Yangtze and Hoang Ho rivers in Southeast Asia and China, have proved to be rich soils where human population can thrive by growing sufficient amounts of crops.

1.3.2 Gelisols

This is the newest soil order added to the US Soil Taxonomy to cover soils in the tundra region of Alaska developed under the influence of a frigid and pergelic soil temperature regime. The FAO-UN and Canadian soil classification systems have already recognized such soils under the taxa gelosols and cryosols, respectively. The choice to name these soils gelisols violates the rules set up by the US Soil Taxonomy itself. As discussed earlier, Greek terms are supposed to be connected with the vowel "o", whereas the connecting vowel "i" is reserved for use with Latin terms. The name gelisols is derived from the Greek term gelid = extremely cold, hence the proper name for these soils should be gelosols. It is hard to understand why objections should be raised in using the name gelosols in the US Soil Taxonomy, because this name was already used by the FAO system.

Not much is known yet about gelisols and the information currently available indicates that they are shallow soils with dark organic surface layers on top of mineral layers underlaid by permafrost. The organic layer is presumably composed of tundra peat, derived from moss, lichens, and/or other tundra vegetation. The permafrost is commonly within the 100 cm depth. Though in general the soils have massive soil structures due to freeze-thaw pressures and desiccation processes, granular, platy and vesicular structures are often present in the surface horizons. Because of mixing due to *cryoperturbation*, the soil profile consists of irregular broken horizons, in which organic matter and stones are incorporated (Miller and Gardiner, 1998). The dominant suborders are histels, orthels and turbels.

1.3.3 Inceptisols

Inceptisols mark the beginning of a mature soil and are characterized by A/Bw/C profiles. The B horizons are in the stage of formation and are called *cambic* horizons (Bw). As such, these soils are in a more advanced stage of development than are entisols. Whether these kind of soils can be equated with the FAO soil taxa *cambisols* is in the author's opinion debatable. Though the name cambisols has also been coined from the Latin term *cambiare*=change, the FAO's interpretation is that of a far more mature soil than would be reflected by the US inceptisols.

Large areas of inceptisols are found in the United States in the Blue Ridge Mountains and the Piedmont Plateau. Inceptisols are reported to be widely distributed in the world and rank third in area of the soil orders. Such information raises some questions, because many soils are erroneously counted as inceptisols, e.g., lithosols (Brady, 1984) and paddy soils (Miller and Gardiner, 1998). Waterlogging, required for growing paddy rice, has provided redox conditions for the development of soil horizons, characteristic for *paddy soils* (Tan, 1968; 1998). Globally, inceptisols are in fact less known and less understood, especially in the tropics, where the warm and humid climate favors the process of drastic weathering in soils, which is not conducive to formation of a cambic horizon. An exception to this is the development of a Bw horizon in andosols. The natural fertility of inceptisols also varies widely. Inceptisols in New York and Pennsylvania are low in fertility, whereas those in the U.S. Pacific Northwest are quite fertile (Brady, 1990).

1.3.4 Mollisols

Mollisols are mature soils characterized by A/Bk/C profiles. The A horizon is typically a *mollic epipedon*, whereas the B horizon is usually a *calcic B* horizon, which carries the symbols Bk (k for accumulation of calcium carbonates). Mollisols develop by a soil formation process, called *calcification*, under semihumid climates and tall grass vegetation. They are important *grassland soils* and occur extensively in the Great Plains west of the Mississippi River. Large areas of mollisols are also found in the great plains of Canada, the steppes of Mongolia, the pampas of Argentina and in the Ukraine, where they are called *chernozems,* from the Russian terms chern = black, and zemlja = earth. It was Dokuchaiev who proposed this name for the first time for the black fertile soils in the steppe of the Ukraine. His studies, indicating the intimate influence of environmental factors on the development of the mollisol profile, started the pedologic concept in soil science. The name chernozem has been adopted as the official name by the FAO's soil classification system. Mollisols in the dryer region of the Great Plains in the United States are called *ustolls* at the suborder level, which perhaps can be equated with the FAO's soil taxa *kastanozems* (L. castaneo = chestnut, and Russian *zemlja* = earth).

Mollisols are very fertile and are considered to be among the world's most important agricultural soils. The environmental factors affecting formation of these soils, especially the tall grass vegetation and the semihumid condition, insure an annual turnover of abundant amounts of organic matter and a small degree of leaching. Hence, among the twelve soil orders, mollisols have the highest organic matter content, consequently also the highest nitrogen content. Only the andisols will match them in soil organic matter content. These characteristics together with a base saturation of $\geq 50\%$ (mostly occupied by Ca) give them the best chemical properties a soil can ask for. The physical properties, including a strong structure and friable consistence, are also considered excellent. Therefore, these soils have been highly productive for decades for growing corn and wheat and have contributed to the development of the cornbelt in the United States, and the wheatbelts in Canada and Russia.

1.3.5 Spodosols

Spodosols are mature soils with profiles characterized by a sequence of A/E/Bh or Bhs/C horizons. They are formed by a soil formation process called podzolization, typically in cool humid regions under a coniferous or mixed conifer-hardwood vegetation. Exceptions to the above may be present, like the occurrence of spodosols in Florida and in the plains of the Amazon River in Brazil. Under the influence of acid leaching, Al and Fe compounds and/or humus are translocated to the B horizon, creating a *spodic B* horizon. When this B horizon is enriched mainly with humus, a Bh horizon is formed. This type of spodosol is called a *humod* (hum = humus, and od = formative elements from spodosol) in the US Soil Taxonomy. On the other hand, when a mixture of Al and Fe compounds and humus is accumulated in the B horizon, a Bhs horizon is formed. Under certain conditions, Fe is the dominant illuvial constituent, and in this case a Bs horizon is formed. A spodosol with a Bs horizon is called a *ferrod* in the US Soil Taxonomy. The FAO equivalent taxon for spodosols is *podzols* (Russian pod = under, and zola = white layer of ash).

Large areas of spodosols are found in the northeastern part of the United States and Canada, and in northern Europe, Russia, and Siberia. In Europe, these soils are called *podzols* and in Russia they are found underneath the Taiga forest that stretches from east to western Siberia towards the sea of Okhotsk. The soils are very acidic in reaction, and adequate fertilization and liming are required for crop production. In the northern part of the United States they are used for pastures in dairy farming, whereas in the south blue berries are grown without liming (Miller and Gardiner, 1998). Limited spodosol areas in northern Maine are used successfully for potato production (Brady, 1990). The acid condition appears to control the potato-scab disease.

1.3.6 Alfisols

Alfisols are mature soils with profiles characterized by a sequence of A/E/Bt/C/ horizons. They are formed by a combination of podzolization and laterization processes in cool to warm-temperate humid regions, usually under hardwood forest. These soils are affected by a more drastic leaching process than mollisols and are, therefore, in a more advanced stage of profile development. The surface soil varies in color from gray-brown to reddish brown. Alfisols with gray-brown surfaces were called in the past *gray-brown podzolic soils*. This name is still used in Canada, Australia, and parts of Europe. Because of the eluviation process, the B horizon is enriched with illuvial clay and is called an *argillic* horizon (Bt). This is perhaps one reason why the soils are called *luvisols* (L. luo = to wash, meaning illuvial clay) in the FAO soil classification system. Alfisols are highly productive soils with a percentage base saturation in the subsoil of > 35%, which gives them a medium rank in fertility. However, sometimes these soils are considered unfavorable for crop production, because of the presence of an illuvial clay layer. When the surface horizon is eroded, exposing the argillic horizon to become the surface soil, this clay layer appears to inhibit plant growth. Liming is frequently required to neutralize the moderately acidic reaction of the surface soil in order to get the best results in crop production.

1.3.7 Ultisols

Ultisols are mature soils with A/E/Bt/C profiles. They are formed by a combination of laterization and podzolization, with the emphasis on laterization, in warm humid regions to the humid tropics, where leaching processes are very pronounced. Under these conditions, the soils are highly weathered, and the A horizons may accumulate varying amounts of Fe oxides, which impart their yellow to red colors. Enrichment of the B horizon with illuvial clay has also caused the formation of argillic horizons (Bt). Because of the drastic leaching process, the soils exhibit a very low base status, with a percentage base saturation in the subsoil amounting to < 35%.

Ultisols occur extensively in the southern region of the continental United States, Hawaii, and Puerto Rico, where hardwood mixed with pines is the common natural vegetation cover. In other parts of the world, e.g., northern Australia, these soils are called *yellow podzolic, red-yellow podzolic,* or *red podzolic soils* according to the color of the surface soil. Because of their acidic condition and low base status, these soils have a very low fertility. They also exhibit a low degree of stable aggregation and are therefore sensitive to erosion. Nevertheless, with adequate liming, the addition of organic matter, fertilizer application, and proper management, these soils can become quite productive in the southern region of the United States. The humid climate, providing sufficient amounts of rain, and the long periods of frost-free condition are favorable for crop production. However, pests and diseases may become more serious because of the humid conditions. Cotton, peanuts, soybean, corn and sweet potatoes as well as pine for timber and pulpwood production are today the most common crops.

1.3.8 Oxisols

Oxisols are mature soils with A/B/C profiles. Formed by a laterization process in warm humid and tropical regions, they are typically highly weathered, even more than the ultisols. The B horizons of oxisols are *oxic horizons,* defined earlier as subsurface horizons containing large amounts of hydrous-oxide clays or sesquioxides and 1:1 layer (kaolinitic) types of clays. The oxic horizon is sometimes also called *oxic endopedon* by some soil scientists (Miller and Gardiner,

1998). Plinthite can be present in the subsoil of many oxisols. Because the electrical charges of the clays in oxisols are highly variable, the soils are sometimes referred to as *soils with variable charges* (Theng, 1980). Many alfisols and ultisols are frequently included in this group of soils. Large areas of oxisols occur in Central and South America, and in Africa, Southeast Asia, and Australia. They are highly leached; therefore, they are acidic in reaction and low in bases. Nevertheless, the potential of many of these soils for agricultural production is far in excess of that currently realized, as has been demonstrated in central Africa and Brazil (Brady, 1990). Although they have very high clay contents, these soils have stable granular structures and are frequently nonsticky, loose and friable. Often they can be cultivated even under heavy tropical downpours. However, depending on the iron content, some oxisols can be converted into *laterites*, in which the formation of iron pans may inhibit the growth of plants. Most of the oxisols usually occur under a tropical rainforest, where the fertility of the surface soil is maintained by a process called nutrient cycling. When this native vegetation cover is removed by burning, as is the case in shifting cultivation, this cycle is destroyed and nutrients from the ash may only last for a year or two. The exposed soil will be affected by erosion and may gradually harden to become impossible for further cultivation. Therefore, oxisols, especially those containing plinthite, must be kept under vegetation by growing coffee, tea, rubber, or other tree crops, to prevent them from drying and irreversible hardening. The use of organic mulches and ground- cover crops is recommended. Because daylight in the tropics seldom exceeds 12 hours daily through the year, crop yield is less than yields in the temperate regions where the average daylight is 14 to 15 hours daily during the growing season. According to Miller and Gardiner (1998), the oxisols are highly productive for crops yielding carbohydrates and oil, because the latter are mostly derived from air and water rather than from soil mineral nutrients. The authors also believe that the oxisols are less productive for growing crops producing protein, since these crops require large amounts of mineral nutrients, such as nitrogen and sulfur. Apparently the authors above ignore the fact that large amounts of potassium are required for crops producing carbohydrates.

1.3.9 Aridisols

Aridisols are mature soils with profiles characterized by a sequence of A/Bk, Bn, or Btn/C horizons. They are called *xerosols* (from the Greek xeros = dry areas) or *yermosols* (Gr. yermo = desert area) in the FAO system. Ardisols are formed in arid and semi-arid regions in the world, where the long dry periods favor the accumulation of salts and other compounds in the surface and subsoil. When the B horizon is enriched with illuvial carbonates, it is called a Bk horizon. If sodium is the illuvial soil constituent, it is called a Bn (n = natrium, European for sodium) horizon. When both clay and sodium have accumulated in the subsoil, the horizon is designated by the symbols Btn. Lime cemented hardpans, called *duripans* or *caliche*, may sometimes be present. The soils are covered by a vegetation characteristic of arid regions, composed of scattered desert shrubs, such as short grasses, creosote bush, mesquite, and sagebrush. Since aridisols also occur in cold desert regions, the vegetation here may differ significantly from that in the hot desert regions.

These soils occur in the western part of the United States, where they are sometimes called *white alkali soils*, when a white crust is present on the surface composed of salt crystals, and *black alkali soils*, when black or darker colors are present. Extensive areas of aridisols are also found in the Sahara desert of Africa, in the Gobi and Taklamakan deserts of China, and in the deserts of Turkestan, the Middle East, and Australia. In Russia, the soils are called *solonchaks*, which are the equivalent of white alkali soils, and *solonetz*, the equivalent of black alkali soils. The limiting factor for cultivation of aridisols is frequently water. In areas where irrigation can be provided, such as with the center-pivot irrigation in arid regions of the United States, the soils can be made productive. Since the soil is characterized by a slightly to moderately basic reaction (pH = 7 to 8.5), micronutrient deficiencies are common. Because of the basic reaction, most of the Fe, Cu, Zn and Mn are precipitated in the form of hydroxides, and become unavailable for plant growth. However, these deficiencies are usually controlled by application of the deficient nutrient.

1.3.10 Vertisols

Vertisols are mature soils that are characterized by A/B/C profiles. Their name denotes their high swell-shrink capacity, which has a natural plowing effect on the soil. This effect is caused by the presence of smectite or montmorillonite in the clay fraction, known for its high degree of swelling and shrinking. Wide and deep cracks are formed when the soil shrinks during dry conditions, and surface soil material may then crumble down or slough off into the bottom of the cracks. Upon wetting, the soil swells and subsoil material is worked up again. The overall effect is like the "turning" effect of a plow. During swelling vertical and angled mass movement occurs in the soil, reflected by the presence of a *gilgai* microrelief, smoothed pressure surfaces on peds called *slickensides*, and wedge-shaped structures in the subsoil called *parallel epipeds*.

The formation of vertisols is not limited to a particular type of climate. They are found all over the world, especially where conditions favor the formation of smectites. Large areas of vertisols are present in India, Ethiopia, South Africa, and northern Australia. In addition, these soils are also located in Indonesia, Mexico, Venezuela, Paraguay, and Bolivia. In Indonesia, these soils are known by the names *grumusols* or *rendzinas*. In the United States, vertisols are found in east- central (Houston, TX) and southern Texas, and to a lesser extent in eastern Mississippi and western Alabama.

Physically the soils are very poor, making them less suitable for crops and engineering. In addition to their high swell-shrink capacity, the soils are sticky when wet and very hard when dry. When dry, wide cracks develop due to shrinking of the soil clays, which tend to break plant roots, inhibiting in this way proper plant growth. In wet conditions, the cracks may close and the soil's permeability for water decreases, and aeration is often very poor. These poor physical conditions restrict root growth, hence root penetration is inhibited. Nevertheless, extensive areas of vertisols are cultivated with success in India and Africa for sorghum, millet, cotton and corn production. In Indonesia, sugar cane is grown satisfactorily on vertisols.

Problems in engineering arise during wet conditions, because road beds often break or become deformed by the weight of heavy traffic due to the soil's low support strength. Powerlines and telephone poles have been reported to lean after years of being affected by shrinking and swelling of the soil. Because engineering activities on these soils require extra effort and building costs, vertisols are dubbed as expensive soils (Miller and Gardiner, 1998).

1.3.11 Andisols

Andisols are young soils with A/B/C or A/C profiles (Soil Survey Staff, 1992, 1996), which formerly have been called *andepts*, a*ndo soils* or *andosols* (Soil Survey Staff, 1975; Tan, 1984; 1998). Today they are defined by the US Soil Taxonomy as soils with *andic* properties, which specify among others the presence of <25% organic carbon content, 2% or more acid oxalate extractable Al and Fe, and a P retention of >25% (Soil Survey Staff, 1990; 1992).

The original andosols are volcanic ash soils rich in organic matter, which imparts to the A horizon the distinctive dark to black color from which the name is derived. The B horizon often has the characteristics of a cambic horizon (Bw). The soils, first identified in Japan, were introduced in 1947 by US soil scientists, after surveying black soils in Japan, under the name *ando soils*, from Japanese an = black, and do = soil, though Japanese soil scientists preferred using the name *kurobokudo* (kuro = black, boku = friable, and do = soil). Since then the soil has received world-wide attention, and the name andosol was adopted by many countries, and is used by the FAO-UN as the official name in its world soil map (Tan, 1984; 1998). With the development of the US Soil Taxonomy, these soils were first placed at the suborder level of the inceptisols as *andepts*, and this was later revised to become the *andisol* order. The name andisol, instead of andosol, was chosen because the connecting vowel "o" was to be restricted for use with Greek formative elements. As indicated before, of the twelve orders, only spodosols and histosols carry the connecting vowel "o". The selection was apparently based on just being consistent with the names of the remaining orders, which all carried the connecting vowel "i". The latter was supposed to be restricted for use with Latin formative elements. but entisols, alfisols, oxisols, andisols, and especially gelisols were not derived from Latin terms. As is the case with gelisols, the decision to name the soils andisols violates the rules set up in the US Soil Taxonomy. In many other parts of the world and in the FAO soil classification system, the soils are still known as andosols. Using the new USDA criteria for andic properties, many red soils, that do not resemble andosols, are now "misplaced" in the new andisols order. Since the new name andisol was coined from the old term andosol, the distinctive properties that give the original soil its name should be recognized and maintained, hence the criteria for andic properties adjusted accordingly. If not, it is suggested that the name andisol be deleted completely, and a new order name be created that does not have any association with the original name andosols.

Andosols occupy about one hundred million hectares, or 0.76% of the world land area (Dudal, 1976). However, on a global scale, they often do not occupy large areas in any one place; instead they are associated with the presence of volcanoes, and hence are frequently scattered all around in small areas. Formation of andosols is not restricted to a particular climatic zone, since the soils are found from the subarctic region in Alaska and Iceland to the tropical regions in Indonesia, and from the warm humid tropical lowlands to the subalpine regions high in the mountains. Soil temperature regimes may vary from cryic and pergelic to isohyperthermic and isothermic. The soils are found under forest as well as under grass vegetation. Oak (*Quercus* sp.; *Fagus* sp.) and pine (*Pinus* sp.) are the major forest trees reported to be associated with andosols, whereas pampas grass (*Miscanthus sinensis*) is considered in Japan the reason for the development of andosols with a *melanic epipedon*, representing the central concept of the kuroboku soil.

The presence of large amounts of organic matter – together with amorphous and paracrystalline clays, allophane and imogolite – is the reason for the many unique properties of andosols. These constituents are responsible for the presence of variable charges, extremely high water-holding capacity, and low bulk density. Andosols exhibit high total porosity, friability, and low plasticity and stickiness. When wet, the soil feels greasy and smeary. Amounts of water, equivalent to those at field capacity, are present in andosols at 0.1 bar (10 kPa), whereas large amounts of water are also retained at 15 bars (1500 kPa). Generally the soils yield water when squeezed between the fingers. Such a high water retention and high porosity are believed to be caused by the presence of large amounts of inter- and intraparticle pores of allophane. These physical characteristics are noted to change with soil moisture condition. In Indonesia, it was noticed that drying produced pronounced changes which manifested themselves in a phenomenon called *mountain granulation*, forming *pseudo -sand*. The soil is then very difficult to remoisten and on disturbance produces black dust clouds; hence the name *black dust soil* was used in the past by Dutch scientists in Indonesia. This irreversible drying presents problems in particle size analysis, because of difficulties in obtaining complete dispersion required for proper analysis. In order to obtain complete dispersion, recent evidence suggests the use of ultrasonic methods and – depending on type of allophane and imogolite – a pH adjusted to 4 or 10 (Tan, 1998a).

In Indonesia, andosols are fertile soils, and are a key factor in many successful horticultural and agricultural operations. Tobacco and the best tea plantations are found on andosols, whereas the more rugged part of the country covered by andosols is planted with pine trees. In Japan, andosols also constitute fertile agricultural soils. In a few cases, it has been reported that the soils are either low in phosphates or exhibit high P-fixation capacity (Tan, 1965; 1984; 1998a; Amano, 1981). The latter is reflected in the US Soil Taxonomy, requiring andosols to have a phosphate retention of >85% or >25% as determined by the Blakemore method. However, indications were reported that this method usually yielded exceptionally high values for phosphate fixation compared with the method generally used in Japan with 2.5% $(NH_4)_2HPO_4$. Ligand exchange was considered by several scientists to be responsible for this reaction. However, ligand exchange is in conflict with the chelation process, responsible for humus accumulation in andosols. Accumulation of humus to such an extent that the unique black colors are produced in andosols is made possible by immobilization of the organics through chelation with Al, allophane, and imogolite. Exposed Al and Si on the surfaces of the amorphous minerals, together with free Al and Si ions, are capable of interacting with organic acids, such as humic acids, forming humo-Al, humo-Al-allophane or humo-Si-imogolite complexes or chelates. In the form of chelates, the resistance of humic acids against decomposition is increased. Recent evidence suggests that the increased resistance against decomposition is caused by incorporation of the organics into the structure of the short-range-order minerals, such as allophane (Huang, 1995). Chelation of Al by humic acids would inhibit formation of allophane and imogolite; hence such an interaction is called an *anti-allophanic process*.

If phosphate fixation were to occur by ligand exchange, then from the discussion above it is clear that in order to form insoluble Al-phosphates, the chelated humic acid, replaced by the phosphate ligand, will be liberated into the soil solution. Such a process is very harmful for the preservation of humus. Humus in the free state will be subject to rapid decomposition, and formation of andosols is inhibited, or existing andosols will be destroyed. In addition, basic soil chemistry indicates that phosphate fixation or retention by ligand exchange takes place with difficulty, since the affinity of Al for humic acids is much greater than the affinity of Al for phosphate. Evidence has often demonstrated that phosphate was liberated from insoluble $AlPO_4$, because of chelation of the Al by humic acids. *Co-adsorption* or *Al-bridging* is perhaps a better explanation, because by this inter-action Al forms a bridge between phosphate and humic acid, preserving in this way the humus content in andosols.

1.3.12 Histosols

Histosols are characterized by O/C or O/Ab (b for buried) profiles. The soil's O horizon is relatively thick and contains at least 12% organic carbon (Soil Survey Staff, 1975). It may lie on a mineral C or a buried A horizon. Histosols, therefore, are organic soils, and are typically formed in a water-saturated environment, where conditions are favorable for development of thick peat and muck deposits.

The presence of these soils is not limited to any climatic region. They may occur in environments ranging from the cold tundra to the tropical regions in the Amazon river delta. Large areas of histosols are found in the United States near the Great Lakes and in the coastal areas of the South, especially in Florida. With proper drainage, the histosols in Florida are successfully used in growing vegetables. Because of large expenses for drainage and liming procedures, cultivation of high value crops are favored. In the Great Lakes region of the United States, potatoes and flowers are grown on histosols. In the Netherlands, histosols are the sites for growing tulips and other cut flowers. Of particular concern with cultivated histosols is the problem of *subsidence*, the sinking or falling of the soil's surface below its original level. This phenomenon is caused by the continuous decomposition of the organic matter brought about by the better aeration. The surface of the land has been reported to drop in elevation as much as 4 m after 40 years of cultivation (Miller and Gardiner, 1998). Since histosols exhibit very low bulk density values, for the use of seedbeds, the soil may need to be compacted by "rolling" with 10 metric ton packers in order to improve seed-soil contact.

1.4 RELATION OF SOIL ORDERS WITH THE ENVIRONMENT

As discussed above, several of the soil orders are formed under the dominant influence of climatic and vegetational factors. After formation they are in equilibrium with the prevailing environment. Such soils were once called *zonal soils* (Thorp and Smith, 1949), and their geographic distribution follows climatic and vegetational zones in the world. Moving south on the Northern Hemisphere from the tundra circle to the Equator, a sequence of soils from gelisols, spodosols to oxisols can be noticed (Figure 1.3). Their occurrence reflects the intimate relationship of the soils to the environment.

South of the tundra circle, the climate changes first into a cold-cool humid climate. Such a climate, together with a coniferous vegetation, favors podzolization and the consequent formation of podzols or spodosols. The type of forest, composed of coniferous trees called the *taiga forest* in Russia, stretches from northern Europe, Russia, and Siberia, to Alaska and northern Canada. Here then is the zone of podzols or spodosols. Next to this zone at a lower latitude is a zone with a temperate climate and a mixed stand of coniferous and hard-

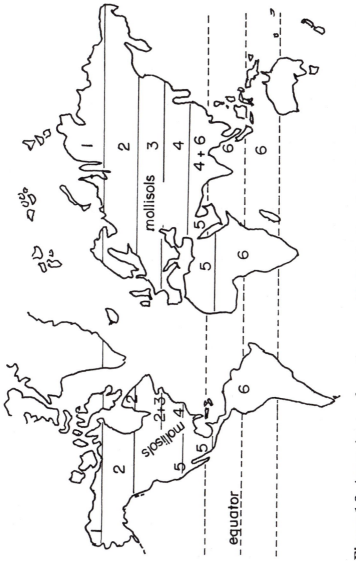

Figure 1.3 Approximate schematic geographic distribution of soils according to climatic and vegetational zones in the world: (1) zone of gelisols (tundra soils); (2) zone of spodosols; (3) zone of alfisols; (4) zone of ultisols; (5) zone of aridisols; and (6) zone of oxisols.

wood forest. This is the zone of gray-brown podzolic soils or alfisols. Further south on a latitudinal basis, a zone exists with a warm-temperate humid climate and a hardwood forest, which favors the formation of red-yellow podzolic soils or ultisols. The next zone to the south, before the Equator, is a desert zone characterized by desert soils or aridisols. On both sides of the Equator lie the tropical zones with warm humid climates and tropical rain forests. These are the zones for laterization, hence the zones of lateritic soils or oxisols. This sequence of soils according to latitudinal zones with changing climates and vegetation is repeated again in the Southern Hemisphere. Such a geographic distribution of soils corresponds with the distribution of soils in the USDA general soil map of the world (Brady, 1984).

A similar sequence of soils according to changing environment can also be noticed from east to west in the continental United States. The eastern seaboard of the United States, characterized by a humid climate, has a mixed stand of conifers and hardwood in the Northeast, and a hardwood forest in the Southeast. The Northeast is the major zone of spodosols and alfisols, whereas the Southeast is the main zone of ultisols. Moving from the eastern seaboard to the Great Plains to the west, a zone will be encountered with a semihumid climate and tall grass vegetation. This is the zone of mollisols, or more specifically the *udolls*. Further westward, the region is characterized by a semiarid climate and short grass vegetation. This is the zone of *ustolls*, or mollisols formed under dryer conditions. Lastly, next to this zone in the Far West lies the zone of aridisols in the arid regions. Such a distribution of soils in the United States, in accordance with the changing environments from east to west, is reflected by the USDA general soils map of the United States (Foth, 1990).

The influence of the environment is not restricted to latitudinal changes in soil formation, but evidence has been reported for changes in soil formation because of altitudinal changes. The latter is especially of importance in the mountainous regions of the tropics. Here, the climate and vegetation change from the tropical lowlands to the top of the mountains (Tan and Van Schuylenborgh, 1959; Tan, 1998a). Rainfall, amounting to 2000 mm/annually in the lowlands, increases to approximately 4000 mm/annually on the slope of the mountains, to decrease again at the *timberline*. The temperature shows a steady decline with increasing altitude towards the top of the mountains. The temperature is noted to decrease 0.6°C for every 1 hm (= 100 m) increase in elevation. These changes in temperature can be calculated with some degree of accuracy by using the formula of *Braak* (Tan and Van Schuylenborgh, 1961; Tan, 1998a):

$$T = 26.3° - (hm \times 0.6°)$$

in which 26.3° is the starting temperature usually occurring in the lowlands, T = temperature in °C, and hm = hectometer (=100 m).

The vegetation will also change in conjunction with the increasing cooler climate with increasing altitude. The tropical rainforest prevailing in the lowland will be replaced by a tropical mountain rainforest. At higher elevation, beard moss growing on the trees due to higher humidity is a characteristic feature, differentiating the mountain rainforest from the rainforest in the lowland. Closer to the timberline, the trees become shorter and stunted. Grass and sub-alpine vegetation prevail at the highest elevation.

These changes bring about changes in soil formation. Laterization occurs usually in the warm humid lowlands when drainage conditions are excellent, forming a zone of *latosols* or oxisols. The environmental conditions at higher elevation favor the occurrence of laterization and podzolization with the consequent formation of a zone of red-yellow podzolic soils or ultisols. Higher up in the mountains, close to the timber line where the climate is cool and humid, a zone of gray brown podzolic soils or alfisols is frequently encountered. Podzols or spodosols may occur at this high elevation when the conditions are favorable for acid leaching necessary in podzolization.

CHAPTER 2

INORGANIC SOIL CONSTITUENTS

2.1 SOIL COMPOSITION

The soil system is composed of three phases: a solid, liquid, and gas phase. The solid phase, a mixture of inorganic and organic material, makes up the skeletal framework of soils. Enclosed within this framework is a system of pores, shared jointly by the liquid and gaseous phase. In *mineral soils*, the inorganic material is present in large amounts, whereas the organic fraction is found in substantially smaller amounts. Mineral soils are defined in *Soil Taxonomy* (Soil Survey Staff, 1990) as soils that contain by weight 80% or more inorganic and 20% or less organic material. On the other hand, in *organic soils*, the amount of organic matter far exceeds that of inorganic material. Organic soils contain by weight 80% or more of organic matter and 20% or less of inorganic material. These soils are formed only where the conditions are favorable for the accumulation of large amounts of organic residue, e.g., low temperatures and/or the presence of excessive amounts of water due to poor drainage. Since mineral soils are the most abundant in nature, the use of the term *soil* usually refers to mineral soils.

Inorganic material, organic matter, water, and air are considered the four major soil constituents (Brady, 1990). Their concentrations

28

may differ from soil to soil, or from horizon to horizon. A soil with a loam texture and that is in optimum condition for plant growth is reported to have a volume composition of 45% inorganic material, 5% organic matter, 25% water, and 25% air (Brady, 1990). As indicated earlier, water and air are present in the pore spaces; therefore, a soil excellent for plant growth is composed of 50% solid space and 50% pore space by volume.

The spatial arrangement in soils of the solid particles and associated pores is called the *soil fabric* (Figure 2.1). In micropedological language the soil pores are called *voids*. Kubiena (1938), who introduced the concept of soil fabric, believed it to be comparable with rock fabric, the arrangement of mineral grains in rocks. The coarse inorganic grains, together with coarse organic fragments (>2 μm),

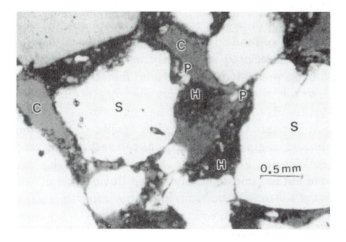

Figure 2.1 A thin section of soil showing a soil fabric, composed of sand grains (S), clay (C), humus (H), and associated pores (P). The sand grains form the skeleton of the soil fabric, whereas the clay and humus are the inorganic and organic plasma, respectively.

form the *soil skeleton* (Stoops and Eswaran, 1986; Brewer and Sleeman, 1988). This can then be distinguished into inorganic and organic skeletons. All other materials smaller than 2 μm constitute the *soil plasma*. Skeleton grains may, of course, weather to produce plasma. Skeleton grains are considered relatively immobile. They will not be readily translocated, except for washing down into cracks. On the other hand, the plasma is the mobile, chemically active part of the soil. It can be translocated and accumulated by soil formation processes. The plasma is called *cutan* when deposited on the surfaces of skeleton grains, on the walls of voids, or on the boundaries of structural units. These cutans are called *argillans* or *clay skins* if they are composed dominantly of clay minerals. The name *soil matrix* refers to the soil plasma (Osmond, 1958).

The pattern produced by the arrangement of these soil constituents in association with pores creates the soil fabric. Many different kinds of soil fabrics can be formed, depending on the various combinations of size and shape of the soil constituents and on their occurrence as single grains or as aggregates. For more details on the different types of soil fabrics reference is made to Brewer and Sleeman (1988). Attempts have been made lately to identify or relate specific soil fabrics with specific kinds of soils. Although some of the soil fabrics have been noted to be associated specifically with particular soils, evidence has also been presented for the same fabrics to occur in different soils. However, the latest information suggests that soil fabrics can serve to identify and define specific soil horizons (Wilding and Flach, 1985). It appears that oxic, argillic, spodic, cambic, and petrocalcic horizons and fragipans and duripans can be differentiated by differences in soil fabrics. For example, the soil fabric of the oxic horizon is plasmic in nature, and does not display illuviated argillans as compared to that of the argillic horizon. Placic horizons, thin black to reddish pans cemented by iron, iron and manganese, or by iron-humic acid complexes, also have distinctive microfabrics, which can be used to identify spodosols (Wilding and Flach, 1985). However, several authors disagree and indicate that placic horizons, occurring below the sola of spodosols, are not formed by a podzolization process, but are the products of redox conditions associated with groundwater fluctuations (De Coninck and McKeague, 1985). In addition, pedogenic properties can be distinguished from lithogenic characteristics by determining differences in soil fabrics.

2.2 INORGANIC CONSTITUENTS

The inorganic fraction in soils is derived from the weathering products of rocks and consists of rock fragments and minerals of varying sizes and composition. Rock fragments are usually not considered soil constituents, but will yield soil constituents upon weathering. Separated according to size, the inorganic soil fraction can be distinguished into three major soil fractions: *sand, silt* and *clay* (Table 2.1), collectively called the soil separates. The particles with a diameter larger than 2 mm are gravel, stones, and boulders, and as indicated above are not considered soil constituents. They are rock fragments which upon weathering may yield sand, silt, and clay. Sand grains are irregular in size and shape, and are not sticky and/or plastic when wet. Their presence in soils promotes a loose and friable condition which allows rapid water and air movement. They are chem-

Table 2.1 Size Limits (Diameter) of Major Soil Separates According to the USDA and International System

Soil separate	USDA system	International system
	-------------- mm ----------------	
Very coarse sand	2.00 - 1.00	
Coarse sand	1.00 - 0.50	2.00 - 0.20
Medium sand	0.50 - 0.25	
Fine sand	0.25 - 0.10	0.20 - 0.02
Very fine sand	0.10 - 0.05	
Silt	0.05 - 0.002	0.02 - 0.002
Clay	< 0.002	< 0.002

Source: Soil Survey Staff (1962).

ically inert and do not carry electrical charges, hence have low water-holding and cation exchange capacities. Silt particles are intermediate in size and possess characteristics between those of sand and clay. Some silt particles may be capped or coated by clay films as a result of the weathering of silt surfaces. Because of this, silt may exhibit some plasticity, stickiness, and adsorptive capacity for water and cations. Clay is the smallest particle in soil and has colloidal properties. It carries a negative charge and is chemically the most active inorganic constituent in soils. The presence of clay gives to the soil a high water-holding and cation exchange capacity. However, clay is sticky and plastic when wet. In the concept of soil fabrics, clay constitutes the inorganic plasma.

The inorganic soil fraction is composed of soil minerals, hence this component is also referred to as the *mineral fraction* of soils. *Minerals* are by definition inorganic substances in nature, possessing definite physical characteristics and chemical composition. In many other textbooks and in medical science, the term mineral refers to nutrient elements, such as N, P, K, Ca, Mg, etc. However, in soil science a clear distinction is made between minerals and nutrient elements. The term *mineral* may be defined as a compound composed of two or more elements, whereas a nutrient element is an element that can be used as food by plants.

The soil minerals can be distinguished into primary and secondary minerals. *Primary minerals* are minerals that have been released by weathering from rocks in a condition that is chemically unchanged. These minerals constitute the sand fraction of soils. *Secondary minerals* are minerals that have been derived from the weathering of primary minerals. They are present in the clay fraction of soils. The use of the terms primary and secondary minerals may frequently create some concern among soil scientists, since secondary mineral deposits are sometimes regarded as primary on a pedological basis.

2.3 MINERAL COMPOSITION

The composition of soil minerals is very variable, and depends on the composition of the rocks. The rocks, from which the minerals originate, are composed mostly of the elements O_2, Si, Al, Fe, Ca, Mg, Na, and K (Table 2.2). The soil minerals are, therefore, made up of these elements. Most of the minerals are either *silicates* or *oxides*. The Si in soil silicates is present in the form of *silica tetrahedrons*, which constitute the basic units of the crystals. On the basis of the arrangement of the SiO_4 tetrahedra in the crystal structure, six types of soil silicates can be distinguished. Listed in alphabetical order, they are

Table 2.2 Chemical Composition of Basaltic and Granitic rocks[*]

	Basalt	**Granite**	**Average**
	----------------- % -----------------		
SiO_2	50.4	69.5	60.0
Al_2O_3	16.9	15.7	16.3
Fe_2O_3	2.8	2.6	2.7
FeO	6.9	0.7	3.8
MgO	7.5	0.9	4.2
CaO	8.2	1.0	4.6
MnO	0.1	--	0.05
K_2O	1.6	2.8	2.2
Na_2O	3.3	2.6	3.0
TiO_2	1.4	0.9	1.2
H_2O+	0.8	1.0	0.9

*Source: Clarke (1924); Mason (1958); Mohr and Van Baren (1960); and Mohr et al. (1972).

(Figure 2.2): cyclosilicates, inosilicates, nesosilicates, phyllosilicates, sorosilicates, and tectosilicates (Tan, 1982, 1993). Most of the minerals in the sand and a major part of the silt fraction are cyclosilicates, inosilicates, nesosilicates, sorosilicates, or tectosilicates. Since they are coarse in size, they have low specific surface areas and do not exhibit

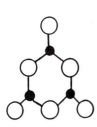

Cyclosilicate
(Benitoite)

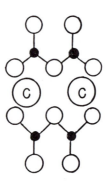

Inosilicate
(Pyroxene)

Nesosilicate
(Olivine)

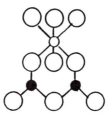

Phyllosilicate
(Kaolinite)

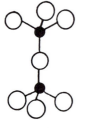

Sorosilicate
(Epidote)

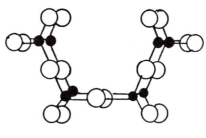

Tectosilicate
(Feldspar)

Figure 2.2 The molecular structure of soil silicates distinguished by the different arrangement of SiO_4 tetrahedra.

colloidal properties. They participate in a number of chemical reactions and exhibit some adsorption, but are not really active in chemical reactions. Most of the minerals in the soil clay fraction are phyllosilicates. Other minerals may also be present, such as quartz and other primary minerals in particle sizes of <2 μm, sesquioxides, talc, sulfides, sulfates, and phosphates.

2.4 PRIMARY MINERALS

Although numerous primary minerals are found in nature, only a few contribute to soil formation. Major primary minerals found in soils are listed in Table 2.3.

2.4.1 Quartz

This mineral, classified as a tectosilicate can occur as a primary or secondary mineral in soils. The latter is often called *pedogenic* quartz. As a primary mineral, it is believed to be the last mineral to crystallize from the magma. It is a major constituent of igneous acidic rocks, such as granite. On the decomposition of such rocks, the quartz minerals persist as detrital grains to accumulate as sand in the soil because of their mechanical and chemical stability. Pedogenic quartz is often an important constituent of the clay fractions of soils. By virtue of its size (<2 μm), it is classified as clay.

The fact that quartz crystallizes at lower temperatures makes it fairly stable at earth temperatures. A quartz crystal does not exhibit cleavage, but will fracture upon impact. Although it is very resistant to weathering, eventually quartz grains can become rounded in nature.

Quartz belongs to the group of silica (SiO_2) minerals within the tectosilicates. The other members in the group are tridymite, cristobalite, and opal. The presence of cristobalite in soils usually indicates its origin from volcanic ash. Depending on the temperature, each of these minerals can exist in α and β form. The α form is the low tem-

Table 2.3 Common Primary Minerals in Soils

Primary mineral	Chemical composition
1. Quartz	SiO_2
2. Feldspar	
Orthoclase, microcline	$KAlSi_3O_8$
Albite (plagioclase)	$NaAlSi_3O_8$
3. Mica	
Muscovite	$H_2KAl_3Si_3O_{12}$
Biotite	$(H,K)_2(Mg,Fe)_2(Al,Fe)_2Si_3O_{12}$
4. Ferromagnesians	
Hornblende	$Ca(Fe,Mg)_2Si_4O_{12}$
Olivine	$(Mg,Fe)_2SiO_4$
5. Magnesium silicate	
Serpentine	$H_4Mg_3Si_2O_9$
6. Phosphate	
Apatite	$(Ca_3(PO_4)_2)_3.Ca(F,Cl)_2$
7. Carbonates	
Calcite	$CaCO_3$
Dolomite	$CaMg(CO_3)_2$

perature variety, whereas the β modification is the high temperature form (Tan, 1998). Quartz itself can exist in soils in many different varieties, e.g., rock crystal, amethyst, citrine, rose quartz, smoky quartz, and milky quartz. Flint, chert, onyx, and agate are also grouped as quartz varieties. Wood that has been petrified by agate, known as silicified or agatized wood, also belongs to the quartz group (Hurlbut and Klein, 1977).

Environmental and Industrial Importance of Quartz. – The presence of quartz gives to the soil a loose and friable consistence. Permeability, drainage, and aeration are enhanced. However, because of its chemical composition, quartz does not contribute towards improving soil fertility. Hence quartz provides to the soil a good physical, but a poor chemical condition. Nevertheless, it is very important as a diluent and as a growth medium in hydroponics. Aside from also being important as gemstones, this mineral has found many applications in industry and scientific instruments. Quartz sand is an important ingredient for the production of mortar and concrete. In powdered form, quartz is of importance for the manufacture of fine crystal porcelain and sand paper. Because of its transparency, quartz crystals are the basic materials for lenses and prisms, essential in optical instruments. The optical activity – the ability to rotate the plane of polarization of light – is a fundamental property that makes quartz useful for the production of monochromators, which separate ordinary light into monochromatic light of different wavelengths. Quartz wedges are key accessories in polarizing microscopes, whereas tiny quartz plates serve as oscillators in digital quartz watches and for controlling radio frequencies.

2.4.2 Feldspars

Feldspars are also members of the tectosilicates. The name is derived from the German term *feld* = field. They can be divided into two groups: (1) potash feldspars, e.g., orthoclase and microcline with a general composition of $KAlSi_3O_8$ as indicated in Table 2.3, and (2) plagioclase, e.g., albite and anorthite ($CaAl_2Si_2O_6$). Microcline can be distinguished from orthoclase only by petrographic microscopy with which the presence of *twinning*, characteristic for microcline, can be seen. The name microcline is derived from the Greek words referring to *little* and *inclined*, whereas orthoclase is derived from the right-angle cleavage exhibited by the mineral (Hurlbut and Klein, 1977). Feldspar minerals are considered solid solution minerals with a framework of silica tetrahedra, in which the cavities are occupied by K, Na, and Ca. The minerals are almost as hard as quartz, but due to the presence of non-framework ions, they weather very easily.

Environmental and Industrial Importance of Feldspars. – Orthoclase and microcline are important minerals of igneous rocks, such as granites, granodiorites and syenites. However, orthoclase is believed to have cooled at a fast rate, whereas microcline was formed by a slower cooling process. The plagioclase minerals are considered more abundant than the potash feldspar minerals, and make up the mineral suite of igneous and metamorphic rocks.

The presence of potash feldspars in soils usually indicates the soils to be potentially rich in potassium. Weathering of feldspars will release the K, improving soil fertility, and at the same time clay will be formed from the Si and Al liberated from the mineral. Since such a weathering process is usually a very slow process, only perennial crops, instead of annual crops, can take advantage of this K supply. Decomposition of albite will also produce clay, but Na will be released from the mineral, which may have a harmful effect on the physical condition of the soil. Since Na is a dispersion agent, the release of large amounts of Na tends to disperse the soil particles. A dispersed soil will cake upon drying. Another mineral of the plagioclase series is *anorthite*, $CaAl_2Si_2O_8$, which upon weathering can enrich the soil with Ca and clay.

In industry, potash feldspar is a major ingredient in the manufacture of porcelain. When ground and mixed with kaolinite, it fuses upon heating and cements the mixture together. Fused K-feldspar produces the nice glaze of porcelain. Albite, known commercially as soda (Na) feldspar, is also used in the ceramic industry as is K-feldspar. Coarse minerals of both K- and Na-feldspars when polished are very valuable as gemstones.

2.4.3 Mica

The mica group belongs to the phyllosilicates and exhibits a sheet structure similar to that of clay minerals. The name mica is believed to have been derived from the Latin term *micare* = to shine, whereas the name muscovite is reported to originate from the name *muscovy-glass*, because of its use as a substitute for glass in Russia, nicknamed *muscovy* in the past (Hurlbut and Klein, 1977). As indicated in Table 2.3, two types of mica minerals, muscovite and biotite, are recognized. Because of their sheet structure, these minerals exhibit cleavage and are easily attacked by water. Because of the presence of Fe and Mg in biotite, this mineral is less resistant to weathering than muscovite.

Environmental and Industrial Importance of Mica. – Muscovite is a rock-forming mineral in igneous rocks, such as granites, and in metamorphic rocks, e.g., mica-schists, whereas biotite appears to be formed in a greater variety of geologic environments, hence is more widespread. Biotites are found in the mineral suite of granites, diorites, and gabbros, in lavas, and in metamorphic rocks. Both muscovite and biotite contribute to enriching the soil with plant nutrients and to formation of clay. Soils containing muscovite and biotite minerals are potentially rich in K, Mg, and Fe. Since muscovite is more resistant against weathering than biotite, the release of plant nutrients occurs more easily from the decomposition of biotite.

Because of such a resistance to weathering, and due to its high dielectric constant and heat resistance, muscovite finds application as an insulating material in electrical appliances. Sheet mica is frequently used in irons, furnaces and stove doors. Ground mica is used for the production of wallpapers with a shiny luster. It is also used as a filler for fireproof material. When mixed with oil, it serves as a lubricant.

2.4.4 Ferromagnesians

Ferromagnesians, the dark-colored minerals, are classified as inosilicates or nesosilicates. The dark colors are due to their high Mg and Fe contents. Two important groups of minerals in the inosilicates are pyroxenes, with single-chain Si tetrahedra, and amphiboles, with double-chain Si tetrahedra. The most common mineral species in the pyroxene group are augite, $Ca(Mg,FA,Al)(Al,Si)_2O_6$, and hypersthene, $(Mg,Fe)SiO_3$, whereas that in the amphibole group is hornblende. Olivine, $(Mg,Fe)_2SiO_4$, another ferromagnesian mineral, belongs to the nesosilicates.

Environmental and Industrial Importance of Ferromagnesians. – The ferromagnesian minerals are usually present in the dark-colored ig-

neous rocks, such as gabbro, basalt, basaltic lava, andesite, and peri-
dotite. The presence of nonframework Ca-O, Mg-O, and Fe-O bonds
forms weak spots in the crystal structure, hence this group of dark-
colored minerals is listed at the top of the weathering sequence as the
least resistant minerals against weathering. They will decompose very
rapidly, enriching the soil with Ca, Mg, and Fe. The Si and Al
liberated from their crystal structure are used for the formation of
clays, which will increase the cation exchange capacity of soils. Hence,
ferromagnesians contribute to the development of good chemical, but
poor physical conditions in soils. The latter is caused by the high clay
content, because soils high in clays tend to be sticky and plastic when
wet, and hard when dry. These soils are subject to rapid dispersion,
forming puddled soils which will cake upon drying.

When cut and polished, the ferromagnesian minerals make nice
gemstones. A famous Na containing pyroxene is *jadeite*, the mineral
constituent of *jade*, a stone characterized by a green color. In Asia,
and especially in China, jade is made into beautiful high-priced
ornaments. In the old days it was also used for weapons and utensils
by primitive people.

2.4.5 Magnesium Silicates

These minerals are correctly called *serpentine* minerals. They
belong to the phyllosilicates (Hurlbut and Klein, 1977). Major
minerals in the group are antigorite and chrysotile, both characterized
by the composition $H_4Mg_3Si_2O_9$ as listed in Table 2.3 for serpentine.
The name *serpentine* refers to the green, serpent-like clouding of the
minerals, whereas the name *antigorite* is derived from antigorio in
Italian, and chrysotile from the Greek terms for golden and fiber.

Environmental and Industrial Importance of Serpentine Minerals. –
The serpentine group minerals – antigorite and chrysotile – are found
in igneous and metamorphic rocks. They are recognized by their
discrete markings of green colors and greasy luster, or by their fibrous
structure. Used as a rock name, serpentine refers to a rock composed
mainly of antigorite.

Upon weathering, these minerals will provide Mg to the soils and may contribute toward formation of clay. The silicon released, when not leached from the soil, may perhaps be used in the formation of pedogenic quartz, which is usually of a particle size <2 µm. Serpentine rocks are often used as ornaments. When present and found in nature as a mixture with marble, this rock is sometimes called *verde* (green) *antique marble*. It is often used as building material, because of its beautiful variegated green coloring. Chrysotile is mostly fibrous and is the major source for asbestos. Because it is not combustible and has a slow heat conductivity, it has been used in the past for the production of fire and heat resistant materials, and as insulation against electricity.

2.4.6 Phosphate Minerals

This very large group of minerals does not belong to the silicates. Most of their members are so rare that they are seldom mentioned. Of the many types of phosphate minerals, perhaps only the apatite group is common in soils. Three types of apatite minerals can be distinguished: *fluorapatite, chlorapatite*, and *hydroxylapatite*, containing F, Cl, and OH, respectively. Of the three, fluorapatite is the most common. The ions F, Cl, and OH can replace each other by isomorphous substitution, producing the *isomorphous series* of apatites. Isomorphous substitution of (CO_3, OH) for PO_4 gives the *carbonate-apatite* series.

Environmental and Industrial Importance of Apatites. – Apatite is found as an accessory mineral in a variety of rocks, e.g., igneous, metamorphic, and sedimentary rocks. Present in soils, apatite contributes only in enriching the soils with plant nutrients. It does not contribute to clay formation, since its composition lacks the major elements Al and Si, necessary for formation of clay.

Phosphates in bones and teeth are also compounds belonging to the apatite group, which are mostly calcium hydroxyapatite. The use of toothpaste containing F for cleaning and polishing teeth is supposed to replace by isomorphous substitution the OH group with F, forming fluorapatite, a more resistant form of the mineral against chemical and physical degradation. Ground animal bone, called *bone meal*, was used in the past as a nutrient supplement for human health in developing countries. Today it finds application as a fertilizer for acid soils, where phosphate fixation is a problem, and in horticultural operations and nurseries. Bone meal works as a slow-release fertilizer, allowing plant roots to take up the phosphate released before it is fixed by Al in the acid soil.

The apatite mineral is used in the manufacture of phosphate fertilizers. To render the P more soluble, the mineral is treated with sulfuric acid, H_2SO_4, and converted into the fertilizer *superphosphate*. The residue from the production of superphosphate contains large amounts of gypsum, $CaSO_4$, which is a valuable source of liming material. Currently, attempts have been made to enhance the solubility of apatite and phosphate rocks by adding microorganisms, instead of using H_2SO_4. Microorganisms such as *Rhizobia* sp. and *Mycorrizha* sp., were frequently used with the assumption that compounds secreted by the microorganisms have a similar solubilizing effect as H_2SO_4 on the phosphate rock. It is a well-known fact that organic secretions have the capacity to chelate the phosphate group in the rock, forming soluble organo-phosphates readily available for plant growth (Tan, 1998). These products, distributed in the market under the name of *biofertilizers* or *bio-phosphate-fertilizers*, require further investigations as to their capacity as fertilizers.

Apatite is usually too soft to be used as gems. In soils it weathers very rapidly, and ground apatite is sometimes used as a fertilizer. An exception is perhaps *turquoise*, $CuAl_6(PO_4)_4(OH)_8.4H_2O$, a mineral possessing a characteristic blue color, because of its Cu content. The name turquoise is French for *turkish*, because the first stones imported in Europe came from Persia through Turkey. The stone is usually cut in oval form, and is favored for use as an ornament among the Native Americans of the American West.

2.4.7 Carbonates

Carbonates can be distinguished into three groups: *calcite, aragon-*

ite, and *dolomite* group. They are not silicates, and are rather compounds composed of triangular groups of metal-carbonate,-$(CO_3)^{2-}$, complexes. Calcite is one of the most common minerals in soil formation. The name *calcite* comes from *calx*, the Latin term for burnt lime. Aragonite is less common because it is less stable than calcite under atmospheric conditions. It is usually formed at higher temperatures than calcite. In experimental conditions, aragonite is precipitated from warm carbonated water, whereas calcite will precipitate in cold carbonated water (Hurlbut and Klein, 1977). For this reason, aragonite occurs near hot springs, and is also a major component of coral reefs and the pearly layers of oyster shells. Aragonite was first discovered in Aragon, Spain, hence its name. Dolomite is reported to have been formed from limestone, $CaCO_3$, by the substitution of some of the Ca for Mg. The mineral is named in honor of Dolomieu (1750-1801), a French chemist.

Environmental and Industrial Importance of Calcite. – Calcite and dolomite are common minerals in sedimentary rocks, especially limestone rocks. Limestone is converted into marble by a process called *metamorphism*. Chalk is a pulverized form of calcium carbonate. Limestone is usually deposited at seabottoms from the remains of shells and skeletons of sea animals. Occasionally, it can also be formed by terrestrial precipitation of calcium carbonate. Water containing $CaCO_3$ in solution and dripping from cave ceilings often deposits the $CaCO_3$ in the form of *stalactites* or *stalagmites*.

Calcite and dolomite contribute toward enriching the soil with Ca and Mg. They do not contribute toward formation of clay, because their chemical composition lacks Al and Si needed for clay formation. They are the main mineral constituents of limestones, which are used as liming materials to increase the pH of acid soils, thereby making the plant nutrients available for plant growth. Chemically, $CaCO_3$ is insoluble in water, but can be made soluble by water containing CO_2. This process, called *carbonation*, can be represented by the following reaction:

$$CaCO_3 + H_2O + CO_2 \rightarrow Ca(HCO_3)_2$$

The calcium bicarbonate formed is a weak acid and is soluble in water. The CO_2 required in the reaction is supplied by the respiration of plant roots and microorganisms in abundant amounts.

Calcite as a pure mineral is very important for use in optical instruments. The *nicol prism*, cut from calcite, is an essential part of the petrographic microscope, which produces polarized light. In the form of limestone, this rock composed of calcite finds many applications in industry. It is the source for the production of *quick lime*, CaO, which is formed by heating limestone in a kiln at high temperature (900°C). The reaction is usually written as follows:

$$CaCO_3 \rightarrow CaO + CO_2^{\uparrow}$$

When mixed with water CaO forms $Ca(OH)_2$, called *slaked lime* or *hydrated lime*, while giving off a whole lot of energy. Slaked lime is used as *whiting material* or *whitewash*. Quicklime can also be mixed with sand to form mortar. Limestone is usually used for the production of cement. Small amounts of Si, Al, $MgCO_3$ and Fe oxide must be present for the production of proper quality cement. In their absence, these ingredients can be added in the form of clay or shale and mixed with the limestone before burning. When water is added to cement, hydrates of Ca-silicates and Ca-aluminates are formed, which harden upon drying. Calcium carbonate finds also application in the steel industry and as material in the *wet-limestone scrubbing* process for removing S and SO_2 from burning coal (see section 10.3.4). When polished, limestone slabs and especially marble slabs are used as building materials, pillars for houses, table tops, and other kitchen wares.

2.5 WEATHERING OF PRIMARY MINERALS

2.5.1 Abiotic Weathering of Primary Minerals

Primary minerals are affected by weathering processes. *Weather-*

ing is defined as the disintegration and alteration of rocks and minerals by physical, chemical and biological processes.

Abiotic Physical Weathering. – Abiotic physical weathering involves the breakdown of rocks and minerals into smaller particles without the assistance of biotic factors. It was also called in the past *mechanical weathering*, and today it is often designated as *disintegration*, because forces responsible for this weathering process bring about a decrease in particle size without changing the chemical composition. As such this type of weathering can perhaps be considered as destructive. Major environmental factors of importance in physical weathering are (1) temperature, (2) water, (3) ice, and (4) wind. The temperature has a pronounced influence on contraction and expansion of rocks and minerals. Rocks are aggregate of minerals, and each mineral within the rock exhibits different expansion and contraction upon being heated or cooled. When heated by the sun during the day, the expansion of a mineral exerts pressures on the neighboring, also expanding, minerals in the rock, creating thereby stresses at the mineral borders. At night the rock often cools off below the air temperature, and a simultaneous contraction due to cooling of the minerals yields a pulling effect at the boundaries. With every temperature change, occurring day and night, differential stresses are set up which may eventually produce cracks and rifts at the boundaries of the minerals, encouraging mechanical breakdown. This facilitates flow of water in the cracks, for easy access to the minerals, which then can be attacked by hydrolysis and dissolution. In winter, water in cracks freezes into ice and may split the rocks or minerals apart. The formation of ice causes water to expand in volume by about 9%, and the force created by freezing water, due to this volume expansion, is approximately 150 kg/cm^2 or 150 tons per square foot (Tan, 1998). Therefore, when water freezes in soils and plants, the expansion may also cause the soil structure to change and the plant cells to rupture. Fine mineral or rock particles may be picked up by the wind or flowing water, and during their transport collision and abrasion of one particle against another take place, causing the particles to disintegrate even further. When blown against a large rock, dust particles are capable of chipping off particles from rocks. The rounded rock remnants in arid regions of the American West are products of wind action. The force of a volcanic duststorm is known to be able to *sandblast* the paint from automobiles within minutes.

Rocks are known to exhibit a slow *heat conduction*, which causes the outer layer to be at a different temperature than the inner protected portion. By day the periphery of the rock is often higher in temperature than the inside of the rock. At night, the outside of the rock cools off more rapidly than the inside, causing the periphery to be cooler than the inside. This differential heating and cooling tends to produce *lateral stresses*, which may cause the surface layers of the rock to peel apart from the parent mass, a weathering phenomenon called *exfoliation*.

Abiotic Chemical Weathering. – Abiotic chemical weathering is the decomposition and alteration of rocks and minerals without the help of biotic factors into new materials with a completely different chemical composition, such as the secondary minerals or clay minerals. Soluble materials are also released, and a further decrease in particle size occurs. Clay produced in chemical weathering is the smallest particle in soils. Several of the elements released have an important bearing in soil formation:

1. Si, Al, and Fe contribute to clay formation
2. Fe and Mn are essential in reduction and oxidation processes
3. K and Na are dispersing agents in soils, and
4. Ca and Mg have strong flocculating powers, hence contribute to formation and stabilization of soil structure.

Therefore, in contrast to physical weathering, chemical weathering can be considered a constructive process, because of the production of constituents important for building up the soil. The chemical reactions in chemical weathering are hydrolysis, hydration-dehydration, oxidation-reduction, carbonation and dissolution (Brady, 1990):

Hydrolysis. – Hydrolysis is the attack of the mineral by water (hydro = water, and lysis from analysis), and is a major decomposition process of feldspars and ferromagnesians. The weathering processes can be represented by the following reactions:

$$KAlSi_3O_8 \quad + \quad H^+ \quad \rightarrow \quad HAlSi_3O_8 \quad + \quad K^+$$
orthoclase $\qquad\qquad\qquad$ clay

$$NaAlSi_3O_8 \quad + \quad 2H^+ \quad \rightarrow \quad HAlSi_3O_8 \quad + \quad Na^+$$
albite $\qquad\qquad\qquad$ clay

$$3MgFeSiO_4 \quad + \quad 2H_2O \quad \rightarrow \quad H_4Mg_3Si_2O_9 \quad + \quad SiO_2 \quad + \quad 3FeO$$
olivine $\qquad\qquad\qquad$ serpentine

The hydrolysis of feldspar above indicates that a precursor of clay is formed, which is a new material, while at the same time soluble components are released (K^+ and Na^+ ions). Weathering of olivine produces clay, though its composition lacks Al. The mineral contains Fe and Si, which yield SiO_2 and FeO. Because they are formed in a colloidal state, both SiO_2 and FeO are clays by definition.

The H^+ ions needed in the reactions above are provided by either the dissociation of water molecules or the ionization of H_2CO_3. The latter is formed by CO_2, produced in the respiration process of microorganisms and plant roots, reacting with soil water. In this case, such a hydrolysis reaction is not purely an abiotic process, because environmental or biotic processes are involved for the production of carbonic acid. A more complicated reaction of hydrolysis of orthoclase is as follows:

$$2KAlSi_3O_8 \quad + \quad 3H_2O \quad \rightarrow \quad H_4Al_2Si_2O_9 \quad + \quad KOH \quad + 4SiO_2$$
orthoclase $\qquad\qquad\qquad$ kaolinite

The kaolinite formed is mostly in the colloidal state and is important to soil fertility. Because it is electronegatively charged, kaolinite can adsorb and exchange plant nutrients. However, kaolinite itself is composed of Al, Si, H, and O, and consequently does not supply any of the essential plant nutrients.

Hydration and Dehydration. – Hydration is the rigid attachment of water to the crystal structure of the mineral, whereas dehydration is loss of water from the mineral. As is the case with hydrolysis, hydration and dehydration are interdependent on environmental factors. Both hydrolysis and hydration require abundant amounts of water, which is possible only in humid regions. Dehydration needs in addition a warm and dry condition. These reactions are major decomposition processes of micas and iron minerals. As micas become hydrated, water moves between the plate-like layers, and in doing so swelling occurs, which tends to expand the crystal. This process makes the mica crystals soft, which facilitates other decomposition reactions. Hydration of iron minerals increases their solubility and frequently affects their colors. A good example of hydration of iron minerals is the formation of limonite from hematite:

$$2Fe_2O_3 \quad + \quad H_2O \quad \rightarrow \quad 2Fe_2O_3.3H_2O$$
hematite (red) \qquad\qquad limonite (yellow)

During wet periods hydration takes place, whereas during dry periods dehydration prevails. This is one reason for the changing colors in the soils of the southern region of the United States, where they look bright red in summer and yellowish during winter time.

Oxidation and Reduction. – Oxidation and reduction reactions are also called *redox* reactions. Oxidation refers to a loss of electrons or negative charges, and in the older theory it is considered the addition of oxygens. In contrast, reduction is the gain of electrons. Oxidation and reduction in soils happen in particular with iron compounds. The process can be represented by the following reactions:

$$Fe^{3+} \quad + \quad e^- \quad \rightleftarrows \quad Fe^{2+}$$

$$4FeO \quad + \quad O_2 \quad \rightleftarrows \quad 2Fe_2O_3$$
ferrous oxide \qquad\qquad hematite

If the oxidation to the ferric ion takes place while the iron is still part of the crystal lattice, other ionic adjustment must be made since the Fe(III) ion is smaller in size than the Fe(II) ion, and because a trivalent ion is replacing a divalent ion. These adjustments result in a less stable crystal. The reactions are again interdependent on environmental factors. Oxidation reactions are prevalent in well-aerated soils, whereas reduction processes are strongly developed in wet conditions, where aeration is inhibited. Bacteria living in reduced environments may obtain their oxygen from ferric oxide. As a result, the ferric oxide is reduced into ferrous oxide. In soils where reduction processes have been active, gray colors usually predominate in contrast to the red colors of soils where oxidation is active. Reduction tends, especially with iron and manganese compounds, to increase solubility and mobility of the ions.

Carbonation. – Carbonation is usually defined as the process involving the combination of carbonic acid with bases to form carbonates. It is an important weathering process of limestone and dolomitic limestone. According to chemistry, limestone is insoluble in pure water, but can be made soluble when converted into a bicarbonate. The reaction, requiring the participation of CO_2, is as follows:

$$CaCO_3 \quad + \quad H_2O + CO_2 \quad \rightarrow \quad Ca(HCO_3)_2$$
limestone (calcite)   calcium bicarbonate

The CO_2 required by the reaction is supplied by the respiration process of microorganisms and plant roots. Calcium bicarbonate is soluble in water. This is a weathering process by which limestone is decomposed in nature. When lime is applied to acid soils, the acidity in acid soils may already be sufficient to dissolve part of the ground limestone ($CaCO_3$). By nature limestone does not exist in acid soils, but the carbonation process as shown in the reaction above may be in effect and assist in making the applied liming material more soluble.

2.5.2 Biological Weathering

Biological processes have been known for a long time to play a major role in rock and mineral weathering, and in soil genesis, but only with recent advances in soil organics and technology have these environmental processes gained prominence. The main problem was that the effect of biotic processes is often overshadowed by that of the abiotic weathering processes. Microscopic studies are frequently required to determine the weathering effects especially occurring at the interfaces between live organisms and soil minerals. At first considered very difficult and tedious, these studies today are becoming less complicated with the availability of electron microscopy.

Evidence has been reported and accumulated that live organisms and dead soil organic matter have a significant effect on weathering. The information currently available indicates that the degree of weathering induced by soil organisms and soil organic compounds may be more important than that brought about by abiotic chemical reactions alone. To recognize the increasing importance of organisms and organic matter in weathering and soil formation two types of biological weathering processes are distinguished today: (1) *biological weathering*, and (2) *biochemical weathering* (Robert and Berthelin, 1986). Biochemical weathering is related to reactions with chemical compounds excreted or secreted by organisms, and to interactions with the decomposition products of organic matter. Rock and mineral dissolution through chelation by organic acids, such as humic acids, is presumed today to be more effective than abiotic hydrolysis (Tan, 1986; Birkeland, 1974). Lately, the production of extracellular polysaccharides (EPS) by a variety of microorganisms has received much attention, because of its possible involvement in the formation of organomineral complexes in soils (Chenu, 1995). Biological weathering, on the other hand, is assisted by the presence of live organisms, and includes biochemical weathering and other processes related to the presence and activity of live organisms. Often a sharp distinction cannot be made between biological and biochemical weathering. Physical disintegration of rocks by growing roots requires the presence of live organisms, but rock and mineral dissolution by chelation can take place in the absence and/or presence of live organisms.

Biological Weathering. – Biological weathering includes physical and chemical weathering of rocks and minerals under the participation of live organisms. It has been known for a long time that live roots growing into cracks of rocks can exert considerable pressure and thereby contribute to physical weathering. Roots 10 cm in diameter and 1 m in length have been reported to raise a mass having a weight of 30,000 to 50,000 kg. When the roots die and decay, they leave bigger channels favoring easy penetration of water and permitting surface material to fill in. However, less is known on swelling gelatinous or thallial tissue of microorganisms exerting appreciable pressure on the substratum. Hyphae, penetrating cracks deep into the rock, may separate or detach the minerals from rock surfaces. Tiny cracks in crystals, only observable by microscopy, can be penetrated by fungal hyphae causing the mineral to disintegrate into microcrystals. Surface adhesion and hyphal penetration are assumed to induce microdivision of minerals (Robert and Berthelin, 1986). Rock and mineral dissolution by live organisms is believed to be of general occurrence and represents phenomena of high local intensity. Bacteria, algae, fungal and lichen hyphae are virtually glued to mineral surfaces, because of the presence of a polymer interface, composed of polysaccharide gels, on their bodies. These polymers, recently called *extracellular polysaccharides* (EPS), occur as a capsule or as *mucigel* (slime) on the cell surfaces, and constitute the interface between the cells and the soil solution, or between the cells and soil minerals. Beside providing protection against microbial attack, EPS is believed to also interact with surrounding soil constituents (Chenu, 1995), facilitating dissolution of minerals and rocks. Under the encrusting hyphae or thallus, dissolution patterns were identified on mica, plagioclase, and quartz surfaces. The dissolution of silicate and phosphate minerals by microorganisms has received considerable research attention. *Pisolithus tinctorius*, an ectomycorrhizal root fungus, is known for its potential in the dissolution of insoluble phosphate compounds. When inoculated into Caribbean pine, *Pinus caribea*, the fungus increases P-uptake by the pine, grown in soils low in P content. *Glomus* sp., an endomycorrhizal fungus, is reported to increase K-uptake by soybean plants, grown in a biotite medium. The fungus is believed to cause increased weathering of biotite (Mojallali and Weed, 1978).

The weathering of soil minerals by soil organisms is apparently not limited to physical disintegration and to dissolution reactions only. Evidence has been presented that the root system of many plants contributes toward transformation of clay minerals. Biotite was reported to be converted into vermiculite (Mortland et al., 1956) and kaolinite (Spyridakis et al., 1967) by a process called *neoformation* induced by plant roots. Calcite microcrystals have been detected at the surface of hyphae and roots growing on carbonate rocks. These crystals are even detected as fine needles inside the hyphae and the epidermal root cells. After the organisms die, the calcite crystals may give rise to formation of cytomorphic sands (Robert and Berthelin, 1986).

Biochemical Weathering. – As defined earlier, biochemical weathering involves the interaction of rocks and minerals with organic compounds produced during the decomposition of soil organic matter, or to reactions of the minerals with organic substances secreted by live organisms. A variety of organic compounds are produced by soil organisms, and some of them are released constantly into the soil as root exudates or microbial secretions. They are generally low molecular weight compounds, and many of them are organic acids. However, by decomposition of plant and animal tissue, more complex organic substances and acids are released or synthesized. The organic compounds include carbohydrates, proteins, lignin, etc., whereas the organic acids may range from the nonhumified simple aliphatic, aromatic, and heterocyclic acids to the very complex humified organic acids, such as humic and fulvic acids. Humic acid is a new soil material, synthesized during the decomposition of organic matter. In this case, it is not customary to use the term *weathering* of organic matter. All these compounds make up the humus fraction of soils. Formation of humus is dictated by environmental factors. Many people believe that large amounts of humus can only be formed in a warm humid climate where plant growth is abundant. However, maximum amounts of humus are detected in mollisols formed in semihumid climates under tall grass vegetation. The annual turnover of dead grass makes large accumulation of humus possible. The next highest organic matter content is found in soils developed in humid regions under forest vegetation. The humus content in these soils are less than that in mollisols, because large amounts of organic matter are not recycled into the soils, but are retained in the big tree trunks.

Without humus, the surface of the earth is just a mixture of weathered rocks and minerals. Similar to the moon material, it is just plain *regolith*. When live organisms have set a foothold and humus is incorporated in the weathered mass of rocks and minerals, this inorganic and organic mixture is by definition a soil.

In contrast to biological weathering, biochemical weathering is mostly chemical in nature and commonly involves dissolution of rocks and minerals and transformation of clay minerals. It also contributes to mobilization of the released elements, a process of importance in soil profile development. The effectiveness of these organic acids in mineral dissolution depends on a number of factors, e.g., concentration and chemical reactivity. By sheer concentration, humic and fulvic acids, forming the bulk of soil humus, are more important than the other organic acids. Many of the latter are perhaps as effective as humic acids, but due to their presence in minor concentrations, their effect is often overshadowed by that of humic acids (Tan, 1986). In laboratory analysis, oxalic acids and salicylic acids exhibit the same degree of dissolution of silicate minerals as humic acids. On the basis of chemical reactivity, the organic acids can be distinguished into two groups: (1) organic acids in which the acidic characteristics are attributed only to the presence of carboxyl, -COOH, groups, and (2) organic acids in which the acidic characteristics are attributed to carboxyl and phenolic-OH groups. Examples of the first category of acids are formic, acetic, and oxalic acids. Although they may exhibit a complexing capacity, their effect on mineral weathering is generally through the acidic (H^+ ions) effect. Many display dissociation constants comparable to those of strong acids. The second group of organic acids includes humic and fulvic acids and a number of the more complex nonhumified organic acids. By virtue of the presence of carboxyl and phenolic-OH groups in their molecules, they have the distinct advantage over the simple organic acids of being able to exert an acidic and an interaction effect. The interactions can be in the form of (1) electrostatic attraction, (2) complex reaction or chelation, and (3) coadsorption or water and metal bridging (Tan, 1998). Recent information indicates that the acidity, coadsorption, and chelating capacity bring about the dissolution of elements, hence the degradation of rocks and minerals (Schnitzer and Kodama, 1976; Tan, 1976; 1978; 1980). By prying loose structural Al, Si, and Fe from micas, feldspars, and kaolinite, or any other soil mineral, mineral decomposition is accelerated. The organic chelating agent perhaps reacts with an exposed cation, followed by movement of the complex compound or chelate into solution. The effect of chelation on mineral weathering is believed to exceed that brought about by abiotic hydrolysis (Birkeland, 1974). As a chelate, Al and other metals are rendered soluble over a pH range in which these metals are normally insoluble, hence chelation increases their mobility. The latter has an important bearing in soil formation and nutrient supply to plant roots. Mobilization of Si, Al, Fe, and clay in the form of their organo-chelates is currently the basis for the theory of formation of argillic and spodic horizons (Tan, 1998). Not only will such a mobilization and precipitation of metal chelates result in horizon differentiation, giving rise to different kinds of soil profiles, but depending on stability and pH, such chelates provide the carrier mechanisms for replenishing depleted nutrients at root surfaces.

Lately, indications have been presented that organic acids can enhance or inhibit formation of clay minerals. Formation of imogolite and allophane in laboratory conditions appears to be inhibited by the presence of citric, tannic, malic, tartaric, salicylic, humic and fulvic acids (Huang and Violante, 1986). Hydrolysis and polymerization of aluminum hydroxide compounds are also harmfully affected by citric acid and humic acid. The organic acids occupy coordination sites of monomeric and dimeric aluminum hydroxides, preventing in this way further polymerization and/or hydrolysis of the Al monomers and dimers (Tan, 1998). The latter has a perturbing effect on the crystallization of Al and Fe precipitates. Another explanation is that chelation of Al by humic acids is assumed to inhibit formation of allophane and imogolite in andosols, because the Al necessary for formation of these clay minerals is taken out of the soil solution. Such an interaction is dubbed as an *anti-allophanic* process in Japan (Shoji et al., 1993). Other scientists suggest that complex reactions of organics with Al, Fe, and Si compounds and their subsequent incorporation into the structural network of the mineral colloids promote formation of short-range-order clay minerals (Huang, 1995).

2.5.3 Environmental Factors in Weathering

The weathering of these primary minerals to form secondary minerals is quite complex and depends upon the climate, biotic factors, and a number of other parameters. Climate is perhaps the most important factor in mineral weathering. In arid climates, physical weathering predominates. Chemical alteration is at a minimum due to lack of water, and not much clay is formed. On the other hand, in a warm humid climate, both moisture content and temperature are favorable for chemical weathering, and large amounts of clay are produced. The biotic factors of importance in the weathering of primary minerals include aerobic decomposition of organic matter, respiration of plant roots and microorganisms, biological nitrification, and the presence of humic acids and other organic acids. Both aerobic decomposition of organic matter and respiration of plant roots produce CO_2, which increases the dissolution capacity of soil water. Under warm humid conditions favoring abundant plant growth, respiration is perhaps more important than aerobic decomposition, although the latter does contribute to the production of CO_2 in soil air. The process of respiration is represented by the reaction below:

$$C_6H_{12}O_6 + 6O_2 \rightarrow 6CO_2 + 6H_2O + \text{energy} \qquad (2.1)$$

In this reaction, carbohydrate is oxidized by microorganisms to obtain the energy needed for performing metabolism and growth. The CO_2 is a by-product of the oxidation reaction and must be discarded from the plant body into soil air. This reaction increases the CO_2 content in soil air to ten to several hundred times more than is found in atmospheric air. The CO_2 content in the earth atmosphere is approximately 0.03%, or 300 ppm. The CO_2 released in the soil dissolves in soil water to form carbonic acid, H_2CO_3. The latter, a weak acid, is capable of dissociating its proton, which increases the dissolution power of soil water. The reactions of formation and dissociation of H_2CO_3 are represented as follows:

$$CO_2 + H_2O \rightarrow H_2CO_3 \qquad (2.2)$$

$$H_2CO_3 \quad \rightarrow \quad H^+ + HCO_3^- \hspace{5cm} (2.3)$$

Nitrification also increases the dissolution capacity of soil water by yielding protons during the reaction. Nitrification, the conversion of ammonium, NH_4^+, into nitrate, NO_3^-, takes place in two steps. In the first step, ammonium is converted into nitrite by *Nitrosomonas* bacteria. In the second step, nitrite is transformed into nitrate by *Nitrobacter* bacteria. As indicated by the reactions below, protons are produced during the first step of the process:

$$2NH_4^+ + 3O_2 \quad \rightarrow \quad 2NO_2^- + 2H_2O + 4H^+ + \text{energy} \hspace{2cm} (2.4)$$

$$2NO_2^- + O_2 \quad \rightarrow \quad 2NO_3^- + \text{energy} \hspace{3.5cm} (2.5)$$

Humic acids and a variety of organic acids, formed during the decomposition of soil organic matter, also serve as sources of protons that enhance the dissolution power of soil water. As discussed earlier, hydrolysis can only occur in the presence of sufficient amounts of protons. When abundant plant growth results in the formation of large amounts of plant residue, large amounts of humic acids and other organic and inorganic acids are produced. These acids not only release protons, they also enhance chelation and complex reactions, important forces in the physical and chemical weathering of soil minerals as indicated above.

In addition to the aforementioned climatic and biotic factors, mineral weathering also depends on three general factors: (1) the physical properties of the minerals, (2) their chemical characteristics and crystal chemistry, and (3) the saturation of the environment.

2.5.4 Physical Properties of the Minerals

Crystal size, shape, hardness and crystal imperfection are some of the physical characteristics affecting mineral weathering (Ollier, 1975). Large minerals are more resistant than small minerals, due to the greater surface areas of smaller particles. Larger surface areas

Table 2.4   Relative Stability of Soil Minerals on the Basis of Hardness

Mineral	Scale of hardness (ease of scratching)	Mohs' scale
Talc	Easy to scratch with fingernail (very soft)	1
Gypsum	Just scratches with fingernail (soft)	2
Calcite	Scratched by copper coin, not by fingernail (slightly hard)	3
Fluorite	Easy to scratch by glass, not by copper coin (moderately hard)	4
Apatite	Just scratched by glass (hard)	5
Orthoclase	Mineral scratches glass easily (hard)	6
Quartz	Difficult to scratch by glass; mineral scratches glass very easily (very hard)	7
Topaz	Very difficult to scratch by glass (very hard)	8
Corundum	Very difficult to scratch by glass; mineral cuts glass (very hard)	9
Diamond	Extremely difficult to scratch by glass; mineral cuts glass (extremely hard)	10

Source: Hunt (1972); Tan (1981; 1986)

provide more sites for contact with water and other weathering agents. Therefore, the smaller the particles, the faster the rate of weathering.

Hardness depends upon the bond strength between the elements in the structure of the mineral. Bond strength will be discussed in the next section on the effect of chemical properties on weathering. The degree of hardness of minerals is frequently rated on the *Mohs' scale* developed around 1839 by Friedrich Mohs, a mineralogist. The Mohs' scale ranges from 1 to 10, with 1 indicating the softest mineral, such as talc, and 10 the hardest mineral, such as diamond (Table 2.4). Minerals with a hardness > 7, such as topaz, corundum and diamond, are uncommon in soils. Therefore, the degree of hardness of most soil minerals ranges from 1 to 7. Quartz, with a hardness of 7, is generally the hardest mineral in soils and is very resistant to weathering.

Crystal shape is another important physical property that affects mineral weathering. Platy crystals with perfect geometrical shapes are more resistant than imperfect crystals. Loose bonds are perhaps the cause of crystal imperfections. Porosity and cleavage in crystals are additional physical factors in weathering. Rocks and minerals with high porosity tend to weather more rapidly than do those that are more dense and solid. Both porosity and cleavage allow solutions to penetrate deeply into the crystal structure.

2.5.5 Chemical Characteristics and Crystal Chemistry

As discussed earlier, the fundamental units of silicate minerals are silica tetrahedrons, which can be joined together in several ways. As cited earlier (Figure 2.2), the silica tetrahedrons in quartz are joined together by a mutual sharing of tetrahedral oxygen atoms. This sharing produces a Si-O-Si linkage, called the *siloxane bond*, which is considered the strongest chemical bond in nature (Sticher and Bach, 1966; Tan, 1992). This is the reason why quartz is the hardest mineral in soils. In other minerals, single or several units of silica tetrahedra can be linked together by cations. These cations are nonframework ions, and form weak spots in the mineral structure. For example, in ferromagnesians, double chains of silica tetrahedra are linked together by Ca and Mg atoms (Figure 2.2). These cations can be made soluble by water, resulting in a collapse of the mineral structure. On the basis of a progressive increase of siloxane bonds, Keller (1954) ranked the stability of silicate minerals as follows:

neso- < soro- < ino- < phyllo- < tectosilicates

The implication is then that tectosilicates are more resistant to weathering than phyllosilicates. Tectosilicates include feldspar and quartz, whereas phyllosilicates are mainly silicate clays. The ranking above is perhaps valid for stability assessments, if quartz is compared with silicate clays. However, silicate clays, being the weathering products of feldspar, are expected to be more resistant than feldspar minerals. A ranking using specific mineral species rather than group names is perhaps less subject to arguments. Such a ranking is presented by Birkeland (1974):

olivine < pyroxene < hornblende < biotite < quartz

Not only does the above ranking indicate the presence of an increase in oxygen sharing in the direction of olivine → quartz, but it also agrees with the order of minerals listed in the weathering sequences presented in many standard mineralogy textbooks. For a more detailed treatise on stability of minerals and bond strength, reference is made to Tan (1982, 1992, 1998), Sticher and Bach (1966), and Keller (1954).

2.5.6 Saturation of the Environment

The removal of weathering products by leaching, precipitation, and absorption by plant roots are critical to the process of mineral decomposition. Mineral weathering may stop when the soil solution bathing the minerals becomes saturated as decomposition progresses. For example, dissolution of calcite releases Ca in the soil solution. When the released Ca is not removed from the soil solution, the latter is soon saturated with Ca. The resulting state of equilibrium prevents further weathering of calcite. Stability regions, based on supersaturation, saturation, and under-saturation of soil solutions, have been developed for major soil minerals (Tan, 1993, 1998). Some of the stability and phase relation diagrams are very complex (Garrels and Christ, 1965).

Because of the aforementioned factors, some minerals weather rapidly, whereas others weather slowly. Quartz, a hard mineral, is very resistant to weathering and may accumulate in the soil's sand fraction. Other minerals, such as the ferromagnesians and plagioclase, weather more rapidly. A large number of weathering sequences have been formulated, and one is given below as an example:

most resistant   →   least resistant

quartz > muscovite > feldspars, biotite > plagioclase, ferromagnesians

2.6 SECONDARY MINERALS

As indicated in preceding sections, secondary minerals are formed by the weathering of primary minerals. They are microscopically small (<2 µm) and make up the clay fractions of soils, or the soil inorganic plasma. The primary minerals are broken down and altered into secondary minerals under the influence of climate and biological activity (Ollier, 1975; Birkeland, 1974). The original (primary) minerals can be completely or partially dissolved and transformed into new materials, the clay minerals. The resulting products are generally in equilibrium with the newly imposed physico-chemical environment. As discussed in abiotic weathering one of the chemical reactions important in the breakdown of feldspars was hydrolysis. It was defined earlier as the reaction of the mineral with water (hydro is the Greek word for water), and the process in a generalized or simplified form was written as:

$$KAlSi_3O_8 \; + \; H_2O \; \rightarrow \; HAlSi_3O_8 \; + \; KOH \qquad\qquad (2.6)$$
$$\text{clay} \qquad\qquad \text{soluble material}$$

As shown above, orthoclase is altered by hydrolysis into clay, while at the same time a soluble compound is produced. The formation of clay minerals is, however, very complex and will be affected by many factors, e.g., climatic, and lithologic factors. In temperate regions, hydrolysis of feldspars tends to form more illitic than kaolinitic types of clays. This reaction can be written as:

$$3KAlSi_3O_8 + 2H^+ + 12H_2O \; \rightarrow \; KAl_3Si_3O_{10}(OH)_2 + 6H_4SiO_4 + 2K^+$$
$$\text{illite} \qquad\qquad (2.7)$$

However, in warm humid regions, such as the subtropics and tropical regions, hydrolysis of orthoclase, producing kaolinitic types of clays, is more prevalent:

$$2KAlSi_3O_8 + 2H^+ + 9H_2 \rightarrow H_4Al_2Si_2O_9 + 4H_4SiO_4 + 2K^+ \qquad (2.8)$$
$$\text{kaolinite}$$

Lithologic differences may also produce different types of clays. Duchaufour (1976) postulated that the types of clays produced by weathering of basic rocks differed from those produced by weathering of acid rocks. According to his hypothesis:

Basic rocks → ferromagnesians, feldspars → smectite → kaolinite

Acid rocks → Quartz, feldspars → kaolinite

The minerals that make up the basic rocks are composed of large amounts of ferromagnesians which are rich in bases. These minerals release sufficiently high amounts of bases upon weathering to create the basic environment required for the formation of smectite. In contrast, acid rocks contains minerals that produce relatively low amounts of bases upon weathering. Quartz, a major component of acid rocks, is composed mainly of SiO_2, and is resistant to weathering. On the other hand, the small amounts of K released by feldspar are hardly sufficient to create the basic environment needed to form smectite. Consequently, kaolinite is formed in this relatively more acidic environment.

2.6.1 Major Types of Secondary Minerals

The major secondary minerals present in soils are listed in Table 2.5 (Tan, 1992; Mackenzie, 1975), and each of the minerals will be discussed in more detail below. The best known in the list are the crystalline silicate clays, but with the recent progress in clay mineralogy, many other types of clay minerals have been detected in soils, e.g., amorphous clays, sesquioxide clays, and silica minerals. The amorphous clays are noncrystalline and include a wide variety of

Table 2.5   Major Clay Minerals in Soils

Major minerals	Layer type	CEC cmol(+)/kg	Major occurrence
Amorphous/paracrystalline clays			
Allophane		35	Volcanic ash soils, spodosols
Imogolite		35	Volcanic ash soils, spodosols
Crystalline silicate clays			
Kaolinite	1:1	08	Ultisols, oxisols, alfisols
Halloysite	1:1	10	
Smectite	2:1	70	Vertisols, mollisols, alfisols
Illites	2:1	30	Mollisols, alfisols, spodosols, aridisols
Vermiculite	2:1	100	Accessory mineral in many soils
Chlorite	2:2	000	Accessory mineral in many soils
Sesquioxide clays			
Goethite, α-FeOOH		03	Oxisols, ultisols
Hematite, α-Fe_2O_3		03	Oxisols, ultisols
Gibbsite, $Al(OH)_3$		03	Oxisols, ultisols
Silica minerals			
Quartz, $n(SiO_2)$		–	Accessory mineral in many soils
Crystobalite, $n(SiO_2)$		–	Accessory mineral in volcanic ash soils

materials, such as silica gel, sesquioxides, silicates and phosphates. They are called amorphous because they are amorphous to x-ray diffraction analysis, meaning that they exhibit featureless x-ray diffraction patterns. Many object to the term amorphous, because it may be that it is the method of analysis that is inadequate to detect crystallinity in the so-called amorphous clays. These clays have often been detected in very fine microcrystal forms, hence several soil scientists suggest naming them *noncrystalline* instead of amorphous minerals, but lately the name *short-range-order* (SRO) mineral appears to be favored. The sesquioxide and silica minerals listed in Table 2.5 are mostly crystalline.

Allophane. – Allophane is noncrystalline, and several definitions for this mineral have been given (Tan, 1992, 1998). It was classified as amorphous clay because it is amorphous to x-ray diffraction analysis, whereas today it is considered a short-range-order mineral (Tan, 1998; Huang, 1995). The name allophane was first introduced by Stromeyer and Hausmann in 1861 for a hydrous alumino silicate in nature. Since then the name allophane has found general acceptance for a wide variety of clay minerals amorphous to x-ray diffraction analysis. Its chemical composition is postulated to be between:

$$SiO_2.Al_2O_3.2H_2O \quad - \quad Al_2O_3.2SiO_2.H_2O$$

Therefore, depending on its Si and Al content, an isomorphous series of different types of allophane is recognized in Japan between two end members distinguished by atomic ratios of Al:Si = 1:1 and 2:1, respectively (Shoji et al., 1993). Allophane, characterized by an Al:Si atomic ratio of 1:1, is called *Si-rich allophane*, whereas that with an Al:Si atomic ratio of 2:1 is called *Al-rich allophane*. Of the two end members, Al-rich allophane is the most common variety in andosols, whereas the Si-rich variety is less known. Because the atomic ratio of Al-rich allophane is similar to that of imogolite, another paracrystalline clay, it is believed that this variety is similar to protoimogolite (Shoji et al, 1993). Other allophane varieties with atomic ratios <1 and >2 to 4 have been detected only under laboratory conditions. By selective differential dissolution, using dithionite-citrate and Na_2CO_3

solution, Wada and Greenland (1970) succeeded in separating several types of noncrystalline alumino-silicates, but these can only be considered *allophane-like* constituents. They have not been detected in soils, hence the possibility exists that they are artifacts produced during the dissolution analysis. In the older literature three types of allophane were recognized, e.g., allophane B, AB, and A (Birrell and Fieldes, 1952). Allophane B is believed to be formed first from the weathering of volcanic ash, which is converted by desilicification into allophane A through an intermediate species AB. Many questions have been raised about the occurrence of allophane B, and the opinion exists in Japan that it is just plain opaline silica (Shoji et al., 1993). However, indications are present that allophane B is more likely related to Si-rich allophane.

Allophane is characterized by a very high specific surface area, which is even higher than that of smectite. The mineral also exhibits a high variable charge, and is amphoteric in nature. The cation exchange capacity is around 20-50 cmol/kg, and its anion exchange capacity is between 5-30 cmol/kg. Allophane is reported to have a high phosphate fixation capacity. The ZPC (zero point charge) value of allophane is believed to change with type of allophane, especially with an increase in its Si content. Allophane with an atomic ratio of Al:Si = 1:2 has a ZPC = 5.5, whereas allophane with an atomic ratio of Al:Si = 1:5 exhibits a ZPC = 6.5. The ZPC increases to 6.8 in allophane with an atomic ratio of Al:Si = 1:8 (Clark and McBride, 1984). The ZPC values presented above are close to or even larger than the pH values of andosols in nature. Andosols in Japan exhibit pH values between 5.0 - 5.7 (Shoji et al., 1993), whereas the pH of andosols in Iceland is between 5.6 - 6.7 (Arnalds et al., 1995). In Indonesia the pH in andosols ranges from 4.0 to 6.0. In view of these soil pH values, allophane present in nature should either have no charge at all or should be positively charged. However, andosols are characterized by high CEC values, indicating the presence of high negative charges in allophane. Such a contradiction is perhaps caused by the reported high ZPC values, which can not be applied to allophane present in natural soil condition. In soils, allophane and other amorphous Al hydrates are present in close association with silica and humus, compounds with very low ZPC values. Hence, it is expected that the

ZPC of allophane in soils is going to decrease and is adjusted to values close to the ZPC values of silica and humus (Greenland and Mott, 1978).

Environmental Importance of Allophane. – At first thought to be found only in volcanic ash soils, at present allophane has been detected also in the Bh horizons of spodosols of nonvolcanic origin. Nevertheless, allophane is considered a mineral characteristic for andosols. The presence of allophane gives to these soils the unique properties discussed earlier in Section 1.3.11 on andisols. The mineral interacts with humic matter preserving in this way decomposition of the organic material. The soils are usually characterized by thick black A horizons, rich in organic matter, hence the name andosols or ando soils is given, from the Japanese words *an* and *do* for black and soil. As discussed earlier the US Soil Taxonomy used the name andisol, derived from the term andi. However, the term *andi* is not known in the Japanese language (Shoji et al, 1993), and is perhaps just a nonsense syllabus, similar to alfi and enti. The capacity of andosols to retain exceptionally high amounts of water, while retaining its *biologically dry* nature, is believed to be caused by the structure of allophane in the form of hollow spheres with a diameter of 3 to 4 nm (Wada, 1977). The walls of these hollow spheres look like a sieve with very fine holes, each of a diameter of 0.3 – 0.5 nm. Since a water molecule has a diameter of 0.3 nm (Tan, 1998), water can flow through the holes and accumulate in the hollow spheres of allophane.

Imogolite. – Imogolite was found for the first time in 1962 in weathered volcanic ash or pumice beds in Japan, called *imogo* (Yoshinaga and Aomine, 1962). Since then it has been detected in many other volcanic ash soils, and in the Bh horizons of spodosols, derived from nonvolcanic materials. It has also been synthesized under laboratory conditions (Inoue and Huang, 1984). Frequently, imogolite is closely associated with the presence of allophane in soils. The intermediate phase between allophane and imogolite, or imogolite-like allophane, is called *protoimogolite*. In many respects, the chemical characteristics of imogolite are similar to those of allophane. Its chemical composition is assumed to be:

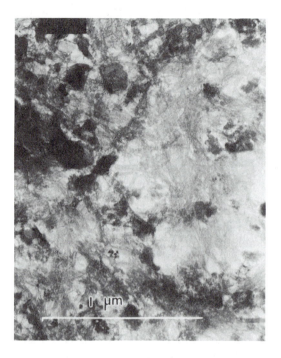

Figure 2.3   Transmission electron micrograph of imogolite in the clay fraction of an andosol derived from volcanic ash in Indonesia.

$SiO_2.Al_2O_3.2.5H_2O$

However, in contrast to allophane, the crystal structure of imogolite is better defined. Electron microscopy shows imogolite to be composed of very fine cylinders (diam. $18.3 - 20.2$ Å) giving it hairlike or spaghetti-like structures (Figure 2.3), hence this mineral is classified as paracrystalline clay. The outer surface of the imogolite cylinders is composed of 10 to 12 unit cells of gibbsit. Wada (1977) believed that high resolution electron microscopy showed indication that allophane exhibits cylindrical structures similar to those of imogolite. However, sheet structures similar to those of kaolinite have also been suggested for allophane by other soil scientists.

Environmental Importance of Imogolite. – Allophane is proposed to be formed by coprecipitation of amorphous alumina and silica hydrates, produced by the weathering of volcanic ash. The mineral tends to be formed especially in the B horizons of andosols, where a high concentration of free Al ions is present. In the A horizons, most of the free Al ions are chelated by humic acids, and consequently are prevented from participating in the formation of allophane. Such a chelation reaction was previously called an anti-allophanic process. Under excellent drainage conditions, desilicification of allophane is expected to produce imogolite, and Japanese scientists consider it an intermediate mineral in the weathering sequence (Shoji et al., 1993). When environmental conditions allow desilicification to proceed further, imogolite will be converted into halloysite and the latter eventually into kaolinite. The relationship between the alteration of allophane into imogolite and/or other clay minerals can be illustrated by the following weathering sequence:

Volcanic glass → amorphous Al and Si hydrates → allophane →
imogolite → halloysite → kaolinite → gibbsite

The hypothesis above is in contrast to the theory suggested by a number of US scientists, who believe kaolinite decomposes into allophane.

Kaolinite. – Kaolinite, a phyllosilicate mineral, is called *kandite* in the British literature. The name is derived from the term *kaolin*, presumably a clay deposit high in kaolinite in the hills or ridges, called *kauling* in Jauchu Fa, China. The name kaolin, or China clay, is still used today. The mineral is characterized by the following general chemical composition per unit cell:

$2SiO_2.Al_2O_3.2H_2O$

Structurally it is a 1:1 lattice type of mineral. As illustrated in Figure 2.4, each crystal unit of kaolinite is composed of one octahedral sheet

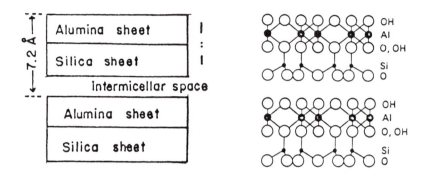

| : | layer type (e.g. kaolinite)

Figure 2.4   The structure of kaolinite, composed of alumina octahedron and silica tetrahedron sheets, stacked into a two–layer crystal unit. This is the reason why kaolinite is grouped as 1:1 layer lattice type of clay.

stacked above one tetrahedron sheet. The two crystal units, making up one kaolinite particle, are held together by hydrogen bonds, and the intermicellar space has, therefore, a fixed dimension. Since the crystal units are rigidly held, resulting in very little expansion and contraction of the intermicellar space, kaolinite displays low plasticity. Little isomorphous substitution occurs in the crystal, and, consequently, its permanent charge is very small. However, kaolinite has a variable or pH dependent negative charge, attributed to the dissociation of protons from exposed OH groups. As a result, its cation exchange capacity may vary with pH. Usually it is in the range of 1 to 10 cmol(+)/kg (= me/100g). Members of the kaolinite group are kaolinite, dickite, and nackrite. Dickite and nackrite are chemically similar to kaolinite, but exhibit a different t-o stacking from that in kaolinite.

Environmental and Industrial Importance of kaolinite. – Kaolinite is an important mineral in the clay fractions of ultisols and oxisols. As discussed previously, it has been formed from the weathering of feldspars. Frequently, it is found mixed with feldspar in rocks that are undergoing decomposition. When rock decomposition is carried out to completion, thick deposits of kaolinite are produced. These deposits are valuable sources for mining kaolinite for use in industry. Red brick, sewer pipes and drain tiles are produced from clay containing kaolinite. Because of its low expansion and contraction capacity, kaolinite is very important in the ceramic industry. Fine porcelain is made from high-grade kaolin or kaolinite with an admixture of smectite or *ball clays*. Clays that are plastic and sticky, but turn white upon burning, are called ball clays. They are composed of three major minerals, e.g., kaolinite, hydrous mica, and quartz (Grimshaw, 1971). Ball clays are added to kaolin to provide strength to the moulded shape while maintaining the white color upon firing. Kaolin or China clay deposits are usually formed *in situ* from the weathering of feldspars. They are the result of hydrothermal reactions in granite rocks, and this process, called *kaolinization*, is reported to occur to great depths in igneous rocks. It is in fact a form of *geochemical weathering*. Some kaolin may have been formed from feldspar under atmospheric weathering conditions, also called *pedochemical weathering*, but this type of kaolin deposit differs in thickness and texture from those of hypogenic origin. The best kaolins for ceramic use comes from the hydrothermally formed deposits. Sedimentary or redeposited kaolins, such as found in the southern United States, are of economic value but are not as pure as kaolin from China rocks. Kaolins formed by pedochemical weathering contain high amounts of Fe minerals and will turn brown, pink or red on firing. Hence, they are not as suitable for some ceramics as hydrothermally formed kaolins. Nowadays, kaolinite is also used for the production of false teeth, for knives exhibiting extreme sharpness, such as surgical knives, and in paper making. Kaolinite finds application in the pharmaceutical industry. Its buffer capacity appears to be useful for the use of kaolin in the manufacture of medicine to remedy stomach-related illnesses.

Halloysite. – Halloysite is also a 1:1 layer type of clay. It has a general composition of $Al_2O_3.2SiO_2.4H_2O$, and is similar in structure to kaolinite. However, in contrast to kaolinite, halloysite exhibits some expansion and contraction, and contains interlayer water. Upon heating, halloysite is irreversibly dehydrated and the mineral is converted into metahalloysite. The mineral is considered a precursor of kaolinite, as can be noticed from the following weathering sequence:

Igneous rocks → smectite → halloysite → metahalloysite → kaolinite

Environmental Importance of Halloysite. – Clays containing the mineral halloysite resemble China clay, but are more plastic because of the presence of water in the intermicellar spaces of the mineral. When used in ceramics, halloysite-bearing clays produce denser fired bodies.

Smectite. – Smectite is an expanding 2:1 layer lattice type of clay, and is one of the most common minerals in the smectite group. Formerly, this mineral was called *montmorillonite*. Two other minerals of importance in this group are *beidellite*, with a high Al content, and *nontronite*, an Fe-rich smectite. Frequently, the term *bentonite* is used for smectite minerals. However, bentonite is a clay deposit from which commercial-grade smectite is produced. The name is presumably derived from Fort Benton, WY, where the deposit was reported in 1896 by W.C. Knight (Grimshaw, 1971). The chemical composition of smectite is variable, but it is usually expressed as $Al_2O_3.4SiO_2.H_2O + xH_2O$. The crystal unit is composed of two silica tetrahedrons sheets with one alumina octahedron sheet sandwiched in between (Figure 2.5). The crystal units, making up a smectite clay particle, are loosely bonded together by water in the intermicellar space. Hence, this space may expand or contract depending on soil moisture conditions. Accordingly, the crystal units can move in either direction, causing the high plasticity of the mineral. When wet, smectite is very sticky and the mineral may exhibit internal swelling.

Environmental Importance of Smectite. – Smectite minerals are especially prevalent in the clay fractions of vertisols, mollisols, and some alfisols. The high plasticity and swell-shrink potential of the mineral make these soils plastic when wet and hard when dry. The dry soil is difficult to till and displays wide cracks. Because of the mineral's high surface area, it finds application in the paper industry

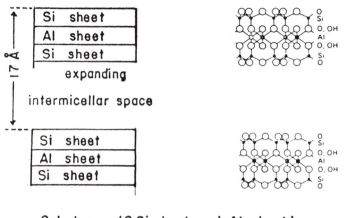

2:1 type (2 Si sheets : 1 Al sheet)
Example : montmorillonite

Figure 2.5 The structure of smectite (montmorillonite) composed of an alumina octahedron sheet sandwiched between two silica tetrahedron sheets, forming a three-layer crystal unit. This is the reason why smectite is grouped as a 2:1 layer lattice type of clay.

for making glossy surfaces on paper. Apparently, smectite also has the capacity to reduce surface tension in liquids, hence it is added to latex in the production of foam rubber. The mineral is reported to be very valuable for its binding power and the strength it imparts to other material. In aqueous solution it disperses easily, forming a suspension, which will change in viscosity, and exhibits *thixotropy* (from Greek *thixis* = touch, and *trope* = change). Upon standing, a smectite suspension turns into a *gel*, but is restored into a *sol* by shaking or stirring. Because of this characteristic, smectite suspensions are employed in drilling fluids. The high cation exchange capacity is the reason for utilizing smectite in the purification of liquids, in bleaching oils, and in the paint industry. In the production of paint, smectite clays are called *fuller's earth*.

Illite. – Illites are micaceous types of clays and are variously identified as *hydrous micas* and *hydrobiotite*. However, because illite is of secondary origin, and because it has a different chemical composition, illite differs from true micas. Illite contains more SiO_2 and less K than muscovite. Structurally, it is similar to smectite, and varies from smectite only because of the presence of interlayer K (Figure 2.6). The crystal units are held strongly together by the K ions, and the mineral, therefore, has a lower swell-shrink potential than smectite. Illite is usually considered a nonexpanding type of clay in the 2:1 group. Its close relationship with smectite can be illustrated by the reaction:

Illite $\rightleftarrows$ Smectite + K^+

Consequently, in soils affected by high precipitation, the mineral tends to be altered into smectite because of leaching of the interlayer K. Illite is an important mineral in the clay fraction of mollisols, alfisols, spodosols, aridisols, inceptisols, and entisols.

Vermiculite. – Vermiculite can be distinguished into: *true vermiculite* and *clay vermiculite*. True vermiculite is a rock-forming mineral that becomes twisted and curled upon heating. After heating, the mineral usually expands to 20 to 30 times its original size. On the other hand, clay vermiculite is of pedogenic origin, and is sometimes called *soil vermiculite, pedogenic vermiculite*, or *14 Å mineral*. It is a 2:1 layer type of clay with Mg in octahedral positions between the two silica tetrahedral sheets. When $Al(OH)_3$ or gibbsite is present in the intermicellar spaces, the mineral is called *hydroxy aluminum interlayer vermiculite* or *HIV*.

As shown in Table 2.5, vermiculite is one of the clay minerals with the largest CEC among the inorganic soil colloids. The mineral is known to exhibit *wedge zones*, attributed to marginal curling of layers on the mineral surface (Tan and McCreery, 1975; Raman and Jackson, 1964). These wedge-shaped zones provide partially enlarged interlayer spacings for entrapment of organic matter (Figure 2.7), or for fixation of K, NH_4^+ and other cations. The high K and NH_4^+ fixation capacity in many soils is attributed more to the presence of

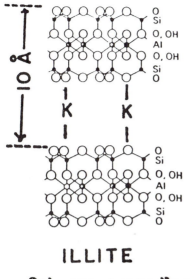

ILLITE
2:1 non-expanding

Figure 2.6 The structure of illite with K in the intermicellar space between the crystal units. Illite is also a 2:1 layer lattice type of clay.

vermiculite than to that of smectite or illite.

Environmental Importance of Vermiculite. – Vermiculite usually occurs as accessory minerals in the clay fractions of ultisols, mollisols and aridisols. It is formed more often in well-drained soils, in contrast to smectite, which requires for its formation a Si-rich environment due to impeded drainage and gley condition. True vermiculite finds application as a valuable packing material, because of its capacity to expand into coarse and fluffy material after heating. Often it is also used as a filler or carrier, and as a potting medium in nurseries.

Chlorites. – Chlorites are hydrated Mg and Al silicates that are similar to mica minerals in appearance. The minerals occur as green-

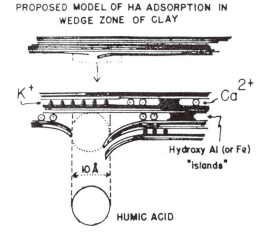

PROPOSED MODEL OF HA ADSORPTION IN
WEDGE ZONE OF CLAY

Figure 2.7 Wedge-zones in clay mineral structures, attributed to curling of layers on mineral surfaces. These wedge-shaped zones provide enlarged interlayer spacings for entrapment of cations and humic acids (Tan and McCreery, 1975).

ish flakes resembling the flaky nature of micas, hence the name chlorite is chosen from a Greek term meaning green. Structurally, it is related to vermiculite, but a recent trend is to use the term 2:2 layer for chlorite. This term reflects the fact that both the octahedral sheet sandwiched between the silica tetrahedral sheets, and the intermicellar space are filled with $Mg(OH)_2$, called *brucite layers*. The replacement of Mg by Al in the brucite layers creates positive charges, which practically neutralize all negative charges. Consequently, chlorite has no charge or only a very small charge, hence, a small CEC. The abundance and occurrence of the mineral are very low. Chlorite is usually detected as accessory minerals in the soil's clay fractions.

Interstratified clays. – Soil clays exist in nature as a mixture of different types of clays stacked together in a packet. The process of stacking is called *interstratification*, and the clays are called *mixed layer* or *interstratified clays*. The different types of clays cannot be separated from the stacks or packet by physical means as is the case with an ordinary physical mixture of clays. Interstratification can occur *randomly* or *regularly (systematically)*. Interstratified clays are present in a large variety of soils in cold, temperate, and tropical regions. In the soils of the humid regions, interstratification often takes place in the sequence smectite-chlorite-mica, or mica-illite. In soils of the subtropical regions of the United States, interstratified clay, composed of smectite-kaolinite or vermiculite-kaolinite, has also been detected.

Sesquioxides. – The sesquioxide clays do not belong to the phyllo-silicates, but are oxides of iron and aluminum and/or their hydrates. They do not contain Si (Table 2.5). Some of them are present in soils in crystalline form, e.g., gibbsite, hematite, goethite, lepidocrocite and ferrihydrite. Others are amorphous in nature, whereas many may exist in soils in the form of short-range-order minerals. Sesquioxide minerals are amphoteric in character and exhibit variable charges. Therefore, these clays, together with amorphous clays, are also known as *clays with variable charges*, whereas the soils are called *soils with variable charges*. The minerals are characterized by a high surface charge density, and a high specific surface area. They are also known for their high phosphate and metal fixing capacity.

Environmental Importance of Sesquioxides. – The sesquioxide mine-rals usually occur in highly weathered soils, e.g., in oxisols and ultisols. The iron oxides in particular determine the color of soils. *Goethite* and *hematite* are responsible for the yellow and red colors of the oxisols and ultisols. Goethite minerals are naturally yellow, whereas hematite is bright red. These colors will be reflected in the soil color depending on the relative abundance of the respective minerals. On the other hand, *gibbsite* is colorless, but its content is used to indicate the degree of weathering, as formulated in the oxidic ratio (Soil Survey Staff, 1975):

oxidic ratio = (%extractable Fe_2O_3 + %gibbsite)/%clay

Soils are considered to be highly weathered if the value of the oxidic ratio ≥ 0.2.

The iron oxide minerals have also been reported to influence the physical properties of soils. Iron oxides can be adsorbed on soil mineral surfaces, inducing a cementation effect that leads to the development of strong aggregation and to the formation of concretion and crust (Baver, 1963). This enhanced stable aggregation causes oxisols in the tropics to be friable, rather than sticky and plastic, as would be expected from their high clay content. The strong structure allows oxisols to be cultivated even in heavy downpours.

With the advanced knowledge of sesquioxides, it is currently believed that many of the minerals may exist in soils as short-range-order minerals. The evidence presented indicate that organic acids interacting with mineral colloids favor the formation of short-range-order minerals. Polymerization of Al and Fe compounds, leading towards formation of long-chain minerals, is inhibited due to chelation of the inorganic colloids with humic acids (Tan, 1998; Huang, 1995). By the attachment of an organic ligand to an Al or Fe monomer or dimer, the hydroxyl bridging mechanism for further polymerization is broken. The amorphous short-range-order mineral varieties often form coatings on the surfaces of crystalline clays or even primary minerals, changing in this way the surface properties of the coated minerals. The ill-defined coatings of sesquioxides present the most reactive surfaces in soils. It can be expected that the high specific surface areas and disordered network of short-range-order colloids create a different surface chemistry.

Silica minerals. – Silica minerals are also not phyllosilicates. They are composed entirely of silica, and structurally they belong to the tectosilicates. The coarse silica minerals are constituents of the silt and sand fractions, whereas the particles ≤ 2 μm belong to the clays. Two groups of silica minerals are usually recognized, *noncrystalline* and *crystalline*. *Opaline silica* belongs to the noncrystalline group. It is of biological origin, formed from the silicification of grasses and parts of deciduous trees. The crystalline group is composed of quartz, tridymite, and crystobalite. As primary minerals in the form of large crystals, quartz has been discussed in section 2.4.1 on page 35.

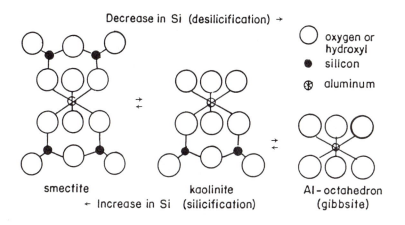

Figure 2.8 Transformation of smectite into kaolinite and gibbsite by a desilicification process. The reverse process, silicification, is the addition of silica to gibbsite and kaolinite to form smectite.

Environmental Importance of Silica Minerals. – The silica minerals occur in a wide variety of soils. Their presence and content are related to parent materials and to the degree of weathering. Inceptisols and entisols can be rich in quartz, but this may be a lithologic effect, or the effect of the parent material. Crystobalite is often volcanic in origin, and its presence is of importance in many volcanic ash soils. As discussed in section 2.4.1 on page 35, silica minerals also find application in the ceramic industry. Although several scientists believe that its importance is only second to kaolin clay, free silica, composed of amorphous and crystalline silica, is very valuable in the glass industry, for the manufacture of so-called crystal or *crystal wares*. It also finds application for making bricks and roof tiles, and in the form of sand it is used for a variety of industrial purposes, such as for making sandpaper.

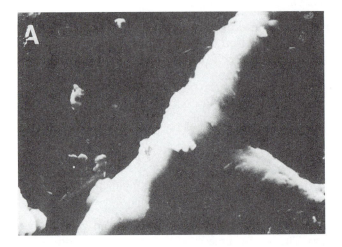

Figure 2.9   Scanning electron micrographs showing the conversion by neoformation of imogolite (A) into kaolinite pseudomorphs (B) stacked into an accordion-like pattern. The clay minerals have been detected in the C horizon of an ultisol derived from volcanic ash in Indonesia (Goenadi and Tan, 1991).

2.6.2   Weathering of Secondary Minerals

Because the clay minerals are the results of weathering processes, one may infer that they are stable or in equilibrium with the prevailing environmental conditions. Nevertheless, these minerals are subject to further weathering. Under changing physico-chemical conditions, clay minerals can be converted from one type to another, a process referred to as *alteration*. Two main processes of alteration can be distinguished: *transformation* and *neoformation*. Transformation is defined by Singer (1979) as an alteration process of clay minerals that occurs without changing the basic layer structure. This process can further be differentiated into *degradation* and *aggradation*. A degradation process involves loss of substances by

destruction, whereas in aggradation, substances are added to the mineral structure to form the new clay minerals (Millot, 1964). Under warm and humid conditions, desilicification and silicification are the two main trends of transformation processes (Figure 2.8).

Desilicification commonly takes place during a laterization process in soils where drainage is not restricted, and Si can be leached. Laterization gives rise to formation of ultisols and oxisols rich in kaolinite and sesquioxides. Over geologic time periods, a continued desilicification process will ultimately transform these minerals into bauxite. Under conditions with a more restricted drainage, silica will not be leached, but will rather be added. This process, called *silicification*, favors formation of smectitic minerals. On the other hand, neoformation is the synthesis of clay minerals from other secondary minerals with different structure, and/or from primary minerals by pseudomorphic replacement, or by direct crystallization from solutions or colloidal gels. Figure 2.9 provides examples of neoformation, showing the conversion of imogolite into kaolinite pseudomorphs, and eventually into discrete kaolinite units (Goenadi and Tan, 1990).

CHAPTER 3

ORGANIC SOIL CONSTITUENTS

3.1 SOIL ORGANIC MATTER

The organic fraction of soils, also called *soil organic matter*, is derived from the soil biomass, and strictly speaking it consists of both living and dead organic matter. Frequently the term is used to indicate the dead organic fraction only, and the live fraction is usually ignored. The dead organic matter is formed by chemical and biological decomposition of organic residues. It can be distinguished into (1) organic matter in various degrees of decomposition, but in which the morphology of the plant material is still visible, and (2) completely decomposed material. The first category is often called *litter* when it lies on the soil surface. In forest and grassland soils, litter is particularly important in the process of nutrient cycling. The second group, the decomposed fraction, consists of numerous organic compounds, but only a few are present in detectable amounts in soils. Some of them are *nonhumified*, whereas others are *humified* compounds (Stevenson, 1967,1982; Flaig, 1971; Tan, 1993). The nonhumified compounds have been released by decay of plant, animal, and microbial tissue in their original or in a slightly modified form. They include carbohydrates, amino acids, proteins, lipids, nucleic acids, lignin, pigments, hormones, and a variety of organic acids. The humified compounds are products that have been synthesized from these nonhumified substances by a process called *humification*. They consist of a group of complex substances such as humic and fulvic acids. The nonhumic and humic material is collectively called *soil humus*, though some scientists prefer to use the name humus for designating soil organic matter as a whole (Waksman, 1936). Many suggest distinguishing the whole or the total organic matter in soils into several *pools*. This concept of pools of organic matter is formulated on the basis of differences in susceptibilities to microbial attack, hence these pools exclude the microbial fraction and are composed only of the dead organic fraction. Three major pools are currently recognized: (1) active, (2) slow, and (3) stable or passive pool (Faustian et al., 1992; Stevenson, 1994; Brady and Weil, 1996). The active pool of soil organic matter is the least resistant fraction and is composed of compounds most readily accessible for food to micro-organisms. Under sustainable agriculture, it is the most important fraction for maintaining soil productivity. Stevenson (1994) includes a *light* fraction in the active (labile) pool, which is defined as plant residues at varying stages of decomposition. Its properties and composition are said to be comparable to plant litter. The difference is that the light organic fraction is mixed within the soil, whereas litter is a plant residue on the soil surface as stated earlier. The stable pool is the most resistant organic fraction against microbial decomposition and is characterized by the highest *mean residence time* among the different types of soil organic matter. The mean residence time (MRT) is a term used to express the age of soil organic matter as determined by ^{14}C dating. Its use is preferred to that of the older term *half-lives*. The MRT of organic matter in US soils varies from 800 to 3000 years (Stevenson, 1994). The passive fraction is usually composed of the humic matter and is responsible for the chemical activity in soils, e.g., cation exchange capacity, interactions, and water-holding capacity. The slow pool of organic matter is inter-mediate in nature between the active and passive pools.

The amount of organic matter varies from soil to soil. Based on standard analysis on organic C content, highest organic matter content, ranging from 5 - 10 % C_{org}, is found in andosols and mollisols, whereas the lowest organic matter content (1 - 3%) is normally noted in oxisols. Both the climate and vegetation are determining factors in soil organic matter content, though Stevenson (1994) believes that vegetation is secondary to the climate and proposes the following decreasing order of importance of soil formation factors in determining organic matter content in soils:

climate > vegetation > topography = parent material > age

3.1.1 Beneficial Effect of Soil Organic Matter

The soil organic fraction as described above affects the physical, chemical, and biological conditions in soils. Physically, it increases organic matter content, imparts darker colors to soils, and decreases bulk density with an increase in organic carbon content. It improves aggregation of soil particles, resulting in the development of stable soil structures. Chemically, it increases the cation exchange capacity, and the water-holding capacity of soils. Its cation exchange capacity, far exceeding that of clay minerals, is the reason for the high soil's buffer capacity. Harmful elements and compounds, such as pesticides toxic to plants and humans, are detoxified by interaction with soil organic matter. Organic matter affects soil fertility by increasing the soil's nutrient content, especially N and S content. It is the main source of N in soils. It can also have a direct effect by stimulating plant and root growth. A variety of growth-promoting substances are said to be secreted by microorganisms or formed as organic residue decomposes. The fact is that growth and yield of crops have been noted to be better when plants are grown in soils rich in organic matter. However, some doubts exist on this matter, allegedly due to lack of scientific data (Stevenson, 1994). Biologically, soil organic matter is the main source of food and energy for soil organisms. The size of fungal and bacterial population increases and decreases in relation to rising and declining organic matter content, respectively. Earthworms especially are strongly influenced by organic matter content. The population of soil organisms will decline with a decrease in organic matter content. In the absence of soil organisms many, if not all, biochemical reactions will come to a standstill.

3.1.2 Allelophaty

In some cases, plant growth is harmfully affected by soil organic matter. The term *allelophaty* is proposed for such an effect by Hans Molisch in 1937 (Miller and Gardiner, 1998; Brady and Weil, 1996). Though only the harmful effect is emphasized by most scientists, allelophaty can also be very beneficial. The production of growth-promoting substances, such as auxin and gibberellin, is one example. By definition allelophaty is the beneficial and detrimental effect of phytochemicals on soil organisms. These compounds, called *allelochemicals* or *allelophatic chemicals*, are produced within the plant tissue or microbial cells, and are released as exudates and/or secretions by roots, leaves and microbial cells. This is the reason why no undergrowth is present underneath the foliage of some trees. The phytotoxins can also be formed by enzymatic decomposition or by hydrolysis of organic residues. An allelochemical which is of considerable concern in animal and human health is *aflatoxin,* produced by the fungus *Aspergillus flavus*. The organism thrives on corn and peanut plants grown in wet conditions and especially on grains stored in wet conditions. This mycotoxin is suspected to be carcinogenic, hence very harmful for animal and human health. Poultry fed with contaminated grain is noted to develop a disease in the gizzards, known as *gizzard erosion*. When the liver is affected, the toxin may cause instant death. In 1974, it was reported from Hong Kong and Thailand that consumption of grain affected by this toxin-producing mold has caused many people to die and many more to become seriously ill (Miller and Gardiner, 1998). Allelophaty is also assumed to be the reason for the dominance of several types of weeds in the field. An example is the spread of *dyer's woad (Isatis tinctoria)* in the West. In Europe this plant is used to produce blue dye. When the seed pods of the plants fall and decompose in the soil, a toxin is produced killing nearby growing grass species (Young, 1988). Overcrowding the remaining grasses is then the next process. Other examples are Johnson grass, sorghum, and peach, plants that are capable of producing cyanide-containing chemicals which upon hydrolysis are converted into HCN (cyanide) a powerful toxic gas. Such plants are called *cyanogenic* plants (Putnam, 1994). Amygdalin produced by peach roots is broken down by hydrolysis into HCN gas and benzaldehyde, which cause young peach roots to die. Acetic acid is reported to be produced by anaerobic decomposition of wheat straw. This chemical will inhibit the growth of sorghum and barley when plants are mulched with wheat straw in cool and wet conditions or when plants are grown into the wheat stubble in no-till farming (Wallace and Elliott, 1979). Coffee (*Coffea arabica*) and tea (*Camellia sinensis)* are known to contain caffeine and other alkaloids. When the leaves die and decompose in the soils, the chemicals when accumulated in large amounts can allegedly cause degeneration of the coffee and tea plantations (Miller and Gardiner, 1998).

Currently, these phytotoxins have attracted considerable attention in industry for their potential in the production of biological or environmentally friendly pesticides. The glycosides produced by many cereal plants when broken down have antifungal properties, whereas the phytotoxins in dyer's woad can be used as an environmentally friendly herbicide. A success story is the production of insecticides from pyrethrum produced by chrysanthemums. Used in powdered form, pyrethrum is noted to be effective for insect control in vegetables and flowers. Pressed into solids, it is an inexpensive and favored insecticide in Southeast Asia, where the solid sticks or solid coils are burned as incense to combat mosquitoes.

3.2 HUMUS

As defined earlier, the term *humus* or *soil humus* refers to a mixture of decomposed nonhumified and humified compounds. In the past it referred to the total organic matter in soils. In geochemistry exotic names have been proposed for humus developed in different environments. Humus formed from microscopic plants in eutrophic lakes and marshes is called *copropel*, whereas that located in deeper hypolimnetic areas of lakes and bays is called *sapropel. Förna* is a variety of sapropel formed from cellulose-rich pondweed. Allochthonous deposits in dystrophic lakes are called *dy,* and amorphous, gummy deposits within peat bogs are named *dopplerite.* Marine deposits composed of planktonic debris are called *pelogoea* (Swain, 1963). In forestry, the names of *mor* and *mull* are frequently used to designate nondecomposed and decomposed organic matter, respectively. Mor is equivalent to litter, hence does not qualify to be considered humus.

Humus is defined by Stevenson (1994) as the total organic fraction in soils, exclusive of nondecomposed plant and animal material, their partial decomposition products and the soil biomass. Humic compounds, e.g., humic acids and fulvic acids, make up the bulk of humus, though lignin can also be substantial in amounts. The humic matter fraction is called *humus acid* in the European literature. Stevenson (1994) believes that carbohydrates are the next most abundant component of humus. The carbohydrate concentration in humus has been estimated to range from 5 to 25%. Soils rich in nondecomposed litter may contain small amounts of simple sugars, but large amounts of cellulose, which is the main form of carbohydrate in higher plants. Currently the presence of *extracellular polysaccharides (EPS)* has attracted much attention, because of their importance in soil aggregation and interaction with soil clays. These compounds are produced by a variety of organisms, e.g., bacteria, fungi and algae, and occur as a mucous layer around the cells. As such they are the interfaces between the cells and the surrounding soil, and are also assumed to provide protection to the cells. Their amount in soil humus is very small and ranges between 0.1 and 1.5% of soil organic matter (Chenu, 1995). Lipids are also important constituents with concentrations of 2 to 6% in soil humus. They are a diverse group of organic compounds, ranging from fatty acids, wax, resins to sterols, terpenes, and chlorophyll. Free amino acids are also important constituents of soil humus, but their concentration is very difficult to determine due to analytical problems. Their concentration is estimated to be 2 mg/kg of soil, but may be sevenfold higher in the soil rhizosphere, where they are believed to be secreted by live plant roots (Rovira and McDougall, 1967). In addition to the aforementioned compounds, humus may also contain an assortment of water soluble organics and enzymes. Enzymes are proteinaceous compounds, which are extremely difficult to determine directly. They can be determined indirectly through their capacity to transform one compound into another. By definition, enzymes are thermolabile catalysts produced by living tissue but capable of action outside this tissue (Gortner, 1949). Hormones, such as auxins, are excluded since they conduct their physiological functions only in the living tissue. A catalyst is then a substance capable of altering the speed of reaction without appearing as part of the final product. The names of enzymes all end with *ase* and are descriptive for the type of compounds broken down. For example, cellulase is an enzyme that breaks cellulose into its constituent sugar components, whereas protease is important in the splitting of protein into amino acids, and urease breaks down urea, in urine and fertilizers, into ammonia. Some of the enzymes are produced by plant cells as *constitutive enzymes*, whereas others are only produced when a susceptible substrate is present, and such enzymes are called *induced enzymes*. Cellulase is an example of an induced enzyme and urease is an example of a constitutive enzyme. Free enzymes, enzymes not associated with the microbial biomass, are reported to be accumulated in soils by entrapment (fixation) within intermicellar spaces of expanding clays (Paul and Clark, 1989), though the opinion in clay mineralogy is that most enzymes are too large to penetrate the intermicellar spaces of clay minerals. Complex formation and interaction between enzymes and soil colloids are perhaps better reasons for the presence of free enzymes in soil humus. In this way, the compound ATP (adenosine triphosphate), an important coenzyme in all biosynthetic and catabolic cell reactions, can be present in soils, though ATP has not been isolated yet from soils.

3.2.1   Environmental Significance of Humus

As defined above, humus is a complex mixture of decomposed organic residues. Part of the nonhumified fraction is present as free compounds, but some is present in close association with the humified fraction. For example, carbohydrates, especially polysaccharides, amino acids, and lignin are bonded together to form the humic matter. Since these components are colloidal in nature, humus represents the chemically active fraction of soil organic matter. The beneficial effect and the discussion above about allelophaty are especially valid for humus. It is the humus fraction that is responsible for increases in cation exchange capacity, buffer capacity, water-holding capacity, and for the presence of chelation reactions in soils.

Humus content usually declines in soils with the first cropping, but eventually it will reach a new level that is in equilibrium with the environment. At this point the rate of formation equals the rate of decomposition of humus at the prevailing environmental conditions. Stevenson (1994) believes that equilibrium levels are attained after a period of 10 to 30 years. In the southeastern United States humus content appears to level off at 2 - 3% C_{org} after just a couple of years of adding grass clippings (Beaty et al., 1976).

3.3 SOIL BIOMASS

There are several definitions for biomass, and one definition states that the biomass is composed of microbial tissues only. It can be estimated by microbial counts and a knowledge of cell weights as formulated in the following relationship (Stevenson, 1989):

Biomass = number of cells x volume x density

Another definition indicates that the biomass is the total mass (dry weight) of living organisms (Paul and Clark, 1989; Brady, 1990). The latter are composed of the live *soil faunae* and *florae*, each of which consists of a macro and micro group. This definition appears to be used by a great number of people, because the definition shows that all organisms, not only microorganism cells, contribute toward formation and accumulation of organic matter in soils.

The biomass is constantly being generated by growing organisms. When the organisms die, their bodies are decomposed, compounds are released, oxidized, and degraded, and the final products contribute to the group of compounds that polymerizes to form humic matter. Under these conditions, humic matter in soils is slowly but constantly being formed and decomposed at the same time.

3.3.1 Macrofaunae

The macrofaunae are the higher animals. According to Miller and Gardiner (1998) this group is now called *Animalia,* one of the five kingdoms. The other four are *Plantae* (plants), *Fungi*, *Protista* (protozoa), and *Monera* (bacteria and actynomycetes). The importance of the macrofaunae in soils is primarily through (1) the breakdown of organic matter, and (2) the cultivation effect. They are, therefore, essential participants in the natural cycle of organic matter. Dead leaves and twigs are chewed by these animals, and moved from one to another place in the soil. Holes and burrows are made into the soil by some animals, such as earthworms, moles, chipmunks, and prairie dogs. In doing so, surface material is brought down, whereas subsoil material is worked up. This has a similar effect as soil cultivation with the plow, hence the term *cultivation effect.*

Earthworms. - The earthworm is perhaps of special importance in the cultivation effect. Under the tropical rain forest in Nigeria, the earthworm population can amount from 30 to 210 worms per m^2, which translates to 9×10^5 earthworms per hectare (Ghuman and Lal, 1987). By digging their way through the soil, earthworms consume organic matter and soil alike, which are thoroughly mixed and ground in their body. At the same time, this inorganic-organic mixture is subjected to the digestive action of the stomach enzymes. The soil turnover from earthworm activity under a tropical rain forest in Africa is estimated to be 21.5 Mg (tons)/ha per year (Ghuman and Lal, 1987). Earthworm casts are reported to be very high in organic matter, N, P, K, and Ca content, hence can improve soil fertility. The amount of casts produced is substantial, and is estimated to be 250 Mg/ha per year (Brady, 1990). In addition, earthworms may have an important effect on the physical condition of soils. The holes left serve as channels for air and water passage. Most of the earthworms live in the root zone and only few of them are reported to burrow as deep as 5 m. They prefer moist, well-aerated, warm soil (25°C) rich in organic matter, and a pH between 5 and 7. The two common earthworm species of importance in the USA are *Lumbricus terrestris*, a red deep-boring species which was imported from Europe, and *Allo-*

lobophora caliginosa, a pinkish shallow-boring earthworm.

Ants, Termites, and Miscellaneous Animals. - Ants, mites, beetles, flies, and other insects are listed as beneficial animals feeding on decaying vegetation and assisting in soil aeration because of their burrows (Miller and Gardiner, 1998). However, other scientists usually consider mites, beetles and flies as pests. Beetles and flies also do not live by burrowing into the soils, perhaps only their larvae do. Ants are perhaps better examples for cleaning organic debris. Some types of ants will take care of dead bodies of animals, which when left alone will contaminate the environment. Other types of ants may cut live leaves for feeding their fungal cultures. These ants are called *leaf cutter ants*. Ants may burrow holes, but the size and amounts of ant holes are relatively small so that water movement is expected to take place only by capillary action. Moreover, fire-ant mounds are scattered erratically over the area, whereas below the thin layer of single grain soil mass, the ant nest is usually cemented together into a mass more compact than expected. In this case, the effect is a radical change in soil structure, as claimed by several authors. However, the single grain and massive structures that these ants create are not beneficial for agricultural operations. More controversial is the effect of termites. True that they are feeding on dead wood, contributing in this way to organic cycling. However, the mounds they build, located as protruding islands scattered in the landscape, are tall and huge. Because they are very compact and hard, efforts to level the mounds and convert them into croplands require special equipment. Even in Australia, with its advanced techniques and availability of modern equipment, termite mounds have not been reclaimed yet for agricultural lands. It is usually considered a waste of time, since the area of soils created from termite mounds is generally small and also infertile due to the very low organic matter content.

The larger animals that burrow in the soils to make dens and holes, e.g., prairie dogs, woodchucks, moles, and badgers, are perhaps more harmful than beneficial in agricultural operations. If these dens and holes are distributed evenly over the whole area, they become a hazard for grazing animals and horses.

3.3.2 Microfaunae

The microfaunae are the microscopically small animals, and two of the major species in soils are nematodes and protozoa.

Nematodes. - Nematodes are very small threadworms, normally 1 to 2 mm in length, although exceptions may occur. They can be distinguished into (a) saprophytic, (b) predatory, and (c) parasitic nematodes (Brady, 1984). They are all essential in the natural recycling process of soil organic matter. The _saprophytic nematodes_ consume dead organic matter, whereas the _predatory nematodes_ feed on other nematodes, and also on a number of microorganisms, such as bacteria, algae, and protozoa. _Parasitic nematodes_ feed on live plant roots, causing _nematode disease_ in cultivated crops. The parasitic nematodes are attracted to plant roots by a chemical compound secreted by the roots. Therefore, only plants that can produce chemicals attractive to parasitic nematodes can be affected, e.g., soybean, and tomato. If the larvae of parasitic nematodes fail to migrate to the root surface, they will die naturally in soils. As can be noticed in Figure 3.1, the parasitic nematodes have needle-like organs in their mouth to inflict damage and penetrate the root tissue in which they multiply. Although the parasitic nematodes can become harmful for cultivated crops, they are an essential part in the natural recycling process of soil organic matter. They are the major food source of predatory nematodes, whose population will grow or decline with abundance and decrease in parasitic nematodes. Therefore, in nature a certain balance exists between the number of beneficial and harmful organisms. In other words, the population of parasitic nematodes is kept under control by the predatory nematodes. If by human interference, however, conditions are created favorable for the growth of parasitic nematodes, an outbreak of a nematode disease occurs. Rootknot disease caused by infection with the rootknot nematode _Meloidogyne_ sp., can cause heavy damage in soybean plants. Another example is the _golden nematode (Globodera rostochiensis),_ which attacks potato roots (Hays, 1996). Methyl bromide has been used in the past to control nematode 'diseases', until EPA considered it harmful for the environment. A new environmentally friendly approach is to

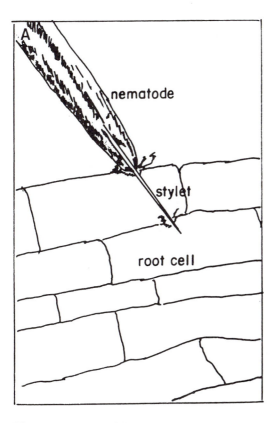

Figure 3.1 (A) Schematic drawing of a parasitic nematode inflicting damage to root cells. (B) Reproductive organs of a rootknot nematode, *Meloidogyne* sp. (J. Plaskowitz, USDA, courtesy Tousimis Research Corp., Rockville, MD).

interplant nematode-sensitive crops with marigold or canola plants, which can produce phytotoxins possessing nematicidal properties. Root exudates of marigold plants are reported to contain thiophenes and benzofurans, allelophatic chemicals which are perhaps toxic to nematodes (Brady and Weil, 1996; Miller and Gardiner, 1998). Today, studies have also been conducted to cultivate the predatory nematodes

in vitro for use in the biological warfare against nematode diseases in cultivated crops. Recently, a beneficial nematode has been reported for control of the larvae of the Japanese beetle, damaging lawns in cemeteries and golf greens. The nematode (*Steinernema glaseri*) enters the white grubs of the beetle through their mouth to release a bacterium in the larvae body which kills the grub overnight (Lyons-Johnson, 1996).

Protozoa. - Protozoa are considered the most elementary form of animal life in soils. They are microscopically small and include *amoeba, ciliates*, and *flagellates*. The distinction is based on the type of locomotion. The amoeba moves by stretching and contracting like a worm. Ciliates are surrounded by hairlike structures that are used for movement. Flagellates, as the name implies, use a *flagellum*, a whip-like tail. Some of them are pathogenic to animals and humans, whereas others feed on organic residue and other microorganisms, e.g., rhizobium bacteria (Jones, 1983). The opinion exists that the population of protozoa in soils is often not sufficiently large to influence the recycling of organic matter (Brady, 1990).

3.3.3 Macroflorae

The macroflorae are the higher plants, including the cultivated crops. All of them affect the physical, chemical and biological properties of soils. Physically, they are capable of increasing organic matter content in soils when left alone in nature to grow and complete their life cycle.

Roots can enhance development of soil structure. They also grow through cracks, and when they die they leave channels for air and water passage. Chemically, plant residues affect the nutrient content of soils, and as such the macroflorae play an important role in nutrient cycling, a process that preserves and maintains soil fertility. In this process, the organic residue decomposes, releasing nutrients into the surface soil where they are taken up again by growing plants. Roots can also absorb from subsoils the nutrients that have been leached from surface soils. These nutrients are accumulated in the plant tissue, and will be released in the surface soil with leaf-fall or when the plants die and decompose. Plants grown deliberately for improving soil fertility, e.g., nutrient content and physical conditions of soils, are called *green manures*. The macroflora are also important for use in the protection of the surface soil against erosion and other destructive forces in the environment. Plants grown to protect soil surfaces against these harmful forces are called *cover crops*.

Plant Roots. - Biologically, higher plants affect the life of almost all organisms, macro and micro alike. In the soil, it is especially the root system that affects the physical, chemical and biological properties of soils. Based on their physiological functions, roots can be distinguished into *feeder roots, root hairs, anchor,* and *taproots.* Depending on their dimension, Brady and Weil (1996) distinguish them into meso- and macroroots. They also consider the roots as soil organisms. Apparently feeder roots in the range of 100 to 400 µm are meso-organisms, whereas root hairs in the size of 10 to 50 µm are microorganisms since according to the authors this is the size of microscopic fungus' strands. Accepting the validity of the concept that root hairs are microorganisms is left to the reader. At the seedling stage, the root system is an absorbing organ, and as the roots thicken with age, the area for absorption becomes restricted at the area right behind the root tips. The root tips themselves are usually protected by *root caps*, and that part of the root below the root cap is the meristematic region characterized by high physiological activity. The root cap does not only shield the meristem, but it is the organ that makes the root respond to gravity, called *geotrophic response.* The newly formed rootlets or the terminal ends of extending roots are called feeder roots (Campbell and Hendrix, 1974). In many plants, feeder roots consist of incompletely differentiated tissues. Their cells, covered only by a thin epidermis, are usually in a very active physiological state. Lateral roots develop from cells within the vascular tissues, pushing through the cortex and epidermis at regular intervals, leading to the characteristic branching of a normal root system. This forms the anchor roots. Taproots, on the other hand, characterized by a high geotrophic response, grow deeper in the soil. These roots also function as anchors for the plants. Root hairs are elongated single epidermal cells. They disappear in many plant roots and are sloughed off as the epidermis thickens in the maturation zone. Generally the fine roots or feeder roots are abundant in the upper 10 cm of soil. This zone is usually called the *root zone*. Anchor and taproots are growing deeper in the soils.

The root system is not only very important in maintaining soil organic matter content, but it affects also directly the physical, chemical and biological conditions in soils. The high organic matter content in mollisols is believed to be caused by the decomposition of residues from the dense root system of the tall grass vegetation rather than from the aboveground biomass. The root mass, remaining in soils after the crop is harvested, is estimated to be 15 to 40% of the aboveground vegetative parts (Brady and Weil, 1996). If 25% can be considered a good average, these authors expect that a good crop of oats, corn and sugar cane would yield 2500, 4500, and 8500 kg/ha of root residues, respectively. A dense root system is known to increase soil aggregation, hence promotes formation of favorable soil structures. Respiration from a large number of roots releases large amounts of CO_2, which has an enormous effect on a variety of chemical reactions. The carbon dioxide dissolves in soil water to form H_2CO_3. Being a weak acid, the carbonic acid dissociates according to the following reaction:

$$H_2CO_3 \rightarrow H^+ + HCO_3^-$$

The H^+ ions, or protons, released in the soil solution increase soil acidity. This is the reason why the pH in the soil rhizosphere is noted to be a unit or two lower than in the bulk soil. Such an increase in soil acidity enhances the dissolution power of water. Since hydrolysis is also dependent on the presence of H^+ ions, such an increase in protons is expected to increase the rate of hydrolysis reactions. A large number of these protons is known to be adsorbed by the root surface and plays an important role in cation exchange. They can be exchanged for nutrient elements free in solution or adsorbed on clays.

Rhizosphere. – From the standpoint of soil microorganisms, a large number of organisms are congregating around the surfaces of roots. They are attracted to the root surface because of chemical compounds secreted by live roots, which are vital sources of food and energy for the microorganisms. The chemicals are called *root exudates* and can be distinguished into three groups (1) *mucigel* or *mucilage*, a mixture of polysaccharides and uronic acids, enveloping especially the root tips, (2) a variety of organic acids, amino acids, and simple sugars excreted by root hairs, and (3) cellular organic substances produced by senescence of root epidermis. The process of formation and accumulation of the compounds above is called *rhizodeposition* by Brady and Weil (1996). Because of these substances, the number of microorganisms is noted to be more abundant near the root surface than in the bulk soil. Nematodes and rhizobia alike are attracted to the root surface because of the chemicals produced by growing roots. However, the population of pathogenic organisms may also increase and may cause the development of root diseases, e.g., nematode disease. This part of the root-soil interface, where biological activity is at a maximum, is called the *soil rhizosphere* (Rovira and Davey, 1974). The root surface itself, providing critical sites for interactions between microorganisms and plant cells, is called the *rhizoplane*, whereas the epidermal and cortical tissue of roots, colonized by microbes, are called *endorhizosphere*. Paul and Clark (1989) prefer to call the latter *histosphere* or *cortosphere* (from histos = tissue, and cortex, respectively). It is in these regions where microorganisms congregate, and values of 1×10^9 bacteria and fungi in 1 g of soil are common for well-drained and fertile soils.

Many of the soil chemical and biochemical reactions, vital for the growth of plants, occur in the soil rhizosphere. Soil pH is generally lower in the rhizosphere, whereas denitrification is believed to be more rapid close to the root surface. This is due to a greater respiration in this zone by the larger population of microorganisms. In turn, an increased respiration produces a greater reducing condition necessary for the denitrification process. The root exudates are also capable of dissolving and chelating metal compounds. Such a reaction is believed to be the process by which roots obtain iron.

3.3.4 Microflorae

The microflorae are the lower plants, or the microscopic small plants. Numerically, they are the most abundant of all the living organisms in soils. They are usually present in large numbers near feeder roots of plants and play a vital role in many dynamic microbial reactions. On the basis of their feeding habit, they can be saprophytic, parasitic, pathogenic, or symbiotic in nature. Most of them are present mainly in the surface soil or the root zone (Table 3.1). The population

Table 3.1 The Number of Microorganisms in a Well-Drained Soil

Soil horizon	Aerobic bacteria	Anaerobic bacteria	Fungi	Actino-mycetes	Algae
	------------------- x 10^3 g/g soil --------------------				
A	7500	2000	150	2000	30
B	2	1	2	tr	–

Source: unpublished data of the author.

of microorganisms decreases rapidly with depth in the soil. In the subsoil and/or C horizon only minimal amounts of these micro-organisms can be detected. Although a large variety of micro-organisms exists in soils, only a few major species will be discussed in this book (Table 3.2).

Bacteria. – Bacteria are the simplest form of plant life in soils and are composed of single cells, usually 5 µm in size. They lack nuclear membranes and their nucleoplasm is, therefore, not separated from the cytoplasm. Hence, they are called *prokaryotic* (Paul and Clark, 1989). The majority of bacteria species are *heterotrophic* or *organo-trophic*. These are bacteria that need organic matter for food and energy, or in scientific terms: the cell carbon is obtained by the bacteria from an organic substrate. Heterotrophic bacteria can be divided into N_2-fixing and nonfixing bacteria. The nitrogen fixers are subdivided into symbiotic and nonsymbiotic bacteria. The symbiotic group lives in symbiosis with legume plants. The importance of both groups in the environment will be discussed in more detail in section 3.4.4 on nitrogen fixation. On the other hand, *autotrophic* or *lithotrophic* bacteria are bacteria that use inorganic matter, such as NH_4^+, Fe^{2+}, SO_4^{2-} and/or CO_2, for food and energy.

Bacterial cells are often distinguished into *gram-positive* (G^+) and

Table 3.2   Some Common Soil Microflorae

I. Bacteria	1. heterotrophic
	2. autotrophic

I. Bacteria   1. heterotrophic
                       2. autotrophic

II. Fungi   1. mold
                       2. yeast
                       3. mycorrhizae

III. Actinomycetes
VI. Algae
VII. Lichen

---

gram-negative (G⁻) cells. The term G⁺ is given to cells that becomes crystal violet after treatment with a *gram staining solution*, containing KI, whereas the term G⁻ refers to cells which destain readily (Paul and Clark, 1989).

Environmental and Allelophatic Importance. – Bacteria take part in many, if not all, organic transactions of importance to soil fertility and the growth of higher plants. They are responsible for a number of enzymatic reactions, e.g., nitrogen fixation and nitrification. These reactions will be discussed in a later section of this chapter. Many autotrophic bacteria can convert by oxidization NH_4^+, S, Mn^{2+}, and Fe^{2+} into NO_3^-, SO_4^{2-}, Mn^{4+}, Fe^{3+}, respectively. This oxidation process is believed to have some side effects by causing rust formation on metal surfaces. Some bacteria are capable of detoxifying CO gas into the more harmless CO_2 gas by a similar oxidation process, whereas others are reported to be able to reduce the CO gas into methane gas by anaerobic reactions. Several scientists believe that these reactions have a significant impact on maintaining the quality of our environment in view of the increased emissions of carbon monoxide amounting to 200 million Mg/year into the atmosphere by the world population (Miller and Gardiner, 1998).

Bacteria can be both beneficial and harmful to other organisms. Several cause serious diseases in plants. The bacterial wilt disease in tomatoes and the bacterial blight disease in rice and soybean are just a few examples. *Escherichia coli* bacillus infection and food poisoning by *Salmonella* are often in the spotlights of the news media for creating serious illnesses in humans, especially in young children and elderly people. However, beneficial allelophatic effects of bacteria, though less dramatic, are also important. Some of the bacterial allelochemicals have been applied in the production of valuable biological control agents. A toxin produced by *Bacillus thuringiensis*, called Bt toxin, has been applied in the manufacture of insecticides. Used in the form of sprays or sticks on the leaves of cabbage and other vegetables, the Bt toxin proves to be very effective in controlling the diamondback moth. The caterpillars consuming the treated leaves are reported to die, because the Bt toxin causes the guts to rupture. The toxin is said to be harmless to other insects, livestock, and humans (Martin, 1996). However, the development by genetic engineering of plants capable of producing the Bt toxin has created a great deal of controversy. Critics fear that growing Bt-transgenic plants can result in the development of insects that are resistant to the Bt toxin. Reports have also circulated of pollen from Bt-transgenic corn plants that is harmful to non-target organisms.

Fungi. – Fungi are also elementary forms of plant life in soils. The individual cells have a nucleus, and in contrast to bacteria, they are *eukaryotic* organisms. The cells may be linked together into filaments called *hyphae*, which are collectively called a *mycelium* (Paul and Clark, 1989).

Environmental and Allelophatic Importance of Fungi. – Fungi must depend for their food and energy on organic matter in soils. As with bacteria, fungi lack *chlorophyll*, the green substance essential for photosynthesis. Because of the absence of chlorophyll, fungi have developed into saprophytes, pathogens, parasites or symbionts in nature. Saprophytic fungi contribute to nutrient cycling in soils. Together with bacteria, they share the responsibility of decomposing organic residues in soils, breaking them down into simpler forms, which then become available again to plants. Rotting of food, paper, wood and household goods is some type of organic and/or nutrient cycling, although this rotting process is considered harmful from a human standpoint. As parasites or pathogens, fungi may cause diseases in plants and animals. Fungal diseases in plants have been known for a long time, but it was the disastrous impact of potato blight in Ireland in the middle of the 19th century that started the science of *plant pathology* (Webster, 1975). Many types of fungi are used today in the food and pharmaceutical industries. Some types of fungi are edible, whereas others are used in the fermentation of wine, bread and cheese. Production of antibiotics and/or other chemicals are additional examples of the usefulness of fungi. Penicillin, an antibiotic medicine, was isolated originally from the fungus *Penicillium notatum* or *P. chrysogenum*). Aflatoxin produced by *Aspergillus flavus* has been discussed earlier as very detrimental for human and animal health. The fungi thrive and develop in humid conditions, and peanuts or corn plants growing in moist years may be affected by the fungus. Similarly peanut stored in moist condition is an excellent medium for the development of the fungi. Though the harmful effect can be avoided by storing peanut and corn at the proper moisture condition, the control of the fungi in the field is much more difficult. Attempts have been made to control the disease in the field by spreading *Aspergillus* strains that do not produce the toxin (Cole, 1996). However, more research has to be conducted as to the viability of crowding out the toxic fungus by nontoxic fungi.

Mycorrhizae or Root Fungi. – Currently, fungi also find application in improving plant growth and soil fertility. The roots of many plants are infected with a type of root fungi, called *mycorrhizae*. Three groups of mycorrhizae are recognized today: (a) ecto-, (b) ectendo-, and (c) endomycorrhizae (Harley, 1969; Marx and Krupa, 1978). *Ectomycorrhizae* is a root fungus (Figure 3.2A) that forms an inter-cellular fungal network in the root cortex, called the *Hartig net*. Often, the fungus forms a sheath around feeder roots, hence the term *ecto* (outside). It is living in symbiosis with pine, spruce, fir, birch, hemlock, beech, and oak trees, and hence is sometimes referred to as

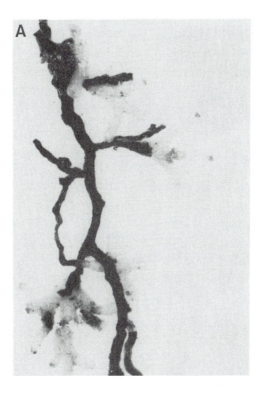

Figure 3.2 (A) Ectomycorrhizae in pine root (courtesy D.H. Marx, USDA-Forestry Science Laboratory, University of Georgia, Athens, GA). (B) Endomycorrhizae (courtesy M.F. Brown, University of Missouri, Columbia, MO).

tree-mycorrhizae. Endomycorrhizae, on the other hand, are living in symbiosis within the root tissue (Figure 3.2B). This group is the more numerous and more widespread of the mycorrhizae, and is found in almost any cultivated crop, to name a few: in roots of wheat, corn, rice, cotton, apple, cacao, coffee, rubber, citrus, maple, and poplar. The walls of root cells in these plants are penetrated by the mycelium, and inside the cells the fungus forms highly branched small structures, called *arbuscules*. This is the reason why this type of fun-

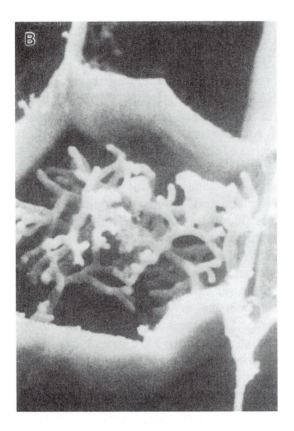

gus is also referred to as *vesicular arbuscular* (VA) mycorrhizae. The *ectendo* group has the characteristics of both the ecto- and endo-mycorrhizae.

The beneficial effect of mycorrhizae on plants is noticed in increased nutrient uptake, and protection against pathogenic attack (Marx, 1969). As can be seen from Figure 3.3 (see also Figure 3.2A), the fungus sends out fine filamentous mycelium into the soil, which translates into increasing substantially the surface area of plant roots. It has been estimated that the active surface area of roots is increased as much as 10 times by the mycelium, which in turn may result in an increase in water and nutrient uptake. The advantage of mycorrhizae is especially noticed in infertile soils, where the fungus can assist the

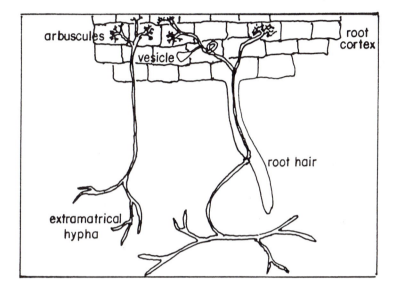

Figure 3.3 Schematic drawing of endomycorrhiza mycelium growing into the soil, increasing in this way the surface area of plant roots.

host plants in nutrient uptake. Some of the fungi are known for their capacity to extract P from soils low in P, or from unavailable-P sources, e.g., *Pisolithus tinctorius*. The latter fungus is well adapted for extracting P from soils low in P, hence allowing the host plant to grow in soils too infertile for other plants.

Actinomycetes. – In many books, *actinomycetes* are defined as lower plants with properties in between those of bacteria and fungi. Like the fungi, actinomycetes have cells that also develop into filamentous hyphae, but their mycelial threads break up into spores resembling bacterial cells. For this reason, they are sometimes referred to as thread bacteria (Brady, 1990). Paul and Clark (1989) classify them within the bacterial group *Streptomyces*, though some authors place them together with bacteria in the *Monera kingdom* (Miller and

Gardiner, 1998). Actinomyces are present in moist, well-drained soils. They are very sensitive to changes in soil reactions, and grow well at pH levels of 6.0 to 7.0, but will disappear in soils with lower pHs.

Environmental and Allelophatic Importance of Actinomycetes. – Most actinomycetes are important in the decomposition of soil organic matter, which can be noticed from the musty smell in decomposing straw piles or freshly plowed land. Some actinomycetes can cause damage to cultivated crops. These parasitic actinomycetes are especially harmful to potato crops by causing potato scab disease. However, since the pathogen is sensitive to low pH levels, this disease is usually controlled by growing the potatoes at lower soil pH by treatment of the soil with powdered S. Many of the actinomycetes are also used in the pharmaceutical industry because of their ability to produce antibiotics, e.g., antiviral, antibacterial, antifungal, anti-protozoal, and antitumor compounds (Paul and Clark, 1989). For example, streptomycin is an antibiotic that was isolated originally from *Streptomyces* by S.A. Waksman in 1942. He was awarded the Nobel Prize for this discovery.

   Lately an actinomycete strain with N_2-fixing capacity was detected living in symbiosis with woody plants. Because of the infection, trees, such as boysenberry, alder, sweet fern, mountain mahoganies, Australian pine, and Russian olive, were noticed to form root nodules (Hensley and Carpenter, 1979). Red alder trees in association with actinomycetes were reported to be able to fix approximately 168 kg of nitrogen per ha (Del Tredici, 1980). A significant feature of such a relationship is that the actinomycetes are capable of forming symbiotic associations with several different plant families, which is in contrast with the *rhizobium*, which is very specific and can only inoculate the *legume* family.

Algae. – Algae are the lower plants that contain chlorophyll. Of all the green plants, they are perhaps the most widely distributed chlorophyll-containing plants on earth. Though mostly aquatic in nature, terrestrial forms of algae, called moss, exist on rocks, soils, stems, barks and leaves. These types of algae grow best in wet conditions.

Aquatic algae include seaweed and plankton, whereas soil algae are composed of *green, blue-green, brown*, and *red algae,* and *diatoms* (Brady, 1990). Diatoms are characterized by stiff cell walls rich in Si, coated by organic matter. A number of microbiologists consider blue-green algae to belong to the *cyanobacteria* (Paul and Clark, 1989). The blue-green algae frequently grow in ponds and inundated paddy soils. They are capable of fixing nitrogen from the air.

Environmental Importance of Algae. – In the tropics, the blue-green algae often grow in close association (symbiosis) with the *Azolla* plant (Figure 3.4), an aquatic fern, commonly found in paddy soils. Such an association is sometimes called a *phototropic* association (Paul and Clark, 1989). They are the reason why the azolla enriches paddy soils with N compounds. As will be discussed in another chapter, algae also find application in the food and health industry. Algae are also known to produce polysaccharides, which when secreted into soils are called extracellular polysaccharides, organic compounds of importance in soil aggregation, as discussed earlier.

<u>Lichens</u>. – Lichens are symbiotic associations of fungi and algae, and according to Webster (1975) belong to the *Lecanorales* group. The fungus and the algae form a single *thallus*, called *lichen thallus* (Paul and Clark, 1989). In most lichen thalli, the algae are confined to a special region, called the *algal zone*, which is interspersed with fungal hyphae. It is believed that the fungus provides mineral nutrients to the algae, which in turn supply the fungus with carbohydrates and, if the algae are blue-green algae also, with nitrogen compounds.

Environmental Importance of Lichens. – Lichens can grow on rock surfaces or infertile soils, and as such they are considered pioneer plants, contributing to soil formation. They also inhabit other sur-faces, such as roof tops, tree trunks and stems, and fence posts. Their growing range is not limited to any climatic condition, since they are found from the Arctic to the humid tropics. In the Arctic region, lichens form the staple food for reindeers and Arctic hares, whereas in the temperate region they constitute part of the winter forage of

Figure 3.4 (A) Algae growth (a) and *Salvinia* sp.(s) mixed with some *Azolla* sp. In a paddy rice field in Bogor, Indonesia. (B) Closeup of *Azolla* sp. plant.

caribous, deers and other grazing animals.

Viruses. – Viruses, whose importance is becoming clear today, are listed here for completeness. According to Miller and Gardiner (1998) the name virus originates from an old Roman term referring to a secretion or poison, whereas in medical science it is used for a disease causing microscopic etiological agent. However, it is not clear yet whether viruses can be called organisms and/or whether they are independent groups of organisms separate from the microflorae. They are either the simplest organisms on earth, or, since they do not have metabolism, are mainly very complex molecules, composed of protein-coated RNA or DNA molecules (Paul and Clark, 1989). Some regard the viruses as an exceptional class of matter which bridges the gap between living, organized matter and nonliving matter. Purified viruses isolated from plants are nucleoproteins, whereas animal vi-

ruses are more complex because of the presence of lipid and enzyme constituents as well as nucleoproteins. The most widely studied plant viruses are the *tobacco mosaic virus* and the *tomato bushy stunt virus,* which are nucleoproteins (Gortner, 1949). Today many regard viruses as *acellular* microorganisms, or organisms without cells that are unable to function physiologically, but can store genetic data that can be replicated in the living cells of a host. Whatever they are, viruses exhibit properties of living matter and grow and multiply only in the cells of a host organism.

Environmental Importance of Viruses. – Not much is known yet how viruses survive and multiply in soils, but fungi, nematodes, or even plant roots can serve as host carriers. It is believed that viruses can also persist in soils as dormant units while retaining their pathogenic properties, though common plant viruses may survive only 4 to 5 weeks in soils. The ability to survive in soils for a long period of time creates concern for degradation of the environment. No chemical control is available yet, and antiviral agents developed to combat viral infections have shown mixed results. Denaturation, as done by boiling protein compounds, inactivates the virus but usually kills the infected organism as well. However, denaturation with urea, a process used in biochemistry to split the tobacco mosaic virus in pieces, may have some promise. Urea, a common fertilizer, may prove to be harmful to the virus, but is expected not to cause injury to the infested crop.

Viruses and viroids have attracted world-wide attention because of their capability to produce diseases in plants, animals and humans. It appears that this pathogenic attribute can be used to our advantage in the biological control of weeds and insects, and in combating plant, animal, and human diseases.

3.4 BIOCHEMICAL REACTIONS OF SOIL MICROORGANISMS

As indicated above, the microorganisms are present in great numbers in the soil. They play an important role in many physiological processes and are responsible for a number of biochemical reactions vital for the life of growing plants. Perhaps one of the most important effects is *carbon cycling*, which includes the biochemical reactions *decomposition* and *mineralization*. Another effect of equal importance is *nitrogen cycling*, including the biochemical reactions *nitrogen fixation, ammonification, nitrification*, and *denitrification*. Many, if not all, of these effects and reactions are interrelated. For example, the N-cycle is in essence part of nutrient cycling. In turn, nutrient cycling is closely related to mineralization. It is not in the scope of this book to discuss all the cycles present in soils. Only the carbon and nitrogen cycles will be presented in the following sections. For the various cycles of the many other nutrients reference is made to Stevenson (1986).

3.4.1 Carbon Cycle

The carbon cycle is the perpetual movement of organic carbon from the air into the soil and back into the air. It is nature's way of cleaning the environment by recycling organic waste, which is mediated by the microorganisms. Many complex diagrams have been developed to depict this cycle, and the overuse of detailed reactions usually tends to cloud the basics of the cycle. In principle, the cycle starts when CO_2 gas in the atmosphere is absorbed by plants and converted into carbohydrates by a process called *photosynthesis*, which can be represented by the reaction:

$$CO_2 + H_2O \rightarrow \text{carbohydrates} + O_2 \tag{3.1}$$

This reaction will take place only in the presence of chlorophyll and sunlight. These carbohydrates are the sources for formation of other organic compounds in the plant body, e.g., protein and lignin. With leaf-fall or when plants die, the vegetative remains are subject to decomposition and mineralization processes, which return the carbon from the soil to the air as CO_2 gas.

3.4.2 Decomposition and Mineralization

Decomposition and mineralization reactions are generally visualized as microbial enzymatic oxidation and reduction reactions. Enzymes are needed to break the bonds within the structure of the chemical substances in the cell tissue. These bonds are considered *activation energy barriers* by Miller and Gardiner (1998), and soil microorganisms can provide the *deactivation energy* through the enzymes. Each bond requires a specific enzyme and after the bond is broken, the chemical compound is released into the soil, and the enzyme can be used again for the same process.

In the decomposition reaction all organic remains in soils are broken down first into their organic constituents, and finally into CO_2 and H_2O. Some nutrients are also released in inorganic forms. The decomposition process can take place in *aerobic* and *anaerobic* conditions.

Aerobic Decomposition. – Aerobic decomposition takes place in well-drained soils, where plenty of oxygen is available. It is characterized by a gradual breakdown of organic waste. The cell tissue is broken down, and at the first instance the cell constituents are released, such as carbohydrates, amino acids, protein, lipid, nucleic acid and lignin. The decomposition process then continues with the less resistant material to be broken down further, whereas the breakdown of the more resistant material, such as lignin and protein, takes place in several stages. Part of the organic compounds released in soils will be converted into *humic matter* by the *humification* process. Eventually humic matter will also be broken down by this process. The end product of aerobic decomposition is CO_2, H_2O, NO_3^-, and SO_4^{2-} (Stevenson, 1986).

Anaerobic Decomposition. – This is a decomposition process in the absence of oxygen, such as occurs in marshes and swampy environments. In poorly drained soils oxygen may rapidly be depleted when water in the soil pores prevents diffusion of O_2 gas from the air. In this case, it is, therefore, difficult to talk about oxidation reactions. Hydrolysis and reduction are more likely the major processes for decomposition. In anaerobic decomposition, anaerobic degradation, called *putrefication*, of organic waste takes place, and additional organic by-products are formed, e.g., methane and other foul-smelling compounds. The formation of methane can be illustrated by the reaction:

$$C_6H_{12}O_6 \;\rightarrow\; 3CH_4 \;+\; 3CO_2 \qquad\qquad (3.2)$$
$$\text{methane}$$

According to Brady and Weil (1996) the production of methane in wet conditions is carried out by *methanogenic bacteria* and a corrected version of the three reactions involved is presented below:

$4C_2H_5COOH + 2H_2O \rightarrow 4CH_3COOH + CO_2 + 3CH_4$
propionic acid   acetic acid   methane

$$CH_3COOH \rightarrow CO_2 + CH_4$$

$$CO_2 + 4H_2 \rightarrow 2H_2O + CH_4$$

Methane gas is often produced in swampy conditions, and in landfills where it becomes a hazard and an environmental issue. Since the reaction is an anaerobic reaction, efforts in creating a more aerobic environment in landfills can perhaps control the situation. In some cases, the decomposition process can also yield ethyl alcohol or methyl alcohol, as illustrated by the following reaction:

$$C_6H_{12}O_6 \rightarrow 2C_2H_5OH + 2CO_2 \qquad\qquad (3.3)$$
ethanol

These reactions form the basis for the production of alternative sources of fuel. In industry, sugars from sugar cane are used for ethanol production, such as in Brazil. Starches from corn and potato, $(C_6H_{10}O_5)_5$, can also be used as inexpensive sources for production of ethanol by fermentation with yeast. In American industrialized cities of the Midwest organic urban waste is burned in the absence of O_2 to produce commercial methane gas.

Mineralization. – Mineralization is defined as the immediate breakdown of organic waste into CO_2 and H_2O, which is accompanied by the direct release of nutrients in the soil in inorganic (mineral) forms. Although the end product is similar to that of aerobic decomposition, a gradual decay with the subsequent release of *intermediate* organic constituents is absent in mineralization. Therefore, humic matter cannot be formed as a result of a mineralization process, but the final process in decomposition reactions can be considered a mineralization process. The inorganic ions released by mineralization are available for ready uptake by plant roots and microorganisms.

Environmental Significance. – From the above, it can be noticed that the soil is a natural producer of polluting gases, CH_4 and CO_2. Both gases are responsible for creating the *greenhouse* effect on Earth. However, the amount of these gases released by a natural decomposition process is very small compared to that produced by the combustion of fossil fuel in automobiles and industry, and by large-scale burning of the tropical rainforest. Most of the CO_2 produced naturally will be *filtered* from the atmosphere by green plants and used again in photosynthesis for the production of carbohydrates, as conditioned by the carbon cycle.

3.4.3 Nitrogen Cycle

Very simply defined, the *nitrogen cycle* is the movement of nitrogen from the atmosphere through the plants into the soil, before it is returned to the atmosphere in its original gaseous state. It is closely associated with the carbon cycle, but can also be considered part of the nutrient cycle. It is believed that a nitrogen cycle as such does not exist in nature, because the soil has an N_2 cycle of its own, called the *inner cycle*, and because the N atom can be transferred from one to another oxidation state in a random fashion (Stevenson, 1986; Paul and Clark, 1989). The cycle is composed of a sequence of biochemical reactions as illustrated in a very generalized diagram in Figure 3.5. The *outer circle* represents the more overall N_2 cycle, whereas the *inner circle* displays the cycle occurring in soils.

3.4.4 Nitrogen Fixation

Nitrogen fixation is a process by which atmospheric N_2 gas is captured by microorganisms and converted into nitrogen compounds available to plants. The atmosphere contains an abundant supply of nitrogen (70%), but in the form of gas it is nonavailable to higher

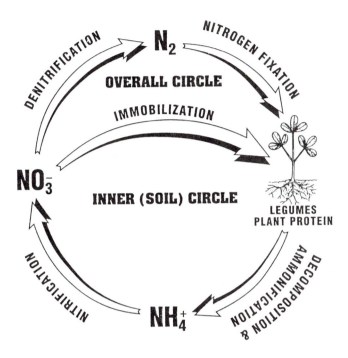

Figure 3.5 Generalized diagram of the nitrogen cycle, showing the overall and inner cycle. Drawing by W.G. Reeves, art coordinator, University of Georgia.

plants. Many people believe that nitrogen fixation is the only process by which soil can be enriched naturally with nitrogen compounds vital for the growth of plants. Other scientists indicate that the plants containing these organisms in fact deplete soil N (LaRue and Patterson, 1981). The N_2-fixation process can be distinguished into a symbiotic and nonsymbiotic nitrogen fixation process.

Symbiotic Nitrogen Fixation. – Symbiotic nitrogen fixation is carried out by the heterotrophic *Rhizobium* bacteria that live in root nodules of legume plants (Nutman, 1965). When a root hair comes into contact with a rhizobium bacterium, curling of the root hair takes place. The cell walls in the root hair tissue are dissolved enzymatically by the bacteria, and an infection thread is formed that invades the root cortex. Once inside, the rhizobium multiplies into *bacteroids*, which are nourished by the host plant with the necessary nutrients for growth. In turn, the host plant receives N compounds produced by the bacteria from N_2 gas. On its own the host plant is unable to obtain the nitrogen for growth. Such an association to the mutual benefit of each other is called *symbiosis*, hence the term symbiotic N fixation. The O_2 requirement of the bacteria is supplied by a substance similar to hemoglobin in blood, called *leghemoglobin*, surrounding the bacteroids in the nodule tissue. The exact reaction for the conversion of N_2 gas into N compounds is not known, and many reaction processes have been proposed, involving complicated electron transfer and conversion of ATP into ADP. Nevertheless, it is believed that the first product is ammonium, NH_4^+, the formation of which can be illustrated by the generalized reaction as follows:

$$N_2 + 8H^+ + 6e^- \rightarrow 2NH_4^+ \tag{3.4}$$

Both poor drainage conditions in soils and the application of nitrogen fertilizers, especially nitrates, are noted to inhibit nodulation and N_2 fixation. Most of the *Rhizobium* is very specific and can inoculate only a specific plant. For example, *R. japonicum, R. trifolii, R. meliloti, R. lupini*, and *R. phaseoli* are strains that are found only in soybean, clover, alfalfa, lupine, and beans, respectively. *Cross inoculation*, (meaning infecting) for example soybeans with *R. trifolii* is very difficult. However, Paul and Clark (1989) reported that the *Rhizobium* in cowpeas can be used to nodulate a variety of legumes without the specificity encountered in the common *Rhizobium* strain of temperate regions.

Recently symbiotic N_2 fixation has also been reported in a number of non-legume plants, e.g., Alnus of the cool temperate regions and Casuarina in the subtropics and tropical regions. The non-legume plants can be nodulated or non-nodulated. The organisms in the nodules of non-legume plants are identified as actinomycetes called *Frankia* (Havelka et al., 1982, Stevenson, 1986). The plants are usually *pioneer* plants growing in very poor soil conditions. They can establish themselves in infertile soils because of their association with the N_2-fixing organisms. As indicated earlier, symbiotic actinomycete association is different from that of rhizobium, because the actinomycete lacks the specificity of rhizobium and can infect several different plant families. The amount of nitrogen fixed by, for example, a red alder tree is substantial and was reported earlier to be approximately 168 kg/ha annually. Another group of bacteria reported to also develop a N-fixing symbiotic relationship with green plants is *Cyanobacteria*. Brady and Weil (1996) indicated that the genus *Nostoc* is capable of fixing 10 to 20 kg N/ha per year. One cyanobacterium strain was noted to produce nodules on the stems of *Gunnera*, an angiosperm growing in marshes in the southern hemisphere.

Nonsymbiotic Nitrogen Fixation. – Nitrogen fixation can also be conducted by microorganisms that live independently in soils. This process, called nonsymbiotic nitrogen fixation, is carried out by two groups of organisms, bacteria and blue-green algae (Jensen, 1965). The bacteria of importance are *Azotobacter* and *Clostridium*, both of which are free-living *organotrophic* organisms. They need, therefore, an organic substrate as a source of energy. The blue-green algae, on the other hand, are *phototrophic* organisms, meaning they can utilize CO_2 from the air and also take up nutrients. The *Azotobacter* lives in aerobic soil conditions, whereas the *Clostridium* is usually found in anaerobic soils. Recently, an *Azotobacter* species has been detected to grow in close association with *Paspalum notatum*, a tropical grass in Brazil, which is called *Azotobacter paspili* (Dobereiner and Day, 1975). This *A. paspali*-grass association is frequently compared with the symbiotic algae-azolla association, which is perhaps somewhat inaccurate. Whereas the algae appear to be an 'integral' part of the *Azolla* tissue, the azotobacter lives in the moist space between the leaf sheath and the stem of the grass, and also between the roots. The similarity is only in the fact that it is also supplying nitrogen compounds to the plant.

Nitrogen Addition by N_2 Fixation. The amounts of N_2 added to soil by biological N_2 fixation are substantial, although they vary from plant to plant (Table 3.3). The data indicate that the symbiotic N_2-fixing organisms are more efficient than the free-living (nonsymbiotic) N_2 fixers in enriching soils with N_2 compounds, especially _Rhizobium_ sp. and _R. meliloti_ in _Leucaena_ and alfalfa, respectively. Because of their nitrogen-fixing capacity, leucaena trees are frequently grown as shade trees in tea, coffee, and cacao plantation. Not only will these trees provide shade controlling sunburn of the crops, but through leaf fall and litter and root exudates, they also build up the nitrogen content of the soil, while at the same time making the soil more favorable for colo-

Table 3.3 Annual Nitrogen Addition by N_2 Fixation

	kg N/ha
Rainfall	1 - 10
Nonsymbiotic	45
Symbiotic:	45 - 300
Legumes:	
Leucaena sp. tree	300
Alfalfa	260
Red clover	170
White clover	130
Kudzu	120
Lupine	75
Peanut	130
Soybean	120
Cowpeas	75
Field peas	54
Nonlegumes:	
Azolla	200
Alder	100
Gunnera	15

Source: Nutman (1965); Jensen (1965); LaRue and
      Peterson (1981); Stevenson (1986); Brady and
      Weil (1996).

nization by other beneficial microbial species. Alfalfa is usually grown in the United States for hay or grazing, but when alfalfa is used as green manure, to be able to obtain the maximum benefit, the plants have to be plowed under before they reach the flowering stage. At this stage, the plants are not woody and are still succulent, so that they will decompose rapidly in soils, releasing in this way all the N_2 compounds that have been produced. Symbiotic nitrogen fixation by nonlegumes, e.g., alder and azolla, appears to be comparable with that of legume plants. Brady and Weil (1996) indicate that *Frankia* is very important for the nitrogen economy of soils undergoing succession, as well as in forested areas and marshes. The benefits of azolla plants are well known in the rice-growing regions in Southeast Asia.

Table 3.3 also lists the annual addition of N_2 from rainfall. The amount is relatively small, and it comes from oxidation of atmospheric N_2 by lightning, photochemical oxidation, and fast-moving meteorites (Hutchinson, 1944). The electrical discharge by lightning, the most significant of the three, oxidizes N_2 gas in the air into NO_2, which dissolves in raindrops to form nitric and nitrous acids. The reactions can be written as follows:

$$N_2 + O_2 \rightarrow 2NO \tag{3.5}$$

$$2NO + O_2 \rightarrow 2NO_2 \tag{3.6}$$

$$2NO_2 + H_2O \rightarrow HNO_3 + HNO_2 \tag{3.7}$$
$$\text{(nitric and nitrous acid)}$$

This phenomenon is a form of *acid rain* that occurs naturally during heavy thunderstorms.

3.4.5 Ammonification

Ammonification is a biological process by which organic nitrogen compounds are converted into ammonia. This is the point where the overall N_2 cycle starts to interface with the inner (soil) N_2 cycle (see

Figure 3.5). Organic residue is broken down by decomposition, and the large molecules of protein, amino acids, amino sugars, and ureases unavailable for plant uptake are recycled into simple inorganic N compounds available to plants. The reaction can be illustrated as follows:

$$R\text{-}NH_2 + H_2O \rightarrow R\text{-}OH + NH_3 + energy \tag{3.8}$$

where $R\text{-}NH_2$ = amino acid. The reaction is an enzymatic reaction, and a wide variety of enzymes are involved, each acting on a specific type of nitrogen compound (Stevenson, 1989; Ladd and Jackson, 1982), e.g., proteinases to break the peptide bonds in proteins, and dehydrogenases and oxidases to break the amino group from the amino acid. The ammonia formed is in gas form, but released in soil water it is rapidly converted into ammonium, NH_4^+, a cation which will be adsorbed by clay and humus.

A number of authors consider ammonification a mineralization process (Stevenson, 1989; Brady and Weil, 1996; Miller and Gardiner, 1998). They believe that ammonia gas is produced by decomposing plant and animal residues and even the foliage of living plants is considered a source of ammonia. Animal manures are given as examples, because of the pungent ammonia odor in barnyards and around poultry houses. The ammonia gas is assumed to be in equilibrium with ammonium ions according to the reaction:

$$NH_{3\,g} + H_2O \rightarrow NH_4^+ + OH^-$$

and the conclusion is drawn that ammonia gas volatilization will be more pronounced as soil pH increases, which is scientifically correct. However, in nature the NH_3 gas formed reacts with soil water to form ammonium hydroxide, which being a weak base ionizes into NH_4^+ and OH^- ions, as indicated by the reaction above. The amount of ammonium hydroxide in solution depends on the partial pressure of ammonia gas, or in plain language "on the amount of NH_3 gas present in the soil." The higher the partial pressure of ammonia gas, the more ammonium will be dissolved in the soil solution. The soils in humid regions are normally characterized by low pH, and basic soil chemistry indicates that acidic condition will tend to preserve the ammonium ions in solution. The OH⁻ ions produced by the ionization reaction are normally neutralized by the soil acidity and the degree of increase in pH is not enough to cause substantial losses of ammonia in humid region soils. Ammonia losses by volatilization will be significant only in arid regions and in alkaline and calcareous soils or when manure and ammonia producing amendments or fertilizers are used.

3.4.6 Nitrification

Nitrification is the conversion of ammonium into nitrate. The process was first discovered in 1856 by J.T. Way, and twenty years later Warington reported that it was a biological reaction. It was Winogradsky who in 1890 finally isolated the bacteria responsible for the reaction (Alexander, 1965; Brady, 1990). Currently, it is known that nitrification occurs in two steps. In the first step, *Nitrosomonas* bacteria are producing nitrite from ammonium, which is immediately followed by the second step, in which the *Nitrobacter* bacteria transform the nitrite into nitrate. The reaction processes have been presented earlier as reactions 2.4 and 2.5, respectively. Nitrification is affected by a number of soil conditions, e.g., aeration, temperature, moisture content, pH, soil fertility, and C/N ratios of organic compounds. Most of these factors should be in optimum conditions for bacterial growth.

Agricultural and Environmental Significance of Nitrification. - The final product, NO_3^-, is the most soluble form of N_2 in soils and has an undesirable effect on the environment. Nitrate is an anion and will not be adsorbed by negatively charged clay minerals, hence will not accumulate in soils. It tends to be leached from soils and may turn up in streams, rivers, lakes, swamps, and groundwater, which is undesirable from an environmental quality standpoint. The concentration limit of nitrate in drinking water in the United States is 45 mg nitrate L^{-1} (50 mg/L in Europe) as set by the EPA. Babies especially are sensitive to high nitrate concentrations. When consumed, the nitrate will be converted into nitrite by certain bacteria in the guts, and prevents the blood from carrying oxygen. Unoxygenated blood causes the skin of infants to turn bluish in color, hence the name *blue baby syndrome* or scientifically *methemo globinemia*, a very rare disease. Consumption of food treated with nitrates, such as bacon, ham and sausage, is suspected to produce carcinogenic compounds, though human fatality from a high nitrate consumption is a very rare occurrence.

In nature, denitrification is an essential process and is part of the nitrogen cycle. If, because of leaching of NO_3^-, denitrification or immobilization of nitrate cannot take place, it may disrupt the N_2 cycle, and nitrogen may be lost from soils. To reduce these losses in agricultural operations, nitrate or nitrification inhibitors have been developed. These are artificial chemicals that inhibit the nitrification process so that most of the N_2 fertilizers stay in the NH_4^+ form. Since ammonium is a cation, it can be adsorbed by negatively charged clay surfaces. The chemicals, such as N-serve (nitrapyrin), DCD (dicyandiamide), and dwell (etridiazol), when mixed with nitrogen fertilizers, inhibit the activity of the *Nitrosomonas* bacteria, but not that of the *Nitrobacter*. However, a new product, ammonium thiosulphate, has been used as a fertilizer and is a nitrate inhibitor. The thiosulphate is reported to act on the *Nitrobacter*, hence preventing the nitrite from being converted into nitrate. Suppressing this second step in nitrification may cause nitrite when formed to accumulate to toxic levels. Therefore, eliminating the first step of the nitrification process is a far more preferable action, because transformation of ammonium into nitrite is prevented. The effect is only temporary, since nitrification may continue when conditions become favorable again. Many assume that nitrification inhibitors are very important in reducing leaching of nitrates in sandy soils, and in decreasing denitrification in poorly drained soils.

Slow-release N_2 fertilizers are also applied to control losses of N in soils because of nitrification. In this case, the fertilizer releases only small amounts of inorganic N at a time, giving time to the plants to absorb it. The fertilizers are usually coated or treated with compounds that inhibit the dissolution or hydrolysis into ammonium. Sulfur-coated urea is an example of a coated nitrogen fertilizer and is usually twice as expensive as plain urea per unit of N. However, this fertilizer finds extensive application in high-value crops and in turf grasses because of the additional effects, such as reduced fertilizer burns and supply of S in soils low in sulfur. In acid soils or soils rich in S, the use of sulfur-coated urea may become a problem. Another example of a coated fertilizer is osmocote, which is a resin-coated nitrogen fertilizer. Urea-formaldehyde is a slow-release fertilizer in which the hydrolysis and subsequent dissolution are decreased by the addition of formaldehyde, a chemical that kills microorganisms responsible for the hydrolysis reactions. Compost and digested sewage sludge can also be considered as slow-release organic fertilizers. A commercial form of these organics is available under the name Milorganite.

3.4.7 Denitrification

Denitrification is the conversion of nitrates into N_2O and N_2 gas, and represents the last step in the overall N_2 cycle. By this process the nitrogen is returned to the atmosphere in the form of gas. The process, sometimes called *enzymatic denitrification*, is a reduction process mediated by anaerobic bacteria, e.g., *Pseudobacter, Achromobacter, Bacillus*, and *Alcaligenes* (Broadbent and Clark, 1965; Paul and Clark, 1989; Brady, 1990). The sequence of reduction can perhaps be presented as follows (Stevenson, 1986):

$$NO_3^- \rightarrow NO_2^- \rightarrow NO \rightarrow N_2O \rightarrow N_2$$

5+   3+   2+   1-   0   ← (Valence of nitrogen)

The amount of NO, N_2O, and N_2 gas produced depends on many factors, e.g., pH, O_2, nitrite and nitrate content. Nitric oxide (NO) is produced in small amounts under acid conditions, whereas under strongly and very strongly acid conditions large amounts of nitrous oxide (N_2O) are formed. High concentrations of nitrite and nitrate are also reported to induce formation of high amounts of N_2O.

Environmental Implications of Denitrification. - The release of nitrous oxide and N_2O gas into the air has caused some concern among environmentalists. It is one of the gases that causes the development of acid rain. In addition, nitrous oxide is reported to contribute in the reactions destroying the ozone, O_3, layer that protects the earth from ultraviolet radiation from the sun. However, as with the issue of CO_2 production, it is the accelerated production of N_2O gas by industry that is more significant than the natural process.

3.4.8 Chemodenitrification

Denitrification can also occur without the help of microbial enzymes, and this nonenzymatic process is called *chemical denitrification* or *chemodenitrification*. This reaction takes place in aerobic conditions, and is sometimes also referred to as *aerobic denitrification*. The final products are NO and NO_2, and the NO gas may in turn be oxidized into NO_2. The latter can dissolve in soil water to form nitric and nitrous acid. These processes can be illustrated by the reactions

$$2NO + O_2 \;\rightarrow\; 2NO_2 \tag{3.9}$$

$$2NO_2 + H_2O \;\rightarrow\; HNO_2 + HNO_3 \tag{3.10}$$

The reactions are similar to those responsible for acid rain (see equations 2.14 and 2.15). It only differs from acid rain in that chemodenitrification causes soil acidity to increase and does not result in the development of acid rain.

Environmental Significance of Chemodenitrification. - The opinion exists that this kind of denitrification is not of too much importance in nature. It will not result in a substantial return of soil N_2 to the atmosphere, hence is insignificant in the last step of the N_2 cycle. The nitrogen remains in the soil because the gases formed are rapidly dissolved in soil water to form the acids as shown above. However, it is reported to be a harmful process in agriculture, because by this process, application of N_2 fertilizers, such as urea, can result in considerable losses of N_2. Consequently the fertilizer loses its efficiency in promoting plant growth. The reaction below illustrates the decomposition of the fertilizer by chemodenitrification.

$$2HNO_2 + CO(NH_2)_2 \rightarrow CO_2 + 3H_2O + 2N_2 \tag{3.11}$$

The large amounts of CO_2 and N_2 gases produced in the decomposition of urea are the reasons for the losses of N_2 from the fertilizer. Soils with low pH values are essential for this reaction (Allison and Doetsch, 1951).

3.4.9 Assimilatory Denitrification

Another type of reduction of NO_3^- is the so-called *assimilatory reduction*, by which NO_3^- is transformed into NH_4^+. The reaction takes place under anaerobic conditions within the plant body for the biosynthesis of amino acids. In contrast to the denitrification process essential to the N_2 cycle, in assimilatory reduction, nitrogen is not lost in the air, but is accumulated in the plant body in the form of amino acids or proteins. The process of incorporating the NH_4^+ as an NH_2 group in the plant tissue is called *assimilation*. Assimilatory reduction is closely related to the immobilization of NO_3^-, which is the topic of the next section.

3.4.10 Immobilization

Immobilization is the uptake of NO_3^- by plants, and its conversion from the inorganic into organic form. Nitrate is a source of food and energy for higher and lower plants and is consumed rapidly by these organisms. Because of the immobilization process some of the NO_3^- in soils will not be denitrified. Immobilization is the point where the overall N_2 cycle is bypassed and converges to become the so-called *inner N_2 cycle* or *internal N_2 cycle*. Because of this process, N_2 is not returned to the atmosphere but remains in the soil, as indicated earlier. This immobilization process is not similar to the process of ammonium fixation by clays as indicated by several authors (Miller and Gardiner, 1998), nor is it the reverse of the process of mineral-ization. It is in essence the uptake of NO_3^- as food by plants and microorganisms, which in their body is transformed into organic nitrogen compounds by metabolic reactions. As such, these reactions are quite different from entrapment reactions by expanding clays. Conversion of nitrate into amino acids as outlined by Brady and Weil (1996) is also an almost impossible reaction without the help of metabolism in the body of the various organisms. Immobilization does not occur in nature to increase or decrease the nitrogen content in soils, but has as its sole purpose the cycling of nitrogen as part of the nitrogen cycle.

3.5   NONHUMIFIED ORGANIC MATTER

The nonhumified organic matter is composed of compounds released during decomposition in the original or slightly modified form. Although numerous organic compounds are present in plant tissue, only a few exist in soils in detectable amounts after their release in soils. They are primarily (1) carbohydrates, (2) amino acids and proteins, (3) lipids, (4) nucleic acids, (5) lignin, and (6) organic acids.

3.5.1 Carbohydrates

Carbohydrates are important constituents of the plant body. Almost 75% of the plant's dry weight is composed of carbohydrates. Of the three major groups of food substances (carbohydrates, fats, and proteins) produced by plants, carbohydrates are formed first directly from CO_2 and water with the help of chlorophyll and sunlight, by a process called *photosynthesis*. They are, therefore, considered the immediate link between the energy of the sun and the energy displayed by living plants and other organisms on earth. Carbohydrates provide the source for the formation of fats, oils, amino acids, proteins, and other organic compounds needed in plant growth.

Carbohydrates are also the ultimate source for formation of cellulose, chitin, pectin, and lignin, important structural components of the plant body. Ascorbic acid (vitamin C) and inositol are related carbohydrate compounds.

The carbohydrates range from simple sugars to polysaccharides or complex carbohydrates, and their properties consequently change with increasing molecular complexity. The sugars are usually crystalline compounds and sweet in taste. They form true solutions in water and are soluble in ethanol. The sugars are further subdivided into (1) monosaccharides or simple sugars, and (2) oligosaccharides, or compound sugars. Polysaccharides are usually amorphous and tasteless, and disperse in water to form colloidal suspensions. They are very complex in structure and have high molecular weights. Polysaccharides are sometimes distinguished into (1) *homopolysaccharides* and (2) *heteropolysaccharides*. Homopolysaccharides are composed of a repeating monosaccharide, whereas heteropolysaccharides are made up of two (or more) different monosaccharides. For a more detailed classification of the carbohydrates, their structure and properties, reference is made to Tan (1993, 1998).

Environmental Importance of Carbohydrates. - Carbohydrates are the principal sources of food and energy for all organisms on earth, including humans, animals and especially soil microorganisms. Much has been said about the vital importance of complex carbohydrates in human nutrition and human health. Used in industry as raw materials for the production of pulp, paper, rayon, and starch, carbohydrates are also renewable resources of primary importance. As such, they are invaluable materials for the manufacture of alternative clean fuel and many other chemical compounds, such as butanol, ethanol, methyl alcohol, acetic acid, citric acid, and lactic acid.

As indicated above, carbohydrates serve as major sources of energy for many biological functions. Sugars are the first targets of microbial attack and are quickly subjected to aerobic and anaerobic decomposition. *Aerobic decomposition* is the complete oxidation of sugars into carbon dioxide and water. Anaerobic decomposition is an incomplete breakdown of carbohydrates in the absence of oxygen. In pure biochemistry, the *anaerobic decomposition* of glucose is called *glycolysis* or *fermentation* of glucose. It is a nonoxidative breakdown and liberates only a small part of the energy of the glucose molecule. The ultimate endproducts of glycolysis are methane, CH_4, and carbon dioxide, CO_2. The reaction processes can be illustrated as follows:

$$C_6H_{12}O_6 \rightarrow 3CH_4 + 3CO_2 \qquad\qquad (3.12)$$
glucose   methane

However, microorganisms vary greatly in their capacity for breaking down carbohydrates. In some fermentation processes, the end products can be alcohols, ketones, and various organic acids mixed with carbon dioxide and/or hydrogen. The type of anaerobic decomposition of glucose leading to formation of ethanol and CO_2 is called *alcoholic fermentation*, and the reaction process can be illustrated as follows:

$$C_6H_{12}O_6 \rightarrow 2C_2H_5OH + 2CO_2 \qquad\qquad (3.13)$$
glucose   ethanol

The formation of both ethyl alcohol and methyl alcohol or both, as depicted by reaction (3.13), is the basis for the production of alternative fuel sources. In practice, sugars from sugar cane or starches from corn or potato, $(C_6H_{10}O_5)_5$, fermented with yeast are the least expensive sources. Most of the 1 billion gallons of ethanol produced annually in the United States is reported to come from corn, a very important crop in the midwestern and eastern states of the United States grown for human food and animal feed. Critics have indicated now that if most of the corn is taken away from food production for the production of ethanol, it may create havoc in the US economy by raising the prices of food and animal feed. They predict that the use of corn for the production of ethanol exceeding 3 billion gallons per year may increase considerably the cost of food and animal feed. Therefore, the search for other and cheaper sources of raw material for the production of ethanol has now been intensified. Currently researchers are pursuing ways to harvest fuel from crops of US farmlands that in the foreseeable future can power cars and airplanes and furnish energy to power stations cheaply. Carbohydrates from grasses and straw are viable alternatives, and rapidly growing bushes and trees can also be planted for the needed cellulose, as is done now for the paper, pulp and rayon industry. The technology to squeeze out ethanol from hay, rice straw, wheat straw, and wood pulp would lower the cost of *biofuel* even more. The possibilities are in fact unlimited, since all kinds of farm refuse, e.g., chaff, corncobs, stalk, and even brewery grain, are valuable sources of cellulose or carbohydrates that can be fermented to yield ethanol. As for the production of methane, many industrialized or metropolitan cities in the Midwest have taken advantage of the huge amounts of city and urban organic waste which are processed by burning in the absence of oxygen. Such a process is in fact not new, but has been applied for decades in the production of cokes and gas from coal.

The production of methane and ethanol is always accompanied by the release of CO_2. This may create some concern about decreasing the quality of air in our atmosphere, since CO_2 is a *greenhouse gas*. However, in the presence of abundant plant growth, a natural production of CO_2 will be "filtered" from the atmosphere through absorption by green plants that use it in photosynthesis for the production of carbohydrates. Therefore, the chances of this CO_2 contributing to global warming due to the greenhouse effect are very small.

In contrast to simple sugars, polysaccharides, and especially soil polysaccharides, exhibit greater resistance to enzymatic degradation due to their size, more complex molecular structure, and their capacity to form complexes with soil inorganic constituents. The intimate association with soil clays particularly does not only slow down chemical decomposition, but also changes the electrochemical properties of clay surfaces important for adsorption of water. The polysaccharides compete with soil water for adsorption sites, and by expelling water reduce wetting and swelling of soil clays. They increase cementation of clay particles, enhancing the development of soil aggregates. The soil structure formed is considered more stable than that produced by fungal mycelium.

3.5.2 Amino Acids and Protein

Amino acids are the building constituents of protein. These compounds contain N in the form of NH_2, called an amino group, attached to the C chain in their molecular structure. The acid part is composed of a terminal C in COOH form, called a carboxyl group. These are the reasons for the name *amino acids*. Twenty-one amino acids are found as protein constituents. Both amino acids and proteins are major sources of nitrogen compounds in soils. They are more difficult to break down than carbohydrates because of the size and complexity of their molecular structure. They are amphoteric in nature, and consequently react with acids and bases. At the isoelectric point, amino acids behave as *zwitterions*, in other words, behave as cations and anions (Tan, 1993). In acid soils (soil pH < isoelectric point), the amino acids are positively charged and behave as cations, whereas in basic soils (soil pH > isoelectric point) they are negatively charged and behave as anions. Proteins are complex combinations of amino acids, and are formed by the linkage of amino acid molecules through the amino and carboxyl groups. The bond linking the two groups is called a *peptide bond,* and the compound formed is called a *peptide*, or protein. By refluxing with 6 *N* HCL for 18 - 24 hours, the protein may be hydrolyzed into its constituent amino acids. Since the N content of most protein is about 16% and since this element is easily determined by the Kjeldahl method, the protein content can be estimated by determination of the N content and by multiplying %N found by 6.25 (=100/16).

Decomposition of Amino Acids and Protein. - Amino acids and proteins are important sources of organic N in soils, and this N can become available again to plants upon decomposition of the amino acids and proteins. The main reaction process for the decomposition of these compounds is *hydrolysis*. Hydrolysis of proteins, brought about by proteinases and peptidases of soil microorganisms, results in cleavage of peptide bonds, releasing in this way the amino acids. The latter compounds are broken down further into NH_3 by the enzymes called amino acid dehydrogenases and oxidases. Schematically the main pathway of decomposition can be represented as follows:

Proteins $\rightarrow$ peptides $\rightarrow$ amino acids $\rightarrow$ NH_3

The decomposition reaction of proteins as described above is frequently called *deamination* or *putrefaction* (Gortner, 1949; Stevenson, 1986). Deamination reactions can take place in aerobic or anaerobic conditions, and are, therefore, also called *oxidative*, and *nonoxidative* deamination, respectively. The reaction for oxidative deamination can be written as follows:

$$R\text{-}CH(NH_2)COOH + O_2 \rightarrow RCOOH + CO_2 + NH_3 \qquad (3.14)$$
amino acid

Anaerobic deamination may result in (1) deamination or reduction, and (2) decarboxylation, as can be seen from the reactions below:

1. deamination or *reduction*:

$$R\text{-}CH(NH_2)COOH + H_2 \rightarrow RCH_2COOH + NH_3 \qquad (3.15)$$
amino acid

2. decarboxylation:

$$R\text{-}CH(NH_2)COOH \rightarrow R\text{-}CH_2NH_2 + CO_2 \qquad (3.16)$$
amine

Reaction (3.15) indicates that deamination is characterized by the destruction of the amino group and its transformation into NH_3. In contrast, decarboxylation (reaction 3.16) is distinguished by the decomposition of the COOH group into CO_2, and the subsequent transformation of the amino acid into an amine compound. The enzyme required for decarboxylation, called *amino acid decarboxylase*, is produced by *Clostridium* bacteria. When formed in animal bodies, some of the amines produced are reported to have important physiological effects. For example, *histidine decarboxylase* in animal tissue can produce *histamine*, an amine substance that can stimulate allergic effects and/or gastric secretion. Another enzyme, *tyrosine decarboxylase*, is an intermediate in the formation of *adrenaline,* an amine

functioning as a *vasoconstrictor*. It is usually released in the blood-
stream when a person or animal is startled or frightened (Conn and
Stumpf, 1967).

3.5.3 Lipids

Lipids are heterogenous compounds of fatty acids, wax, and oils.
The basic component of lipids is glycerol, $C_3H_8O_3$, or other alcohols.
They are classified into three groups: (1) simple lipids, (2) compound
lipids, and (3) derived lipids. The simple lipids include the natural
fats and oils, which can be divided into (a) nonvolatile fats and oils,
and (b) volatile oils, such as turpentine and clove oil. The compound
lipids are more fat-like in nature. They have been formed from fatty
acids in combination with other organic compounds. For example,
phospholipids are lipids in combination with organic phosphorus
compounds, such as nucleic acids, phosphoproteins, and phytin.
Glycolipid is a compound lipid, because the lipid is present in
combination with a carbohydrate (galactose). The derived lipids are
composed of fatty acids, alcohols and sterols. The fatty acids can be
saturated fatty acids, such as palmitic acids, or unsaturated fatty
acids, such as oleic acids. Palm oil and coconut oil are rich in palmitic
acid. Cholesterol is an example of a sterol, which upon UV irradiation
will form vitamin D.

Hydrolysis of fats by saponification with alkalies yields glycerol
and the salts of fatty acids. The metallic salts of the higher fatty acids
are known as soap.

Environmental Importance of Lipids. - Some of the soil lipids find
their origin from higher plants, but many have also been derived from
microbial tissue. Bacterial cells contain 5 to 10% lipids, whereas fungi
may contain 10 to 25% (Stevenson, 1994). These microbial products
are the main sources of the glycerides and phosphatides in soil lipids.
Waxes, covering leaves and fruits of higher plants, are also important
sources of soil lipids, whereas terpenoids are contributed by conifer
plants. Chlorophyll from leaves of green plants is an additional
source. Resistance to decomposition may vary considerably, though

many of the lipids, such as fatty acids, will decompose rapidly in well-drained soils. Others are relatively more resistant, like the waxes, terpenoids, and sterols. However, it is generally assumed that in most soils sufficient amounts of microorganisms are present for a complete decomposition of even the strongest lipid. Soil microorganisms are even available that can attack oil spills, which are lipid-like compounds.

Lipids are known to affect physical properties of soils, though not much information is available on this topic. These compounds are hydrophobic, and therefore, tend to reduce the degree of wetting of soils. A high content of wax in soil humus is expected to make the soil *water repellant*, but no data are available to confirm this scientifically. Relatively more is known on the chemical properties of soil lipids. Many of the high molecular weight organic acids in the lipid group contain hydroxyl and carboxyl groups, e.g., palmitic acid and stearic acid. Organic acids of these kinds can also be present in small quantities in the soil rhizosphere. Depending upon soil pH, the functional groups (OH and COOH) can dissociate their H^+ ions, hence will be negatively charged. Because of these electrical charges, they contribute toward cation exchange reactions, complex formation, and/or chelation reaction with metal cations. The latter can be illustrated by the following reaction:

$2(CH_3(CH_2)_{15}CHOHCOOH) + Fe^{2+} \rightarrow$
(α - hydroxystearic acid)

$$CH_3(CH_2)_{15}CHOHCOO \text{ -Fe -OOCHOHC}(CH_2)_{15}CH_3 + 2H^+$$

These lipoidic acids are, therefore, important agents in the weathering of rocks and minerals, in the dissolution of plant nutrients, and in the mobilization and transportation of elements, critical in plant nutrition and soil formation (Tan, 1998).

3.5.4 Nucleic Acids

All plant and animal cells contain discrete rounded or spherical bodies, called the *nucleus*, which contain nucleic acids. Nucleic acids, first isolated in 1869 by F. Miescher (Tan, 1993), are polymers with high molecular weights. Their repeating unit is a sugar attached to a mononucleotide, rather than an amino acid. Two types of nucleic acids are distinguished: (1) DNA, deoxyribonucleic acid, and (2) RNA, ribonucleic acid.

Currently it is known that nucleic acids can also be formed in other plant organelles: mitochondria and chloroplasts. However, the amount produced in the mitochondrion and chloroplast is much smaller than that produced in the nucleus. They are also smaller in size, and their genetic capabilities are less than the nucleic acids from the nucleus. When released in soils by decomposition of plant residue, nucleic acids are potential sources of P for plant growth.

Environmental Significance of Nucleic Acids. - Not much is known about reactions of nucleic acids in soils. This is one of the organic compounds from living cells that is apparently decomposed rapidly in soils. However, because nucleic acids contain N and P, they are expected to be important sources of N and P. Humus and especially humic substances contain considerable amounts of N that cannot be accounted for in analysis. This unaccountable N content, called *HUN* for hydrolyzable unknown nitrogen or just plain *unknown N*, is believed to be derived from nucleic acids or their derivatives (Schnitzer and Hindle, 1980; Stevenson, 1994).

3.5.5 Lignin

Lignins are highly aromatic polymers and are produced by living plants from carbohydrates by a process called *lignification* (Tan, 1993, 1998). As indicated earlier, the ultimate source for formation of lignin is carbohydrates or intermediate products of photosynthesis related to carbohydrates. In the growth of plants, carbohydrates are synthesized first. The formation of lignin then begins. The process of conversion of nonaromatic carbohydrates into compounds containing phenolic (aromatic) groups, characteristic of lignin, is called *aromatization*. It is believed that dehydration of fructose contributes to the aromatization process. The end products of the aromatization process are pyrogallol, hydroxyhydroquinone, phloroglucinol, or a combination thereof. The spaces existing between the cellulose fibers in the plant body are then gradually filled with these lignified carbohydrates. This process is called *lignification* and serves a number of functions. Lignin cements and anchors polysaccharide fibers in the plant tissue, and gives resistance to the fibers against physical and chemical breakdown. The quantity of lignin increases with plant age, woody tissue, and stem content. A large amount of lignin is also present in the vascular bundles of plant tissue. The purpose is to strengthen and make the xylem vessels more water resistant. By virtue of the presence of larger amounts of vascular bundles, the lignin content of tropical grasses is considerably larger than that of temperate region plants (Tan, 1998). Consequently soils under tropical grasses are expected to have a higher lignin content than soils under temperate region grasses. These differences may have some influence on the nature of humus formed.

Plant lignin can be divided into three types of basic monomers: (1) coniferyl alcohol, derived mostly from softwood or coniferous plants, (2) sinapyl alcohol, derived from hardwood, and (3) coumaryl alcohol, derived from grasses and bamboo. For these reasons, lignin can perhaps be distinguished into softwood lignin, hardwood lignin and grass lignin. These basic monomers form large, complex polymers, and it is common to find a haphazard structure of lignin in many organic chemistry books. A hypothesis of a more systematic arrangement of the basic monomers into lignin is presented by Tan (1993, 1998).

Environmental Significance of Lignin. - Lignin, released in soils by decomposition of plant tissue, is highly resistant to microbial decomposition. Decay of plant material, especially of woody plants, results in an apparent increase in lignin content in soils owing to a preferential decomposition of carbohydrates. Certain fungi, e.g., *Tramitis pini*, have been reported to attack lignin. These organisms are called *lignolitic fungi*, and include white-rot, brown-rot, and soft-rot fungi (Paul and Clark, 1989). In well-aerated soils, the *white-rot fungi* are reported to decompose lignin into CO_2 and H_2O. The *brown-rot fungi* are supposed to be useful for the removal of the methoxyl, $-OCH_3$, groups from lignin, leaving the hydroxyphenols behind, which upon oxidation in the air produce brown and black colors. The *soft-rot fungi* are most active in wet soils and are specifically adapted to decomposing hardwood lignin. The hydroxyphenol units resulting from demethylation of lignin by white-rot fungi can be oxidized to form quinones, which react with amino acids to form humic substances (Flaig , 1975). Lignin itself has the capacity to react with NH_3. This process, called *ammonia fixation*, has been applied in the fertilizer industry to produce nitrogen fertilizers by treatment of lignin, and other organic materials rich in lignin, e.g., peat, sawdust, with NH_3 gas. The exact mechanism of fixation is still not known, but it is believed that the NH_3 reacts with the phenolic functional groups in lignin.

The high resistance of lignin to microbial decomposition is one reason why lignin can accumulate in soils. Depending upon the conditions, accumulation of lignin could result in the formation of peat, which in time can be converted into lignite, leonardite, and coal. Under the influence of high pressure and temperature for geologic time periods, these deposits can ultimately become fossil fuel (oil). In soil science, such a conversion process is called *metamorphism*. In soils, lignin is apparently preserved by incorporation through chelation in the structure of soil humic matter.

3.5.6 Organic Acids

The organic acids may range from simple aliphatic acids to complex aromatic and heterocyclic acids, e.g., formic acid, acetic acid, oxalic acid, amino acid, benzoic acid, and tannic acid. For a partial list of organic acids in soils, reference is made to Tan (1986). Many of them are present in such low concentrations that they can be detected only by thin layer or gas chromatography, and undoubtedly other organic acids may be present in the soils awaiting identification and isolation. The amount of formic acid is usually reported in the range of 0.5 to 0.9 mmol/100g soil, whereas that of acetic acid is between 0.7 and 1.0 mmol/100g soil (Tan, 1986). A large number of these organic acids are intermediate products of plant and microbial metabolism. Some may have been released into the soil as root exudates, whereas others are the result of oxidative degradation of organic matter.

Environmental Significance of Organic Acids. - Several high molecular weight as well as low molecular weight organic acids have been detected or isolated from soils. Some may occur in small amounts in the soil rhizosphere. Low molecular weight organic acids, such as formic acid and oxalic acid, have been found in water extracts of the O and A horizons of entisols, Bt horizons of ultisols, and spodosols Bh horizons. In aerobic environments these acids are believed to be transitory in existence at soil moisture and temperature conditions favorable for microbial activity. Anaerobic conditions are assumed to be more suitable for their accumulation in soils (Stevenson, 1994).

The organic acids may have both favorable and unfavorable effects on plant growth. Under anaerobic conditions carboxylic acid, phenolic acid, and butyric and acetic acids can accumulate to such concentrations that they restrict root growth (Wild, 1993). Accumulation of especially butyric acid in rice fields has been suspected in Japan of causing a disease called *Aki-och* in rice plants (Stevenson, 1994).

Organic acids play a significant role in the dissolution and mobilization of elements from rocks and minerals. Acetic acid and oxalic acid have been used in the study of mineral dissolution. They exhibit some metal-complexing capacity.

3.6 HUMIFIED ORGANIC MATTER

Humified organic matter, or humic matter, is a group of compounds that includes humic acids, fulvic acids, hymatomelanic acids, and humins. This humified soil organic fraction is also known as *humus acids* in the German and Russian literature. Today humic compounds are defined as amorphous, colloidal polydispersed substances of yellow to brown-black color and high molecular weights.

These humic compounds have been formed in soils by a process called *humification*. According to the lignoprotein theory, they are lignoprotein compounds produced by interpolymerization of phenolic compounds, peptides, amino acids and carbohydrates (Flaig, 1975; Stevenson, 1982; Schnitzer and Khan, 1972; Tan, 1993). Lignin and lignin degradation products are the sources of the phenols. The polymers in humic compounds are relatively stable, being relatively resistant to microbial enzymatic attack, although they contain amino acids and amino sugar polymers in their structure.

The distinction into the different types of humic compounds is based on their solubility in alkali and acids or ethyl alcohol. *Fulvic acid* (FA) is soluble in alkali and acid, and is the low molecular weight fraction of humic matter. It is also soluble in water. *Humic acid* (HA), on the other hand, is soluble in alkali, but insoluble in acid and water. It is the high molecular weight fraction of humic matter. *Hymatomelanic acid* is the alcohol soluble part of HA, whereas *humin* is defined as the humic fraction insoluble in water, alkali and acid. However, it can be extracted by hot alkali solutions. As indicated above, in the German and Russian literature, they are collectively referred to as *humus acids* (Orlov, 1985; Tan, 1993), which is equivalent to the names humic matter or humic substances. Within the group, FA and HA are perhaps the most important fractions of humic matter. They are present in soils in relatively large amounts, although their contents may vary considerably from soil to soil (Tan, 1986). The terms humic and fulvic acids have been used for a long time; nevertheless a number of scientists still question the correctness of the usage of these terms. They believe that the terms humic and fulvic acids are valid only as *operational terms* that relate to the methods of extraction of these compounds (Aiken et al., 1985). Many other scientists, on the other hand, believe that the names refer to identifiable humic compounds that are complex in structure (Flaig, 1975; Kononova, 1975; Orlov, 1985; Stevenson, 1986), and numerous intermediate forms may be present between the major humic fractions. Recent evidence, presented by Lobartini et al. (1997), also indicates the presence of a composition in humic acids more homogeneous than expected by the critics. Data of elemental composition, infrared spectra, and electron microscopy show that all the fractions of humic acids, separated by different molecular weights, contain essentially similar components.

3.6.1 Types of Humic Matter

Currently it is known that humic matter exists not only in soils, but also in streams, rivers, lakes, oceans, and their sediments. It can also be found as geologic deposits, e.g., lignite or leonardite, coal, and oil shale. These deposits are the sources for the production of commercial humates, which are used as soil amendments (Lobartini et al., 1992). On the basis of present knowledge, humic matter can be classified into three broad groups:

1. *Terrestrial* humic matter: humic matter in soils
2. *Aquatic* humic matter: humic matter in streams, lakes, oceans, and their sediments
3. *Geologic* humic matter: humic matter in lignite or leonardite, coal, and other geologic deposits.

Terrestrial humic matter can be subdivided again into three categories: softwood, hardwood, and grass humic matter (Tan, 1993). It is composed of humic and fulvic acids, although their contents may differ considerably from soil to soil. Aquatic humic matter is composed mainly of fulvic acid, and can be subdivided into *allochthonous* and *autochthonous* humic matter. The allochthonous humic matter is believed to originate from soil humic matter, but after leaching into streams it has been subjected to further distinct changes (Jackson, 1975). The autochthonous humic matter originates from indigenous aquatic plant material, which has a different composition than terrestrial plants. Especially in the oceans, where plankton, kelp, and other seaweed are major forms of organisms, the nature of humic matter formed is quite different from soil humic matter. Aquatic humic matter consists of carbohydrate-protein complexes as opposed to terrestrial or soil humic matter, which is mainly a ligno-protein complex.

3.6.2 Chemical Composition

In general, humic acid is higher in C and N, but lower in O and S content than fulvic acid (Table 3.4). Comparison between terrestrial, aquatic, and geologic humic acids indicates that geologic humic acid tends to be higher in C, but lower in H, and especially N content.

The total acidity also shows some variations because of origin (Table 3.5). These values are attributed to the sum of the carboxyl and phenolic-OH group content, and indicate the cation exchange and complexing capacities of humic matter. A high total acidity value is indicative of a high CEC and complexing power. The data show that the lowest total acidity is exhibited by geologic humic acid, and the highest by geologic fulvic acid. The data in the literature often show that fulvic acids exhibit higher total acidity values than humic acids.

Table 3.4 Elemental Composition of Humic Matter

Origin	C	H	N	O+S
		---------%---------		
Humic Acid				
Terrestrial	50.6	5.6	4.3	38.5
Aquatic	51.7	4.8	2.3	41.1
Geologic	57.1	3.8	1.6	37.5
Fulvic Acid				
Terrestrial	48.1	4.5	1.7	45.7
Aquatic	47.4	5.0	1.5	46.1
Geologic	45.0	4.2	1.4	49.4

Table 3.5 Aliphatic C, Aromatic C, COOH C Contents, Total Acidity, COOH and Phenolic-OH Group Contents

Origin	Ali-phatic C	Arom-atic C	COOH C	Total acidity	Phenolic COOH	OH
	-----------%-----------			----------mol/kg----------		
Humic Acid						
Terrestrial	48.7	36.4	14.9	7.0	4.3	2.7
Aquatic	50.3	37.2	12.5	7.0	4.4	2.6
Geologic	45.1	43.0	11.9	5.8	3.8	2.0
Fulvic Acid						
Terrestrial	61.0	25.3	13.7	6.8	4.7	2.1
Aquatic	62.6	21.7	15.7	7.0	4.4	2.6
Geologic	-	-	-	9.4	8.2	0.8

Geologic humic acid is also more aromatic, whereas aquatic humic acid tends to be more aliphatic in nature than terrestrial humic acids. In the literature, the tendency can be noticed for fulvic acids to contain more aliphatic compounds than humic acids.

3.6.3 Chemical Reactions

The chemical reactivity of humic matter can be predicted from the total acidity value, which is defined as the sum of carboxyl and phenolic group content. The dissociation of protons from the carboxyl groups, which starts at pH 3.0, creates a negative charge. This negative charge increases in value, and becomes larger when at pH 9.0 the phenolic-OH group also dissociates its proton. The presence of such pH dependent or variable charge enables the humic molecule to take part in many chemical reactions, e.g., adsorption of cations and water, complex formation or chelation of metals, and interaction with clays (Tan, 1993). These reactions will be discussed in more detail in Chapter 6.

3.6.4 Spectral Characteristics

Infrared spectroscopy produces different spectra for terrestrial humic and fulvic acids, though some doubt exists on the usefulness of the spectra. The doubt is caused by the publication of poorly resolved spectra arising from the use of improper samples and improper techniques. The infrared spectra as a whole can be used as a fingerprint to distinguish humic from fulvic acid (Figure 3.6). The infrared spectrum of fulvic acid is characterized by a strong absorption band at 3400 cm^{-1}, a weak band between 2980 and 2920 cm^{-1}, a shoulder at 1720 cm^{-1}, followed by a strong band at 1650 cm^{-1} for vibrations of chemical groups, or in plain language, for the presence of OH, aliphatic C-H, carbonyl (C=O), and carboxyl groups in COO$^-$ form.The strong band at 1000 cm^{-1} is a very distinctive infrared feature of fulvic acid and is not caused by contamination with Si gel, but is attributed to the presence of functional groups only present in fulvic acid. The infrared features as described for fulvic acid resemble closely those of the polysaccharides. Chemical analyses confirm the presence of polysaccharide constituents in fulvic acid (Tan and Clark, 1968). In contrast to fulvic acid, humic acid is characterized by a spectrum with stronger absorption at 2980-2920, 1720, and 1650 cm^{-1}. These stronger absorption bands reflect the differences in nature and amounts of structural components with respect to the C-H, carbonyl, and carboxyl groups in the organic molecule. The striking difference is that the spectrum of humic acid lacks the absorption bands at 1000 cm^{-1}. Hymatomelanic acid spectra differ from those of fulvic and humic acids by possessing very strong bands for C-H and carbonyl vibrations. As indicated above, these stronger absorption intensities reflect the nature and concentrations of structural components in the hymatomelanic acid molecule. Clark and Tan (1969) have determined the presence of ester linkages in hymatolanic acid, which account for the strong bands at 2980-2920 cm^{-1} and at 1720 cm^{-1}. The aliphatic C-H

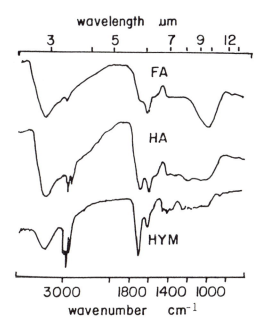

Figure 3.6 Characteristic infrared absorption spectra of fulvic acid (FA), humic acid (HA), and hymatomelanic acid (HYM), respectively. Absorption bands at approximately 2900 and between 1700 and 1600 cm^{-1} are attributed to aliphatic CH and carboxyl group vibrations, respectively.

compounds are linked to the carboxyl group in COO$^-$ form, blocking in this way absorption of infrared at 1650 cm^{-1}.

Currently NMR (nuclear magnetic resonance) spectroscopy is considered an essential tool in the study of humic matter, though NMR instruments are very expensive and the analyses are time-consuming. The NMR spectra of humic matter can be divided into several regions:

1. Aliphatic C region at 0 - 105 ppm chemical shift, with a polysaccharide subregion at 65 - 105 ppm.
2. Aromatic C region at 105 - 165 ppm chemical shift.

3. Carboxyl C region at 165 - 185 ppm chemical shift.

The results of NMR analysis indicate that terrestrial, aquatic, and geologic humic matter are generally composed of aliphatic, aromatic, and carboxylic compounds (Tan, 1993). The NMR spectra demonstrate that soil humic acid is more aromatic in nature than soil fulvic acid. The latter is generally aliphatic in nature, and its composition resembles closely that of aquatic fulvic acid in black water. This suggests that fulvic acids in black water, streams and swamps of the southeastern United States are allochthonous (Tan, 1998).

3.6.5 Effect of Humic Matter on Soils and the Environment

Humic matter is an essential soil component and is known to have a beneficial effect on soil physical, chemical, and biological properties. It may affect plant growth directly and indirectly. Directly, humic acid is capable of improving seed germination, root initiation and elongation, respiration, uptake of plant nutrients, and development of green mass. Indirectly, humic acid can modify the properties of the medium in which plants grow. Humic acid is reported to enhance soil structure formation and increase soil water-holding capacity, aeration and permeability. Chemically humic acids increase the cation exchange capacity, thereby providing to the soil a stronger buffer capacity for resisting sudden drastic chemical changes, which is very important from an environmental standpoint. A lot of material may be added to soils, some hazardous to humans and animals. The presence of humic acids with their huge cation exchange capacity can adsorb and detoxify the toxic compounds. Humic acids can also reduce micronutrient toxicity in acid soils, especially Al toxicity (Tan and Binger, 1986; Ahmad and Tan, 1986). This is attributed to its capacity for complexing or chelating the metals. Unconfirmed reports state that humic acids or at least the degradation products of humic acids can be taken up as food by plant roots (Flaig, 1975). The humic molecules are reported to penetrate the root membranes by way of *depolymerization* or *depolycondensation* (Dell' Agnola and Nardi, 1986; Burns, 1986).

Roots exudates, such as oxalic, fumaric, citric, malic, and succinic acids, are reported to induce the depolycondensation reaction. However, depolycondensation is noted to be a reversible reaction, and the transformation is strictly dependent on low pH values and the presence of the organic acids.

In industry, humic acid finds application as drilling muds for oil wells and as emulsifiers. Its emulsification properties and other attributes of humic acids are attracting the interest of medical science and the pharmaceutical industry. The use of humic acid as an anticoagulant to treat thrombosis has been investigated with success (Kloecking, 1994). It is also a potential antiviral agent, since application of ammonium humate at concentrations of 0.5 mg/L is reported to prevent infection of herpes simplex viruses on surfaces of host cells (Thiel et al., 1977). Clinical tests in Hungary show possibilities for humic acid to be used in cancer therapy and in healing crush, cut, and especially burn wounds (Jurcsik, 1994). Peat has been known for a long time in Europe for its therapeutic properties, which currently is ascribed to humic acid as the dominant component of peat. Peat baths were taken in the old days for therapy of gynecological and rheumatic diseases, and even today mud baths are offered in many European health clinics. In Russia, a medicine prepared from peat, called *torfot,* is used at the Ukranian Institute of Eye Diseases at Odessa as a topical treatment for myopia, opacification of the cornea, and early retinal degeneration (Fuchsman, 1980). Because of its importance in industry and agriculture, humic acid is produced today commercially on a large scale. Extensive deposits of lignite, or leonardite, in Texas, New Mexico, North Dakota, and Idaho, are the sources for the commercial production of *humates*. As such these deposits have an important geologic and economic impact on the US economy. Finally, it must be mentioned that humic matter may affect the environment. The presence of humic and fulvic acids in lakes and streams may stimulate the growth of phytoplankton. However, present in large concentrations, humic acids may reduce the photosynthesis of many aquatic green plants because of the dark-brown color that they impart to the water. Due to its enormous chelation capacity, on the other hand, humic acid is capable of detoxifying lakes that are affected by metal pollution. Suggestions have also been made to use humic acid as a sink of radioactive metals polluting soils and streams. In the form of cations, these radioactive metals can be adsorbed and chelated by humic acids and rendered immobile.

CHAPTER 4

GAS PHASE IN SOILS

4.1 SOIL AIR

The gas phase in soils is called *soil air*. It is located in the pore space, which is defined as the space in soil that is occupied by the gas and liquid phases. This pore space is composed of *macro-* and *micropores*. The macropores are usually the spaces between the soil structural units and as such, they are the main channels for air movement. On the other hand, the micropores are the spaces within the structural units, and are the main spaces for water. The macropores will be filled first with water during a rainfall or by irrigation. This water then moves into the micropores, where it will be held for some time in the soil. Some scientists suggest distinguishing the pores in terms of sizes, e.g., *transmission pores* with an effective diameter >50 µm, *storage pores* with sizes between 0.5 - 50 µm, and *residual pores* with effective diameters <0.5 µm (Wild, 1993). The residual pores then fall in the category of the micropores and will retain soil moisture when the soil is 'dry.' Storage pores have an effective diameter of the size of fine clay and silt and are mixtures of micro- and mesopores, whereas the transmission pores are then the macropores. Wider pores are usually not called soil pores but reveal themselves as *cracks* in the soil. The amount of total pore spaces,

145

called soil *porosity,* differs from soil to soil and from soil horizon to soil horizon. The percentage of pore spaces may range from 25% (by volume) or lower to 50% or higher. The lower number of pore spaces is usually found in the subsoil, where the soil is more compacted and contains more solid space (less organic matter but more mineral matter). A surface soil with a loam texture and in optimum condition for plant growth has a total pore space of 50%, which is filled half with water and half with air (Brady, 1990).

Soil air is composed of the same type of gases commonly found in the atmosphere and differs from atmospheric air only in composition. The major gaseous components and their relative amounts in atmospheric air are listed in Table 4.1. Many other types of gases are present in minute amounts in soil air. For proper plant growth, the pore spaces must contain air in sufficient amounts and in the proper composition. The amount of soil air in the pore spaces is usually controlled by soil water. When the amount of soil water increases, air is pushed out of the pore spaces. When water content in the pore spaces decreases because of uptake by plants and/or evaporation, air content increases by mass flow and diffusion. As indicated earlier, for

Table 4.1 Major Gaseous Components in Air
above the Soil

Gaseous component	Volume %
Nitrogen, N_2	78.0
Oxygen, O_2	21.0
Carbon dioxide, CO_2	0.031
Argon, Ar	0.93
Helium, He	0.0005
Methane, CH_4	0.0002
Hydrogen, H_2	0.00005
Nitrous oxide, NO_2	0.00002

Sources: Taylor and Ashcroft (1972); Manah-
han (1975); Brady (1990).

optimum plant growth, half of the pore space should be filled with air and half with water. The quality of this air is usually measured by its composition in O_2 and CO_2 content. Nitrogen is unavailable to most plants and, therefore, is not taken into consideration in this respect, though it is an important constituent in biological nitrogen fixation. Most biological reactions in soils consume O_2 and produce CO_2; hence, soil air is generally lower in O_2 (<20%), but higher in CO_2 content (>0.03%) than is atmospheric air above the ground. The CO_2 content shows a marked seasonal trend and usually increases with soil depth. Oxygen content in soil air appears to be low in soils saturated with water. It reaches a maximum level at *field capacity,* after which it remains almost constant with decreasing amounts of soil water (Brady, 1984). In general, oxygen deficiency is noted if O_2 content in soil air decreases to 15% or less (Taylor and Ashcroft, 1972). Several biochemical reactions, mediated by soil organisms, are responsible for the O_2-CO_2 balance in soil air, e.g., respiration of plant roots and microorganisms and aerobic decomposition of soil organic matter. Both reactions consume oxygen and produce carbon dioxide (see Tan, 1993). Respiration increases with the presence of abundant plant life; hence, soils under crops generally contain less O_2 and more CO_2 than bare or fallow soils. The application of manure or plant residues may also increase the consumption of O_2 and the production of CO_2. An additional source of CO_2 is the burning of crop residues. Biomass burning is a normal agricultural management process in sugar cane production. The amount of CO_2 generated and its effects on soil and fauna need further investigations.

Unfortunately, no biochemical reactions exist in soils that can reverse the process in favor of the oxygen content. Photosynthesis is the only biological reaction that absorbs CO_2 and produces O_2. However, photosynthesis takes place in green plant parts growing above the soil. Therefore, the O_2-CO_2 balance in soil air can only be corrected by exchange of soil air for atmospheric air, a process called *aeration.*

4.2 SOIL AERATION

The interchange of air between the soil and atmosphere is defined as aeration. It is the only process for removal of excess CO_2 and at the same time for replenishing the depleted O_2 content in soil air. A CO_2 production by respiration to the amount of 7 L/m^2 per day is considered normal (Keen, 1931), and Taylor and Ashcroft (1972) believe that at this rate of carbon dioxide production the soil air has to be replaced every hour to a depth of 20 cm for its composition to remain normal. This hourly replacement of air to a soil depth of 20 cm is called *normal aeration*, and is considered a standard for comparison of relative rates of efficiency in soil aeration. The terms *poor* and *good soil aeration* have been used to refer to soils with poor air circulation and exchange and to those with a less restricted air circulation and air exchange. The restrictions are brought about by at least two soil conditions: (1) poor drainage condition leading to accumulation of excess water, and (2) slow rates of interchange (Taylor and Ashcroft, 1972; Brady and Weil, 1996). Poor soil drainage, brought about by low infiltration rate and slow permeability due to heavy texture, poor structure, and hence low porosity, is usually found in depressions and low-lying areas. Due to topographical conditions, soils in low lying areas can also be inundated with water, forming the so-called *wetlands*. Saturation of the soils with moisture prevents air from moving in and out. Heavy rains can also cause a temporary saturation of well-drained soils, which according to Brady and Weil (1996) can be very harmful for many crops. If proper drainage is not provided for good aeration, many plants are likely to suffer because of lack of oxygen. The rate of gas exchange is affected by *mass flow*, also called *bulk flow*, from the internal spaces in soils, and by *diffusion*. Mass flow or bulk flow of soil air takes place because of expansion and contraction of soil air due to changes in temperature and barometric pressure or because of evaporation. Wind action forcing air into or out of the soil is also an important factor contributing to mass flow. Therefore, mass flow is more likely to be rapid at high wind velocities in very porous soils affected by significant changes in temperatures. Though mass flow can be very important, diffusion is considered the major process for gas exchange. It takes place continuously because of differences in partial pressures of the gases in soil air. Differences in partial pressure or differences in concentration occur continuously due to the biological consumption of O_2 and production of CO_2 by the respiration process. The difference in concentration of CO_2 between soil air and atmospheric air is generally large enough to cause air flow by diffusion. The larger the gradients of carbon dioxide between soil air and atmospheric air, the faster the rate of diffusion. Therefore, slow gas exchange occurs when the differences in concentrations of O_2 and CO_2 in soil air and those in atmospheric air are very small or insignificant. This can only happen in soils without any biological activity due to the absence of plant and microbial growth, even in porous soils. Consequently, where biological activity is present, slow gas exchange will occur only when the amount of pore space is too low for a sufficient channeling of the air flow.

4.2.1 Determination of Soil Air Quality

From the discussion above, it is perhaps clear that the quality of soil air is determined by its O_2 and CO_2 content. The closer the composition of soil air is to the composition of atmospheric air, the better its quality for plant growth. A wide variation has been reported for the O_2 and CO_2 content of soil air. In soils of the temperate regions, O_2 content generally decreases to low values in late fall, winter, and early spring, whereas CO_2 content increases to 12% in summer when moisture condition and temperature are favorable for biological activity (Stolzy, 1974). In the tropics, O_2 content in soil air decreases during the wet season, whereas the CO_2 content increases considerably (Table 4.2). The tropical wet season during December and January is the time for plant growth, hence for maximum biological activity, whereas in the temperate region this is winter with minimum biological activity. Not only are the top layers of soils in the tropics higher in CO_2 content than those of the soils in the US, but the data also indicate that the CO_2 content increases with soil depth in the wet season and winter. Oxygen content in the top layers of the soils in the tropics is lower, but twice to ten times higher in the subsoil than in the US soils, regardless of the higher biological activity in the tropical

Table 4.2 The O_2 and CO_2 Content in Temperate and Tropical Region
Soils During the Winter and Wet Season

Soil depth (cm)	Oxygen (%)		Carbon dioxide (%)	
	Temperate (winter)	Tropics (wet season)	Temperate (winter)	Tropics (wet season)
10 - 30	19.4	13.2	1.2	7.5
30 - 60	11.6	12.5	2.4	9.7
60 - 90	3.5	7.6	6.6	10.0
120	0.7	7.8	9.6	9.6

Sources: Russell and Russell (1950); Wild (1993).

soils. In summer, the CO_2 content in the US soils increases from 2% in the top layer to 9.0% at 120 cm depth. Oxygen content also decreases from a high of 19.8% in the top layer to a low of 14.5% at 120 cm depth in the pedon. Remarkably, CO_2 and O_2 contents in the tropical soils are not different from those of the temperate region soils both in amounts and in the decreasing trend with depth in the pedon. This is perhaps due to the fact that biological activity in the tropics does not come to a standstill during the dry season, whereas biological activity is at a maximum in the US during the summer.

Although high concentrations of CO_2 may cause toxicity, it is deficiency of O_2 that is believed to be more harmful to plant growth by inhibiting respiration. An O_2 content <15% will show a progressive decrease in nutrient uptake, though values of 5 to 10% O_2 in soil air have been reported to sustain growth of existing root tips.

Several methods have been proposed to determine soil air quality, e.g., (1) measurement of O_2 content in soil air, (2) determination of the oxygen diffusion rate or ODR value, and (3) determination of the soil's redox potential. The results of analysis of O_2 content as a means of determining soil air quality have produced mixed reactions. The O_2

content is often only slightly below 20% in the upper layers of friable, well-drained soils. It can decrease to 5% or even <1% in the subsoil, but most roots seldom grow into the subsoils. When O_2 content is low in soil air, a significant amount of O_2 is still present as dissolved oxygen in soil moisture, available to roots and microbes. More attention has been given to the determination of ODR values. As the name, oxygen diffusion rate, implies, this method determines the rate with which oxygen diffuses into the soil, and several authors have reported a definite relationship between ODR values and root growth. The critical ODR value, at which root growth stops, is 0.20 µg cm^{-1} per minute (Stolzy and Letey, 1964). Naturally, tolerance to low ODR values varies from plant to plant, and the following decreasing order in tolerance of plants is reported (Stolzy, 1974):

Rice > corn > Bermuda grass > barley > Newport bluegrass

The third method, the determination of the redox potential in soils, is perhaps the most simple method of the two discussed above. Redox potential determinations are especially useful in the analysis of soils where ODR values have no meaning at all, such as in poorly drained and waterlogged soils in the wetlands and in the rice paddy fields (Ponnamperuma et al., 1966). The redox potential varies with the reduction and oxidation state in soils. As can be noticed in Table 4.3, the value is positive in well-drained soils, and negative or below zero in waterlogged soils. Therefore, the redox potential, E_h, can be used to to indicate the aeration status of soils Drastic changes take place in the physical, chemical, and biological conditions of soils with development of poor aeration or waterlogging. Oxygen content decreases rapidly in poorly aerated soils. Respiration of plant roots and microorganisms will quickly consume the remaining oxygen in soil air. Therefore, oxygen content decreases with a decrease in E_h value. At low E_h values, the dissolved oxygen in soil water will then be used by microorganisms, and at very low E_h values, even the combined oxygen, in the form of ferric oxides, nitrates, and sulfates, will be attacked. A notable chemical effect resulting from water-logging is the conversion of insoluble iron and manganese oxides into soluble Fe(II) and Mn(II), respectively. The reactions can be repre-

Table 4.3   Redox Potential Values at Various Soil Aeration Status.

Soil aeration status	E_h (Volt)
Well-aerated soils	0.4 - 0.7 or higher
Somewhat poorly aerated soils	0.3 - 0.4 or lower
Waterlogged soil	-0.4 or lower

Source: Tan (1998).

sented as follows:

$$Fe_2O_3 + 6H^+ + 2e^- \rightarrow 2Fe^{2+} + 3H_2O$$

$$MnO_2 + 4H^+ + 2e^- \rightarrow Mn^{2+} + 2H_2O$$

Consequently, in the poorly aerated soils where reduction processes prevail, iron is present as Fe^{2+}, manganese as Mn^{2+}, nitrogen as NH_4^+, and sulfur as SO_3^{2-} ions. On the other hand, in well-aerated soils, these ions are present in the oxidized state, e.g., Fe^{3+}, Mn^{4+}, NO_3^-, and SO_4^{2-}. Biologically, a decreasing oxygen content in soils produces drastic changes in the microorganism population. At low oxygen content, *anaerobic* organisms, e.g., *Actinomyces* sp., may prevail over *aerobic* microorganisms. In medical science, the redox potential is applied for the detection of certain diseases. Low values of redox potentials around the gum increase the hazard for occurrence of *gingivitis*, a gum disease.

4.3 BIOCHEMICAL EFFECT OF AERATION

Aeration has a significant effect on many biochemical reactions. As discussed in Chapter 3, many of the decomposition reactions of soil organic matter are affected by the aeration status in soils. When adequate amounts of air (O_2) are present, aerobic reactions prevail. On the other hand, where aeration is poor (lack of O_2) anaerobic reactions predominate. In other words, aerobic reactions (e.g., oxidation) take place with adequate aeration, whereas anaerobic reactions, e.g., reduction, occur under poorly aerated conditions. As will be discussed in the next section, the properties of soils in an oxidized state are markedly different from those of soils in a reduced state. The solubility of many elements changes with a change in the oxidation-reduction state.

4.3.1 Biochemistry in Aerobic Conditions and Its Role in Environmental Quality

In well-aerated soils, organic matter will decompose into CO_2 and H_2O, and the nutrient elements are released to the benefit of plant growth. In such soils, most of the nutrient elements are present in the oxidation state. As indicated in the preceding section, iron is present as Fe^{3+}, manganese as Mn^{4+}, sulfur as SO_4^{2-}, and nitrogen as NO_3^-. Iron in the oxidized state (ferric iron) is more stable but less soluble than iron in the reduced state (Fe^{2+} or ferrous iron). Ferric iron has a red to reddish brown color, and soils rich in Fe^{3+} are, therefore, red to reddish brown. Mobilization of Fe and Mn, due to redox conditions, has been reported in rice paddy soils to form *iron-B* followed by *manganese-B* horizons (Tan, 1968). In tidal flood water areas, reduction processes play an important role in the formation of sulfur-rich soils. These soils, called *cat clays* or sometimes also *acid sulfate soils*, are usually characterized by a pH <3.0, hence have a very strongly acid in reaction (Tan, 1998). If pockets with anaerobic conditions are present in a well-aerated soil, the iron in these pockets is present in the reduced state. Ferrous iron is yellowish or greenish to grayish in color. Therefore, these pockets are yellowish or grayish in color, which produces in the soil pedon a color pattern called *mottling*. The presence of a mottled horizon indicates an inadequately drained soil horizon. Both reduction and oxidation conditions can occur simultaneously in the pedon. While the surface horizon of the pedon is in an oxidized state, the subsoil horizons may be in a reducing condition owing to a fluctuating groundwater level. This may lead to *pseudogley* formation or to *plinthization* (mottling) in the subsoil.

The carbon dioxide produced by aerobic decomposition of organic matter will react with soil water to form carbonic acid, which affects soil acidity. The increase in soil acidity increases the dissolution of soil minerals. Such a dissolution of primary minerals is important in soil formation, soil fertility and plant nutrition. For example, calcium carbonate, $CaCO_3$, the main component of limestone, is insoluble in pure H_2O. However, in water containing CO_2, calcium carbonate is made soluble by its conversion into calcium bicarbonate. The reaction, called *carbonation,* can be written as follows:

$$H_2O + CO_2 \rightarrow H_2CO_3 \tag{4.1}$$

$$H_2CO_3 + CaCO_3 \rightarrow Ca(HCO_3)_2 \tag{4.2}$$
$$\text{calcium bicarbonate}$$

The calcium bicarbonate that is formed is soluble in water. Carbonation causes liming materials to react in soils. By releasing its Ca^{2+} in this way, lime reduces soil acidity. The rate of formation of calcium bicarbonate depends on the partial pressure of carbon dioxide. In other words, the solubility of calcium carbonate increases as the partial pressure of CO_2 increases (Taylor and Ashcroft, 1972). Dissolved carbon dioxide also increases the solubility of apatite, a natural phosphate mineral. In fact, almost any mineral in soils is affected by carbonic acid.

Other acids that are produced by aerobic decomposition of organic matter, such as humic acids and the inorganic acids, nitric and sulfuric acids, also contribute to the solubilization of soil minerals.

Soil acidity also increases the oxidation of sulfur compounds, which produces sulfuric acid. The presence of extremely acid soils in the coastal regions is caused by these processes. These soils, frequently called *cat-clays*, are nonproductive; fortunately, they occur only in limited areas. Of more practical significance is the increased acidity of soils contaminated with *acid mine spoil* due to oxidation of S-compounds. The residue from coal mines, called mine spoil, contains large amounts of pyrite, FeS_2, which upon bio-oxidation produces sulfuric acid, as can be noticed from the reaction

$$2FeS_2 + 2H_2O + 7O_2 \rightarrow 2Fe^{2+} + 4H^+ + 4SO_4^{2-} \qquad (4.3)$$

Acid mine spoil is a harmful contaminant of soils and streams. The extremely high acidity created by acid mine spoil is toxic and has many undesirable effects on soils and the environment. The iron released remains soluble in the acid environment. Although many attempts have been made to solve the problem, the reclamation of soils containing acid mine spoil is facing many difficulties. One of the reclamation methods involves the application of lime, which neutralizes the acidity by converting the sulfuric acid into calcium sulfate. The reaction is written

$$CaCO_3 + H_2SO_4 \rightarrow CaSO_4 + H_2O + CO_2 \qquad (4.4)$$

The carbon dioxide formed is lost in the air. However, the decrease in soil acidity also creates other problems. The large amounts of Fe^{2+} released in reaction (4.3) are rapidly oxidized into Fe^{3+}. In the slightly acid to neutral environment, as a result of liming, ferric iron precipitates as $Fe(OH)_3$. These processes can be illustrated by the following reactions:

$$Fe^{2+} \rightleftharpoons Fe^{3+} + e^- \qquad (4.5)$$

$$Fe^{3+} + 3H_2O \rightarrow Fe(OH)_3 + 3H^+ \qquad (4.6)$$

The $Fe(OH)_3$ produced is often deposited as an amorphous colloidal material or may form coatings around the liming particles. The latter occurrence inhibits any further reaction from the liming material taking place.

4.3.2 Biochemistry in Anaerobic Conditions and Its Role in Environmental Quality

Reduction processes prevailing in anaerobic conditions bring about the reduction of many nutrient elements in soils. Because of this, iron is reduced to Fe^{2+}, manganese to Mn^{2+}, sulfur to SO_3^{2-}, and nitrate to NH_4^+. Ferrous iron, Fe^{2+}, is usually gray or greenish gray in color; hence, a poorly drained soil exhibits gray colors. Pockets with aerobic conditions present in such a soil contain ferric iron, creating reddish to rusty colors. As discussed earlier, this produces mottling in the pedon.

In anaerobic soil condition, nitrate tends to be reduced into nitrite, NO_2^-. The reaction, called *nitrate reduction*, can be illustrated as follows:

$$2NO_3^- + CH_2O \rightarrow 2NO_2^- + H_2O + CO_2 \qquad (4.7)$$

This reaction indicates that nitrate acts as an electron acceptor. Such a process finds application in the reclamation of sewage sludge lagoons. The lagoons, which are typically deficient in O_2, are treated (fertilized) with $NaNO_3$, which provides an emergency source of O_2 for reaction (4.7) to take place. This kind of treatment may reestablish normal bacterial growth (Manahan, 1975; Moore and Moore, 1976). The nitrite, NO_2^-, formed is, however, toxic, and when it is allowed to build up to sufficiently high concentrations, it may inhibit further bacterial growth.

Nitrate reduction may ultimately end in a denitrification process, a process by which the nitrogen compound is reduced into N_2 gas, according to the reaction:

$$4NO_3^- + 5CH_2O + 4H^+ \rightarrow 2N_2 + 5CO_2 + 7H_2O \qquad (4.8)$$

This reaction involves the transfer of five electrons. The denitrification reaction is often applied in water treatment procedures to remove nitrates, which are harmful pollutants. By treating water with small amounts of methanol, the nitrate is reduced under anaerobic condition into N_2 gas. The reaction is written

$$5CH_3OH + 6NO_3^- + 6H^+ \rightarrow 5CO_2 + 13H_2O + 3N_2 \qquad (4.9)$$
methanol

The nitrogen gas is nonionic and is lost in the air.

4.4 BIOLOGICAL EFFECT OF AERATION

4.4.1 Effect on Plant Growth

One of the most obvious biological effects of an imbalanced oxygen and carbon dioxide content in soil air is its impact on plant growth. It is an established fact that improper cultural practices, resulting in poor aeration, reduce plant growth. Water and nutrient uptake are inhibited. Under inundated or swampy conditions, toxic compounds, such as H_2S, are formed.

Aeration affects the growth and yield of crops as well as the growth of roots. However, roots are perhaps more affected by the aeration status of soils than the aboveground parts of plants. Root growth is inhibited by poor aeration in soils, and if roots are not growing adequately, the plant parts above the ground will also not grow well. Root growth of most plants usually ceases at a concentration of 1% O_2 in soil air. Only *hydrophytes*, plants growing in water (hydro = water), such as rice plants and aquatic plants, can grow roots in anaerobic conditions.

It appears that roots can grow and function at oxygen concentrations far below the normal O_2 concentration of atmospheric air. Root growth was recorded at oxygen concentrations as low as 2.2% (Taylor and Ashcroft, 1972). The term *critical value* was introduced to show the lowest limit of oxygen content required for growing roots, and growth is limited by aeration (Wiegand and Lemon, 1958). This critical value has no fixed limits and seems to be a function of the soil moisture content as expressed in terms of the soil matric potential (ψ_m). The critical value decreases from 4 µg O_2/cm^3 to 1 µg O_2/cm^3 as soil matric potential increases from -30 Joules/kg to -10 Joules/kg

(Wiegand and Lemon, 1958). The increase in matric potential means that soil moisture content increases. Plants are less affected by moisture stress at higher soil moisture content. Because of this, they can function with smaller amounts of oxygen. The definition and concept of soil matric potential will be discussed in Chapter 5 (see also Tan, 1993, 1998).

Plants grown in well-aerated soils usually develop long, well-branched fibrous roots. When aeration becomes restricted, these well-developed roots slowly deteriorate and a new root system is produced by undergoing morphological changes as an adaptation to poor aeration. Though plants may vary considerably in their response to poor aeration, root elongation is generally reported to be reduced at 10% O_2 or lower in soil air. In anaerobic conditions, the roots are frequently short and stubby and often grow superficially to take advantage of the proximity of atmospheric air. A good example is the roots of cypress trees, called *cypress knees*, or roots of mangrove trees in swamps that grow out of the water into the air. Other plants, such as in a rice paddy, can adapt themselves to inundated conditions by developing tissues, especially in the roots, filled with cavities, called *aerenchyma,* for the internal transport of air (O_2). The development of aerenchymas is reported to be caused by the presence of ethylene (C_2H_4) gas. This gas is produced by microbial decomposition of organic matter under anaerobic conditions. At very low concentrations, C_2H_4 stimulates root growth, but at concentrations >2 mg/L the gas can be toxic to roots.

Carbon Dioxide Toxicity and Oxygen Deficiency. - As discussed before, when aeration is restricted, several of the biological processes in soils consume oxygen and produce carbon dioxide. The net effect is that the O_2 content in soil air decreases, whereas the CO_2 content simultaneously increases considerably. Two possible reasons have been suggested for the restricted root growth because of the changing O_2-CO_2 composition in soil air. One hypothesis postulates that high amounts of CO_2 are toxic to plant roots, especially to root cell protoplasm. According to this theory, excessive amounts of CO_2 temporarily suppress the functional activity of root cells, inhibiting root growth (Hunter and Rich, 1925). However, the most accepted hypothesis is that O_2 deficiency in soil air reduces respiration of plant roots. An adequate supply of O_2 is necessary to maintain respiration and permeability of plant roots, factors that influence water and nutrient uptake.

Under normal field conditions, oxygen deficiency is considered more serious to root growth than excess CO_2 in soil air. At high oxygen concentrations, root performance, e.g., water uptake, is not affected by the presence of high CO_2 concentrations (Taylor and Ashcroft, 1972).

However, no definite answers have been given yet as to the reasons for the harmful effect of poor aeration on root growth. Is it because of the reduced concentration of O_2 or can it be attributed to increased concentrations of CO_2? Perhaps it is the result of the interaction of decreased O_2 and increased CO_2 content.

Plant Species. - The problem discussed above is complicated by the fact that different plants may react differently to the variation in O_2 and CO_2 concentrations in soil air. The factor of plant species can be illustrated by the following examples. Tomato plants are very sensitive to the aeration status of soils. Maximum growth of tomato plants is reported to occur only when the O_2 concentration of the air is at 21%. Sugar beets require proper aeration, in contrast to ladino clover, which can grow in poorly aerated soils. Barley is another plant that will grow at low concentrations of oxygen. Red pines also grow well in poorly drained soils, but root crops are affected harmfully by poor aeration. Abnormally shaped carrots and sugar beets develop when these plants are grown in compacted, poorly aerated soils (Taylor and Ashcroft, 1972; Brady, 1984).

In addition to respiration and aerobic decomposition, the aeration status in soils is affected by many other factors, such as *soil compaction, soil moisture content,* and *soil temperature*.

Soil Compaction. - Soil compaction affects the pore spaces, which constitute the main venues for air movement and storage. When the pores, especially macropores, are destroyed due to compaction of soils, not only will air be deficient, but also the soil becomes dense and hard. A compacted soil has an adverse effect on root growth by (1) increasing the mechanical impedance to root growth, and (2) decreasing and altering the configuration of pore spaces (Taylor and Ashcroft, 1972). It is believed that the amount of O_2 needed by roots depends on the amount of work that a root must do in expanding against the mechanical impedance present in soils. In a loose and friable soil, only a small amount of energy will be used by roots to penetrate the soil, hence low concentrations of O_2 will give satisfactory growth. On the other hand, in a compacted soil, the root has to use more energy to grow in the soil, and as the amount of energy required for growth increases, the amount of O_2 necessary for growing also increases. In simulated conditions, it was noted that at an oxygen concentration of 20%, root growth was normal in a sand medium at normal pressure (no pressure applied). However, the growth rate of these roots was only one-half as fast when a pressure of 69.0×10^3 Pascal (10 lb/in^2) was applied to the medium (Gill and Miller, 1956).

Soil Moisture Content. - Soil moisture content is another factor that affects soil aeration. As will be discussed in Chapter 5, water is held in the pore spaces by varying forces of attraction. These forces can be expressed in various ways, e.g., in terms of _water potential_ (ψ_w) or soil moisture tension (see Chapter 5). When the soil is saturated with water, the water potential has the highest value (0 Joules/ kg). The value of the water potential becomes increasingly smaller (becomes more negative in value) with decreasing amounts of water. At _field capacity_ the water potential equals -30 J/kg, whereas at wilting point the water potential is -1500 J/kg. At high soil water potentials, there is less or no tension in soils, hence the soil moisture tension is low. Similarly at low water potentials, the tension in soils increases, consequently the soil moisture tension is high. These conditions affect root growth.

Since the amount of water in the pore spaces usually determines the air content in soils, low soil moisture tension corresponds to low amounts of air or, in other words, translates to poor aeration status. In temperate region soils, this condition occurs in winter and spring when soil moisture content is generally highest. The air content, hence O_2 content, reaches a maximum at _field capacity_ (0.3 bars or -3 $\times 10^5$ ergs/g), after which it remains constant with increased drying of the soil (Figure 4.1). The curve indicates that at field capacity, the O_2 content in soil air is approximately 15%, which is rather low. However, when soil moisture content is not a limiting factor, this amount of oxygen still appears to be sufficient for normal root growth. Root growth will be inhibited when the O_2 concentration in soil air decreases to 10.5% (Gingrich and Russell, 1956). In the presence of adequate amounts of O_2, root growth is generally better at low moisture tension than at high soil moisture tension. At high soil moisture tension, the soil is too dry for plants to grow although the oxygen content may be adequate.

Soil Temperature. -   Soil temperature is also a factor in oxygen content in soil air. Cooler temperatures usually decrease the O_2 concentration in air. Consequently, decreasing temperatures may be

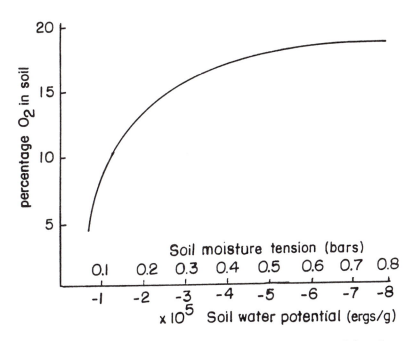

Figure 4.1. The relationship between soil water potential and oxygen content in soil air (adapted from Meek et al., 1980; Brady, 1984).

harmful for root growth. Maximum rate of root growth at 21% oxygen was reported to occur at 30°C. However, this growth rate was noted to decrease almost 50-60% when plants were grown at temperatures of 15° to 20°C. In the presence of 2.2% O_2 the growth rate appears to be 50-70% less than that at 21% O_2 over the range of 15° to 30°C (Taylor and Ashcroft, 1972).

4.4.2 Water Uptake

Root cells consist of (1) cell walls, which are rigid but exhibit some elasticity; (2) protoplasm, which acts as a semipermeable membrane through which water and ions move; and (3) vacuoles that contain cell sap, which is rich in solutes and colloids. The high solute and colloid concentrations in the cell sap attract water from the soil into the plant cell. The forces responsible for the movement of water into the root cells are called *osmotic forces,* and the energy related to these forces is called *osmotic potential* (Tan, 1993, 1998). Such a movement, considered a *passive movement*, requires good permeability of the root cells. Permeability of cell protoplasm is reported to increase by an adequate oxygen supply (Taylor and Ashcroft, 1972). Water can also be transported by *active movement*, especially in the transport of water upwards from cell to cell in the plant body. The energy required for such an active transport is provided by respiration. Active movement is responsible for *exudation* of water from a root or stem cut (Taylor and Ashcroft, 1972). Since adequate O_2 supply is required for respiration, proper aeration is essential for active water transport.

4.4.3 Nutrient Uptake

Movement of plant nutrients from soils into root cells also occurs by *passive* and *active* transport. The passive transport, e.g., mass flow and diffusion, obeys physico-chemical laws (Tan, 1993), whereas active transport requires the expenditure of energy, which is provided by the respiration process. However, the production of CO_2 by respiration and its consequent increase in soil air is reported to decrease nutrient uptake by roots in the following order (Chang and Loomis, 1945):

K > N > P > Ca > Mg

The lyotropic series above indicates that the decrease in ion uptake is decreasing in the direction of Mg. The decrease in K uptake is larger than the decrease in N uptake, and so on, and the smallest decrease in uptake is noticed with Mg.

The decrease in K and Fe uptake may result in K and Fe deficiency in plants. The deficiency symptoms disappear after the O_2 content in soil air is replenished artificially. The latter are reported to increase absorption of N, P, Ca, and Mg at the same time (Lawton, 1945).

4.4.4 Microorganisms

In most soils, respiration by microorganisms is more important than that by higher plants. As indicated in Chapter 2, microorganisms are numerically the most abundant of all soil organisms. They are even more numerous than plant roots. Therefore, their respiration process requires more O_2 than that needed by plant roots, and a deficiency of O_2 in soil air significantly affects all microbial activity, roots included. In anaerobic conditions, where aeration is very poor, microbial decomposition of organic matter is very slow or inhibited, and organic residue tends to accumulate. In such a condition consumption and hence depletion of O_2 are relatively low. Different redox conditions also affect the development of different types of microorganisms, e.g., aerobic and anaerobic bacteria. *Aerobic bacteria* need oxygen from soil air, whereas *anaerobic bacteria* supply their oxygen needs from bonded oxygen. When O_2 is deficient in soil air, the anaerobic bacteria use the O_2 in nitrate, sulfate and carbon dioxide. Between the aerobic and anaerobic bacteria is another bacterial group, called *facultative bacteria*. Facultative bacteria use O_2 when it is available, or bonded O_2 when O_2 is deficient in soil air. Soils in an oxidized state contain predominantly aerobic bacteria, whereas soils in reduced conditions contain mostly anaerobic bacteria. Aerobic bacteria are responsible for the oxidation of Fe, S, and N. The oxidation reactions for these elements into ferric, sulfate, and nitrate compounds have been presented above. The basic metabolism of aerobic bacteria differs from that of anaerobic bacteria; the first produces CO_2, whereas the second yields H_2, H_2S, and CH_4 as ultimate end products. In addition, biological oxidation yields hydrogen or electrons, with O_2 acting as the electron acceptor. In anaerobic reactions, other compounds become electron acceptors, e.g., CO_2, NO_3^-, and SO_4^{2-} (Stevenson, 1986).

4.4.5 Decomposition of Organic Matter and Its Effect on Environmental Quality

In the absence of air, organic matter decomposition follows a different pathway. Anaerobic decomposition results in an incomplete breakdown of organic matter, yielding hydrogen, methane, and other foul-smelling intermediate organic compounds, such as H_2S, indoles, and mercaptans.

The formation of methane, CH_4, is mediated by methane bacteria, which use oxygen from carbon dioxide. Expressed in scientific terms, methane bacteria utilize CO_2 as an electron acceptor. The organic compounds attacked by the bacteria are usually organic acids and alcohols (Manahan, 1975; Moore and Moore, 1976; Stevenson, 1986). The reactions can be illustrated as follows:

$$CH_3COOH + 2H_2O \rightarrow 2CO_2 + 8H^+ + 8e^- \qquad (4.10)$$
acetic acid

$$CO_2 + 8H^+ + 8e^- \rightarrow CH_4 + 2H_2O \qquad (4.11)$$
methane

The production of methane is important in the decomposition of organic matter in an anaerobic environment. It usually occurs at the bottom of lakes and swamps and in coastal regions affected by the tide. It also takes place in landfills and old garbage dumps, where the methane by-product becomes a hazardous pollutant. Methane is a good fuel source, and its presence increases the chances of fire and explosions. It is considered hazardous to human health and is a pollutant gas that contributes to the greenhouse effect responsible for the global warming process. Because methane production is a biological reaction under anaerobic conditions, proper aeration of landfills may inhibit the anaerobic bacterial activity and hence stop the production of methane.

Today, attempts have been made to apply the process of methane production to the biological destruction of municipal waste and sewage sludge, and the methane produced is used as fuel to power these plants. Unconfirmed reports indicate that the amount of methane produced is more than sufficient to power the treatment plants.

In another process, bonded O_2 is used in the anaerobic decomposition of organic matter for the reduction of sulfate into H_2S. Sulfur bacteria are capable of decomposing organic matter in anaerobic condition by using the oxygen obtained from sulfate. The reaction can be illustrated as follows:

$$2CH_2O + SO_4^{2-} + 2H^+ \rightarrow H_2S + 2CO_2 + 2H_2O \tag{4.12}$$

Formation of H_2S occurs especially in swamps and coastal plain soils affected by sea water. Sea water contains high amounts of sulfates, which are retained by the soil and coastal sediments. The reduction of this residual sulfate into large amounts of H_2S creates a potential hazard for pollution in coastal regions. Because of the presence of FeS, formed by the reaction of H_2S with ferrous iron, soil and coastal sediments are frequently black.

The H_2S, CH_4, and other organic compounds produced during anaerobic decomposition are oxidized by adequate aeration of the soil. The oxidation products are often beneficial, but can sometimes be harmful to the environment, soil conditions and plant growth. A beneficial effect is the oxidation of H_2S. When hydrogen sulfide, H_2S, is allowed to accumulate, it not only is toxic to soil organisms, but it also creates pollution problems. Fortunately, a group of bacteria present in soils is capable of oxidizing the H_2S into elemental S and SO_4 compounds. The reactions can be written

$$2H_2S + O_2 \rightarrow 2S + 2H_2O \tag{4.13}$$

$$2S + 2H_2O + 3O_2 \rightarrow 2H_2SO_4 \tag{4.14}$$

Unfortunately the S formed is oxidized further into sulfuric acid, as shown by the second reaction above. This is then the reason for formation of strongly acid soils known under the name of *cat clays*. However, the oxidation process of S is applied in the development of acid conditions for growing *acid-loving* plants and in the reclamation of alkaline soils. Azaleas and tea plants require acid soils, and powdered S is frequently supplied to the soil in which they are grown. In alkaline soils, the soil pH is usually too high for adequate crop growth and needs to be decreased to a suitable level. The latter is also attained by applying adequate amounts of powdered S to the soil. The sulfuric acid produced by the oxidation of elemental S by the sulfur bacteria increases soil acidity.

4.5 CHEMICAL EFFECT OF AERATION

As discussed previously, in the presence of sufficient amounts of air (O_2) oxidation processes prevail over reduction reactions. Such a condition will generally be found in well-drained or well-aerated soils. In well-aerated soils, most of the soil constituents are in an oxidized state, which usually provides a physical and chemical environment suitable for plant growth. Under poor aeration, the soil constituents are in a reduced state, which creates a less favorable condition for plant growth.

Biological reactions in aerobic conditions produce CO_2 and, as discussed before, this reacts with soil water to form H_2CO_3. Other acids are also produced by the aerobic decomposition of soil organic matter, e.g., nitric acid, sulfuric acid, and oxalic acid. All these acids increase the dissolution power of soil water and play an important role in the weathering and dissolution of rocks and minerals. Most of the soil minerals are attacked by these inorganic and organic acids and especially by carbonic acid. The presence of H_2CO_3 is considered essential in the weathering of limestone. Calcium carbonate or calcite, $CaCO_3$, the major component of limestone, is insoluble in water, but will dissolve in water containing H_2CO_3 (see reaction 4.2). The stable solid phase of Ca in a soil system containing $CaO-H_2O-CO_2$ is either $Ca(OH)_2$, $CaCO_3$, $Ca(HCO_3)_2$, or a combination of these. At CO_2 pressures normally present in soils, the stable solid phase is $CaCO_3$. However, $CaCO_3$ is insoluble, but its solubility increases as the partial pressure of CO_2 increases from 0.01 to 0.1 bars (Frear and Johnston, 1929).

Carbon dioxide in water will also affect the solubility of natural phosphate minerals in soils. These minerals are converted into various forms of calcium phosphates, e.g., apatites. The redox state of several elements, serving as plant nutrients, has been discussed earlier in relation to aerobic and anaerobic conditions in soils. In an anaerobic environment, nitrogen exists in ammonium, NH_4^+, form. Under these conditions, ammonium nitrogen is taken up by plants more readily than nitrate (NO_3^-) nitrogen. Nitrate, when present and absorbed, is apparently broken down in the plant tissue by *anaerobic respiration*, to supply part of the oxygen requirement of plants (Taylor and Ashcroft, 1972). Iron exists as ferrous ions in poorly aerated soils, whereas Mn is present as manganous ions. Because these forms of Fe and Mn are the most soluble forms of iron and manganese, their accumulation may increase their concentrations and can create Fe and/or Mn toxicity to plants.

CHAPTER 5

LIQUID PHASE

5.1 SOIL SOLUTION

The liquid phase, also called *soil water* or *soil solution*, is composed of water, dissolved substances, and colloidal materials. The dissolved substances are both solids and gases. The water itself is a renewable resource and is part of the global hydrologic cycle. When this water is present in the soil, it can be distinguished into (1) *surface water*, water above the soil, such as in streams, runoff, lakes, and ponds; (2) *ground water*, which is water present underground; and (3) *soil water*, water in the soil. The chemistry of surface water differs from that of ground or soil water. Surface water may contain high amounts of nutrients, resulting in excessive growth of algae. It may also have high levels of dissolved organic material, which promote bacterial growth. These factors affecting the quality of water are less serious in ground and soil water. Ground water, which percolates through deposits, may contain dissolved substances. It may also accumulate such substances through leaching, but most dissolved inorganic and organic compounds are gradually filtered as ground water moves through the soil. Therefore, chemically, ground water is generally of better quality than surface water. Soil water is usually in between the two in chemical characteristics and water quality.

Many of the chemical properties of water are attributed to its dipolar molecular structure. Water exhibits a high dielectric constant, which affects the dissociation of many compounds in water. Because of these two factors, water is an excellent solvent for a number of compounds. It is the most important transporting agent for nutrient elements and waste products in soils. Water stabilizes the body temperature of organisms and is also known to stabilize the climate. This is due to the fact that a large amount of energy is required to increase the temperature of a unit mass of water. One calorie of energy is needed to raise the temperature of one g of water $1°C$, whereas more energy is required to evaporate water. On the average, 585 calories are needed to evaporate one g of water. These factors prevent sudden and large fluctuations in temperature from taking place in water, and their stabilizing effect protects aquatic organisms in lakes from the shock of abrupt temperature changes.

5.1.1 Forces Attracting Water in Soils

The term *soil moisture* is frequently used instead of soil water or soil solution. Soil moisture is located in the macro- and micropores. When the soil is saturated with water, all the pores are filled with water. After excess water has drained by gravity, water is present in the macropores in the form of a thin film around solid particles, whereas the micropores are still filled with water. Evaporation and consumption of water by soil organisms cause the water content in soils to decrease. At low water content, water exists as thin films and as wedges on the contact points of soil particles.

Water is held in the pore spaces by forces of attraction exerted by the soil matrix, by attraction to the ions, and by surface tension in the capillaries. The two major reactions of importance for holding water in soils are *adhesion* and *cohesion*. Adhesion is the attraction of solid surfaces to water, hence it is sometimes also called *interfacial attraction*. On the other hand, cohesion is the mutual attraction between water molecules. The soil matrix is composed of soil solids whose surfaces are capable of holding water by adhesion. When the solid surface is electronegatively charged, water adhesion is caused by *hydrogen bonding.* Since the clay surface is the soil matrix, the force of such an adsorption is in fact closely related to the matric potential. With noncharged surfaces, e.g., surfaces of sand and silt particles, adhesion of water is made possible by *van der Waals* forces. However, the amount of water held in soil by the latter is substantially lower than that held by the charged solid surfaces. The ions are either negatively or positively charged, hence adhesion of water on their surfaces is caused more by electrostatic attraction. The combined effect of adhesion and cohesion results in thick water films held by the soil matrix or in formation of thick hydration shells around colloids and ions. The colloid surrounded by its hydration shell is called a *micelle*, and soil water which is not part of the micelle is called by Taylor and Ashcroft (1972) *intermicellar* water. This is perhaps quite correct, but very confusing since in Soil Science and Clay Mineralogy intermicellar water is recognized as water held within the intermicellar spaces of clay minerals. Because the solid particles are responsible for retention of water, it follows that the larger the amounts of solid particles, especially the amounts of charged clay and humic particles, the greater will be the amount of water held in soils. Additional water is held in soil capillary pores by hydrogen bonding and surface tension. Adhesion of water on the walls of the capillaries is made possible by hydrogen bonding, because these walls are often made up of negatively charged clay surfaces. Surface tension provides for retention of water by cohesion, which was defined earlier as the attraction of water molecules for each other. This water retained in the capillaries is often called *capillary water.* The water held in the soil by the forces discussed above constitutes the reserve water supply for plant use between additions of water by rain or irrigation. These forces, expressed in terms of energy levels, such as matric (ψ_m), osmotic (ψ_o), and pressure potential (ψ_p), are collectively called the *water potential* (ψ_w) (Tan, 1993, 1998). The definition of the water potential can thus be written as follows:

$$\psi_w = \psi_m + \psi_o + \psi_p \tag{5.1}$$

The water potential is in essence the energy level of water in relation to doing work. Differences in water potential from one to another location in soils tend to create a spontaneous movement of soil water in the direction from high to low potentials. Work is performed as water moves from one to another location and its energy level or the water potential is decreased. As can be noticed from equation (5.1), the water potential is composed of different types of energy levels or potentials. The matric potential is the force or energy level by which soil water is attracted and retained by the soil matrix (soil solids). The osmotic potential is the energy level with which water flows to soil solutes. The force of attraction is called the osmotic force. The pressure potential results from pressure differences in soils, and is a reflection of the total effect caused by retention of water in soil pores (capillary forces) and on the surfaces of soil particles (adsorption).

The water potential (ψ_w) has a negative value because the attraction by matric and osmotic forces reduces the free energy level of soil water. In simple terms, soil water held by the soil matrix and solutes cannot move freely. The larger the negative value of ψ_w, the smaller the amount of water present in soils. At field capacity (the maximum amount of water available for plants) $\psi_w = -30$ joules/kg, whereas at wilting points $\psi_w = -1500$ joules/kg. Many plants can live below the wilting point by adjusting in a number of ways. The most common method is by shedding their leaves during periods of moisture stress, thereby decreasing the intake of moisture. Plants may also adjust to water stress by decreasing their intake of CO_2 through closing or regulating their stomata openings. Closing the stomata opening will reduce their rate of metabolism to the minimum required for staying alive. Intake of moisture is then also reduced to a minimum. These processes are called *osmotic adjustments*, adjustments of the osmotic potential so that only the minimum amount of water is absorbed for sustaining life functions.

These water potentials can also be expressed in terms of positive values. The latter represent the opposite force, or suction force, which is called *soil moisture tension*. The tensional force can be expressed in a variety of ways, pounds/square inch, cm of water, cm of Hg, atmospheres, bars, or pF. The units *cm of water, bars*, and *pF* are the most common units used in soil science. The basic unit is the cm of water, which is defined as the force equal to the weight of the height of a water column in cm. It is interpreted in the same way that the cm

Hg for barometric pressures is interpreted. The values for soil moisture tension in cm of water range from 0 to 10 000 000 cm. A moisture tension of 0 cm water indicates that there is no tension, which is attributed to the presence of excessive amounts of water in soil. A soil moisture tension of 10 000 000 cm means that water is held in soils with a very large force, in other words there is not much water in the soil (the soil is dry). The unit cm H_2O can be converted into bars by dividing by 1000. Hence,

$$0 \text{ cm } H_2O \; = \; 0/1000 \text{ bar} = \; 0 \text{ bar}$$
$$10 \; 000 \; 000 \text{ cm } H_2O \; = \; 10 \; 000 \; 000/1000 = 10 \; 000 \text{ bars}$$

The pF unit is also derived from the soil moisture tension in cm H_2O by taking the logarithm as follows:

$$pF = \log \text{cm } H_2O \tag{5.2}$$

The minimum value for cm H_2O to be used in the log equation is then a soil moisture tension of 1: $pF = \log 1 = 0$. The maximum value is pF $= \log 10 \; 000 \; 000 = 7$.

The pF is unique to soil moisture expression, although atmospheres or bars are also used. As indicated above, water can exist in soil under a tension varying from $pF = 0$ (no tension) to $pF = 7.0$ (very high tension). The pF is zero when the soil is saturated with water. At field capacity, the soil moisture tension equals 0.333 bars (33.3 kPa), which equals a pF of 2.54 (=log 333 cm H_2O). This is the point at which the soil contains the maximum amount of water readily available to plants.

5.1.2 Classification of Soil Water

Over the years several soil moisture values, frequently called *soil moisture constants*, have been distinguished based on the relative degree of soil moisture tension. The soil moisture constants range from saturation to oven dry condition (Table 5.1). These moisture constants are defined as follows.

Maximum retentive capacity: The amount of water held by soils at saturation. All pore spaces are filled with water.

Field capacity: The amount of water held by soils after excess water has drained by gravity and the downward movement has materially ceased.

Moisture equivalent: The amount of water held in soils after excess water has been removed by centrifugation.

Table 5.1. Soil Moisture Constants

Soil moisture constant	cm H_2O	bars	pF
Maximum retentive capacity	0	0	0
Field capacity	346	0.3	2.54
Moisture equivalent	1 000	1	3.0
Wilting point	15 849	15	4.2
Hygroscopic coefficient	31 623	31	4.5
Air dry soil	1×10^6	1 000	6.0
Oven dry soil	1×10^7	10 000	7.0

Wilting point: The amount of water held in soils at which plants start to wilt. The moisture content is not sufficient for plants to maintain their turgor.

Hygroscopic coefficient: The amount of water adsorbed by soils from an atmosphere of known relative humidity. The exact amounts of water are variable depending upon the method and relative humidity at which they are determined. Most scientists in the United States use 3.3% H_2SO_4, which gives about 98% relative humidity.

Air dry: The moisture content of an air dry soil or a soil at equilibrium with the atmosphere.

Oven dry: The moisture content remaining in the soil after the soil has been dried at 105-110°C until no more water is lost.

Physical Classification of Soil Moisture. - These soil moisture constants and the relative degree of tension have been used to classify soil water. Such a classification of the many forms of soil water may be undertaken from a purely physical standpoint relating it to the degree of tension only, or from a biological standpoint relating it also to plant response. The following three types of soil water are distinguished in the physical classification:

Free water: Water held between pF 0 and 2.54. A saturated soil contains free water.

Capillary water: Water held between pF 2.54 and 4.5, or water held between the field capacity and the hygroscopic coefficient.

Hygroscopic water: Water held at the hygroscopic coefficient (pF = 4.5).

Biological Classification of Soil Moisture. - The biological classification defines the types of soil water somewhat differently. The distinctions are similar in some respects to those listed in the physical classification. Three types of soil water are also recognized in the biological classification:

Superfluous water: Water in a saturated soil that is of no benefit to ordinary plants. It is related to free water.

Available water: The total amount of water present in soils between the field capacity and wilting point (or between pF 2.45 and 4.2). This amount is available to plants.

Unavailable water: Water held above a pF of 4.2, or the amount of water below the wilting point. This water is nonavailable to most ordinary plants.

5.1.3   Environmental Significance of Soil Moisture Constants

As indicated above at field capacity, the soil contains the maximum amount of water that is available to plants. Plants use little gravitational water since this water drains quickly to a depth below the root zone. As the soil dries and the water content decreases, the remaining water is present as thin films around the soil particles. These films become increasingly thinner upon continued drying of the soil. The forces holding the remaining water then become increasingly larger (pF value increases). Consequently, the force that the plant has to exert to remove the remaining water increases as the soil dries. Eventually a point is reached at which the force exerted by plants is not sufficient for sustaining growth. At this point, known as the wilting point, the plant starts to wilt. At wilting point the soil moisture tension equals 15 bars (1500 kPa), which is proportional to a pF of 4.2 (=log 15000 cm H_2O). The unit kPa (= kilopascal) is the energy unit in the SI system and 1 kPa = 0.01 bar. The amount of water held between the field capacity and the wilting point is called *available water*.

Although the moisture constants are very useful in crop production, a number of microbiologists believe that these constants and potentials are insufficient in predicting microbial activity, since these organisms can remain active at very low values of ψ_w, where by definition no available water is present. Microbial activity has been reported at a water potential of -500 bars, an unbelievably low value. However, this extremely low value corresponds to a relative humidity of 70% (Papendick and Campbell, 1981). Therefore, most microbiologists believe that relative humidity is more relevant to microbial life than water potentials and soil moisture constants. The topic of relative humidity in soil air will be discussed in more detail in Section 5.2.4 on movement of soil water as water vapor.

5.2 MOVEMENT OF SOIL WATER

5.2.1 Movement of Liquid Water

Water can move in the soil in any direction. Generally water tends to move downwards due to the pull of gravity. In terms of potentials, this force is called the *gravity or gravitational potential* (ψ_g). However, due to differences in water potentials sideways, water can also move sideways, upwards, or diagonally in the pedon. The force causing water to flow is caused by a *water potential difference*. This water potential difference divided by the distance is called *water potential gradient*, which reflects a continuously changing function of potential in a flow medium (Taylor and Ashcroft, 1972). The flow of water continues until the water potential difference between the two locations is zero. At this point an equilibrium condition is reached.

Other terms used to describe movement of soil water are flux, conductivity, infiltration, and percolation. *Flux* refers to the rate of flow, whereas *conductivity* expresses the resistance of water to flow, or in other words the ease with which water is transferred in the soil. *Infiltration* is used for the entry of water into the soil, and *percolation* refers to the downward movement of water in the pedon. The presence of large amounts of macropores and a low water potential are reasons for a high infiltration rate. Wild (1993) believes that the infiltration rate will usually decrease when soil structure collapses and water potential increases as the soil becomes moist. At low infiltration rates a steady state is reached, which Wild calls the *percolation rate*. The infiltration rate is very important in preventing loss of water by *runoff*, especially in regions characterized by abundant rainfalls. When the amount of water supplied by rains exceeds the rate of infiltration, the excess rain water will run off the soil surface by a process called runoff, as stated above. Runoff can be decreased by maintaining a vegetation cover, a strong stable soil structure containing many pores, or drainage channels.

A number of authors have attempted to distinguish movement of water in soils into the following three categories: (1) unsaturated flow, (2) saturated flow, and (3) water vapor movement (Brady and Weil,

1996; Miller and Gardiner, 1998; Newman, 1972; Van Wijk and De Vries, 1963).

5.2.2 Unsaturated Flow

Unsaturated flow is the movement of water in a dry soil, where the macropores are generally filled with air. In an unsaturated soil, water can move downwards and upwards or in other directions in its effort to fill the soil macropores when rain is falling on a dry soil. The force causing water to flow downwards is usually the water potential difference and the gravitational potential. Water that moves in an unsaturated soil may come from rain, snow, seepage water, or irrigation. The surface soil can become entirely wet and water will move only downwards. Wetting the entire surface of the soil creates problems in the removal of air when water pushes further into the pores. This in turn may decrease the infiltration rate, which then limits further entry of water into the soil. However, in case only part of the surface soil becomes wet, water can move downwards and sideways. The rate of water flow sideways may be as great as the downward flow. The amount of water that enters the soil is generally greater in dry than in moist soil. However, the rate of advance of water in a dry soil is noted to be lower than that in a moist soil. When the flow of water advances in soils, it moves as a front of moisture, which is sharply delineated in dry soils. This front of water is called the *wetting front*. Behind the wetting front, water is moving quite rapidly because of the high *hydraulic conductivity*. When the wetting front reaches the dry areas in the soil, it is as if movement stops here and water is piling up until the soil reaches saturation; then water suddenly flows with a jerky movement to the next grain or region. The dry parts of the soil are characterized by lower hydraulic conductivity values.

Capillary rise is perhaps also an important factor in the flow of water in unsaturated soils and occurs generally when the subsoil is still wet while the surface soil is dry. It can also take place in the presence of fluctuating ground water tables that make the subsoil wet. As indicated earlier, the driving forces in capillary flow are adhesion, involving hydrogen bonding, and surface tension as a result of cohesion. The negatively charged surfaces of clay minerals, forming the walls of the capillaries, are attracting water molecules by hydrogen bonding and water moves up along the walls of the fine pores. The adsorbed water molecules in turn are attracting other water molecules by cohesion, filling the capillaries. The height of capillary rise is usually greater in fine-textured than in coarse-textured soils. Although capillary flow of water amounts to only a few centimeters, it is considered an important factor in providing water to plant roots, especially in periods of low moisture content prevailing during hot and dry weather conditions.

5.2.3 Saturated Flow

This is the water flow in a wet (saturated) soil. All the pores are filled with water. Water entering such a soil will flow downwards by a process called *percolation*. Water can be noticed moving down in the soil during a downpour of rain. The percolating water may pass the subsoil and reach the underlying zone of *capillary fringe* before flowing into the ground water. The name capillary fringe is used for indicating the region of a constantly saturated layer located above the water table (free water surface). The water content of the capillary fringe is similar to that in the water table. The zone above the capillary fringe is known as the *vadose zone*, which is usually considered to contain less moisture than the capillary fringe. Brady and Weil (1996) indicate that the vadose zone even includes the pedon. If the percolating water reaches an impervious rock, the water flow may be diverted and may surface as seepage water that can run into rivers and lakes. The driving force of percolation is the combined forces of the water potential and gravitational potential, defined as the *total water potential,* characterized by the following equation (Tan, 1998):

$$\psi_t = \psi_w + \psi_g$$

where t = total, w = water, g = gravity, and ψ = potential.

The movement of water flowing in a saturated soil obeys *Darcy's law*, which can be formulated in several ways. Some of the original equations are very simple, whereas others, especially the modern equations, are very complex. The message that this law conveys is often lost due to the use of complicated mathematics, and all that Darcy's law is meant to convey is that the flow of water in soils (or porous media) is proportional to the hydraulic conductivity, or is dependent on the differences in hydrostatic pressures between two locations in the soil. A simple equation is as follows (Newman, 1974):

$$Q = k \, \Delta P$$

where Q = rate of water flow, k = hydraulic conductivity (cm/sec), Δ = difference, and P = hydrostatic pressure. For a soil system, the unit of Q is vol/unit area, per unit time, e.g., cm^3/cm^2 per second. The value of ΔP is taken as the difference in hydrostatic pressures between the two ends of the soil system. According to Van Wijk and De Vries (1963) the common form of Darcy's law is

$$v = Ki$$

where v = transmission velocity in cm^3/sec per cm^2, K = hydraulic conductivity, and i = hydraulic gradient, a dimensionless unit. These authors believe that the movement of water in an unsaturated soil also obeys Darcy's law. However, the unsaturated permeability and unsaturated hydraulic conductivity are reported to depend on the distribution of water within the pore spaces, hence on the water content. The value of k, the hydraulic conductivity, decreases rapidly with decreasing moisture content. On the other hand, Miller and Gardiner (1998) indicate that Darcy's formula describes a vertical flow of water, whereas Taylor and Ashcroft (1972) suggest it for explaining a steady flow of soil moisture. A steady flow of water occurs usually in saturated soils.

5.2.4 Movement of Water as Water Vapor

Though most of the water in soils is in liquid form, some of it is present as water vapor in equilibrium with liquid water. The amount of water vapor in the pores is related to its partial pressure, and this amount determines the relative humidity of soil air. Since it is in gas form, its movement obeys the rules of movement of gas in soils as discussed in Chapter 4, Gas Phase in Soils. Therefore, water vapor moves from one to another location in soils by mass flow and diffusion. As discussed before, mass flow is caused by expansion and contraction of the water vapor mass due to changes in temperature and barometric pressure. A high temperature at one location in the soil causes the water vapor to expand in volume, whereas a low temperature at another location in the soil will result in contraction of the water vapor volume. This may cause the water vapor to flow from the warmer to the cooler location in soils. This type of movement is not restricted to water vapor only; liquid water is also affected by such a movement, which is called *thermal flow* of soil water (Taylor and Ashcroft, 1972). The flow of liquid water under the effect of a thermal gradient is generally from a warm to a cool spot in soils. On the other hand, diffusion of water vapor will occur because of concentration differences in water vapor at one and another location in soils. Therefore, water vapor will move spontaneously from a moist to a dry region in the soil. The relative humidity of soil air = 100% in the wet location, meaning soil air is saturated with water vapor, and its vapor partial pressure is high. In the dry parts of the soil, the relative humidity of soil air is low, hence the soil air is undersaturated with water vapor. Consequently, this concentration difference forces the water vapor to move to the region with lower concentrations in response to nature's rule for attaining an equilibrium condition.

Though several authors consider the amount of water vapor in the soil as insignificantly small, being 10 kg versus 600,000 kg of liquid water in the same volume of soil (Brady and Weil, 1996), nevertheless it plays a significant role in determining the humidity of soil air, which is very important for biological activity in soils. As is the case with atmospheric air, the amount of water vapor in soil air can be expressed in several ways and one of those is expressing it in terms of *relative humidity* (Tan, 1998). The relative humidity of soil air is defined as the ratio of partial pressure of water vapor, P_w, to its saturation pressure at temperature t, P_w^s. This definition can be formulated as follows:

Relative humidity = $(P_w)/(P_w^s)$

Therefore, the use of the term *low* relative humidity only indicates that the partial pressure of water vapor, P_w, is smaller than its saturation pressure, P_w^s, meaning: in a nonsaturated condition, soil air contains lower amounts of water vapor than it does in a saturated condition. A high relative humidity then means that P_w approaches the value of P_w^s at temperature t. It is customary to express the value of relative humidity as a percentage, which can be calculated using the formula:

% Relative humidity = $[(P_w)/(P_w^s)]$ x 100%

The equation above indicates that at saturation, $P_w = P_w^s$, and the relative humidity is 100%. The temperature at which $P_w = P_w^s$ is called the *dew point temperature*. When the air is cooled below its dew point, dew formation (condensation of water vapor) occurs. This has important environmental implications. In many soils, especially in those of arid and semiarid regions, dew is an important source of water supply for plants and microbial life.

5.2.5 Environmental Significance of Water Movement

The movement of soil water in both the liquid and vapor form is very important in the soil ecosystem. It is the means for replenishing the depleted water content in the root zone, and it is the main vehicle of transport for plant nutrients. Capillary movement contributes very much in providing water to plants growing in dry and hot regions. Water moving via percolation contributes toward soil formation, because of the transport of many elements that are used for the development of different soil horizons. Leaching of materials, though often considered to result in the loss of many important nutrient elements for plant and microbial growth, is also very important because of the dilution effect it causes in the presence of toxic concentrations. Much more can be said of the environmental significance of water movement.

Because soil pores are in the form of irregular tubes, the force required to remove water, either by uptake by plant roots or by evaporation, is then determined by the narrowest parts of the pores. Here, water is most strongly held, but as soon as water flows from these narrow channels, water will automatically flow from the wider parts of the pores. When water is added to the soil by rain or irrigation, it is as if the wider parts of the pores set limits for the movement of water. It takes time to fill the wider pores first before water can be pushed through the narrow pores. The forces for filling are usually lower than that for emptying. As a result of drying and re-wetting, the soil retains more water at the same force during drying than during wetting. This difference in behavior of soil water is known as *hysteresis*. Hysteresis is a common occurrence in the capillary fringe because of the falling (= drying) and rising (=re-wetting) water table. The fluctuating groundwater table determines the thickness of the capillary fringe. The latter is thicker due to falling than to rising water table. A falling water table is a drying process which results in higher amounts of water retained that are used for the development of a thicker saturated zone which is known as the capillary fringe.

5.3   WATER LOSSES

Soil water can also move out of the soil and return to the atmosphere. It is not an absolute loss, but a temporary loss from the soil or the surface of the earth, since water is a renewable resource that belongs to the gigantic cycle called the *hydrologic cycle*. It is evaporated from the surfaces of soils, rivers, lakes, and the sea into the atmosphere. The energy required for evaporation is supplied by the sun. In the atmosphere water is transported by the wind in the form of vapor, until it is finally returned to the earth in some form of precipitation.

Loss of water from soils can be distinguished into (1) vapor loss and (2) liquid loss.

5.3.1 Vapor Loss

Water is lost from soils in the form of water vapor due to (1) evaporation, and (2) evapotranspiration. In physics and chemistry, *evaporation* is defined as the transformation of water from the solid and liquid phase into the vapor phase. When the vapor phase is carried away by any means, this can be interpreted as the loss of water in the vapor phase from any surface. Evaporation of water occurs then from the surfaces of soils and plants, and as such is a passive process. On the other hand, transpiration is usually construed of as the loss of water by plants and microorganisms due to an active metabolism process. The breakdown of sugars by respiration not only yields CO_2, but also equal amounts of H_2O and a lot of energy, as can be noticed from the following reaction:

$$C_6H_{12}O_6 + 6O_2 \rightarrow 6CO_2 + 6H_2O + \text{energy}$$

The energy formed is heat, and in order to maintain a constant temperature for normal growth much water is "evaporated" through the stomata. This loss of water by the active process of plants and soil organisms is called *transpiration*. Transpiration is very critical for the organisms as a means of cooling, for maintenance of turgor, and for performing other active processes for growth. The same process in human beings is called *perspiration*, which has a similar function of maintaining body temperature at almost constant levels suitable for living. The combined effect of evaporation and transpiration is now called *evapotranspiration*. Brady and Weil (1996) assume that evapotranspiration is responsible for the loss of 75 to 100 cm of water from alfalfa plants grown in arid region soils. One hundred cm of water is the equivalent of the amount of water received from 40 inches of rainfall. Such an enormous amount of water lost by evapotranspiration appears to be normal among plants, since a good crop of potatoes, grown in the Netherlands, is noted to have lost 4300 Mg/ha of water, whereas crops with poor yields may have lost half of that by transpiration (Van Wijk and Borghorst, 1963). The fact that enormous quantities of water are lost by transpiration is also supported by reports from other parts in Europe indicating that 10 mm of water per day are lost by evapotranspiration from a complete canopy of plants growing in wet soil (Wild,1993). These losses vary, of course, according to the factors affecting evapotranspiration, such as temperature, radiant energy, vapor pressure, moisture content, wind action, and plant species.

Control of Vapor Loss. - Lost of water by evapotranspiration is usually controlled by the use of mulch, and by cultivation and weed control. *Mulch* is considered the most important material for decreasing water vapor loss. It is defined as any material used on the surface of soils to prevent loss of water by evaporation. Sawdust, manure, straw, leaves, paper and plastic mulch, and stubble mulch have been used to decrease evaporation. In general, organic matter is an excellent source of mulching material. In addition to its purpose of reducing water loss by evaporation, its side effects on improving the physical and chemical properties of soils are beneficial factors that are not provided by the other materials. Paper mulch has been employed with considerable success in the pineapple culture in Hawaii, whereas plastic mulch has found extensive application in the vegetable-growing areas in south Florida and California. Strawberries are grown in California using plastic mulch not only to reduce evaporation but also for producing high-quality strawberries. The berries are less likely to become affected by rot when they mature on the plastic sheet used as mulch, and are not touching the soil. Stubble mulch has been used mostly in semihumid and semiarid regions of the United States. In this case, the mulching materials are composed of the stubble from the previous crops and the new crop is seeded between the dead stubble. However, the chances of allelophaty to develop are high. The growth of sorghum and barley plants can be inhibited when they are seeded and grown according to the no-till method in the wheat stubble of the previous wheat crop. As indicated in Section 3.1.2, allelophatic chemicals may be produced by the products of the decomposition of wheat straw, such as acetic acid, toxic to the new crops. Also worth mentioning perhaps is *soil mulch*, which is formed by granulating the surface soil into a very friable consistence to trap air in the numerous pores created. The contention is that the trapped air functions as an air barrier for water vapor flow into the atmosphere. This idea has received mixed reactions from several soil scientists (Brady and Weil, 1996).

5.3.2 Liquid Loss

Soil water can also be lost as liquid water. As discussed earlier, when precipitation far exceeds the infiltration rate of soils, abundant amounts of water may be lost in the form of runoff. Runoff is considered a very harmful process, not only because large amounts of water are lost from the soil, but also because large quantities of soil constituents can be carried away, resulting in serious erosion and denudation of the soil surface. Two other forms of loss of soil water as liquid water, percolation and leaching, are not that critical, though in certain conditions they can also be harmful. As discussed above, percolation is the flow of water in a saturated soil. When the percolating water runs through the pedon into streams and lakes, this water is considered lost from the soil. A great number of soluble compounds, including plant nutrient elements, are dissolved by the percolating water, and are carried away from the pedon. The loss of nutrient elements and other soluble compounds with the percolating water is called *leaching*. The type and nature of the soluble substances are the topic of Section 5.4.

From an environmental standpoint, it can be argued that percolation and leaching in temperate region soils are noted to be highest when evapotranspiration is at a minimum. Hence, percolation and leaching will be minimal when evapotranspiration is at a maximum. In the humid region of the United States, characterized by an annual precipitation of more than 100 cm, the infiltration rate of soils with good physical properties usually exceeds the rate of evapotranspiration, hence large amounts of water can percolate through the pedon and be lost from the soil. It is assumed that maximum percolation takes place during winter, because evapotranspiration is at a minimum due to dormant biological activity. Most of the plants have shed their leaves in the cold season and transpiration is drastically reduced. In the spring, plants resume their biological activity, and evapotranspiration starts to increase and reaches maximum rates in the summer. Regardless of the high rainfall, Brady and Weil (196) assume that the rate of percolation is offset because most of the water is lost by evapotranspiration during the summer.

Control of Loss of Water in the Form of Liquid. - Loss of water from runoff is usually serious in hilly lands or in regions with steep topography. The common method for reducing runoff and the erosion that it brings is by *terracing*, a process of building terraces that run horizontally across the slope. Water may be captured in the furrows of the terraces and is given time to penetrate the soil, where it is stored in amounts determined by the soil's adsorptive capacity, or water-holding capacity. In less steep areas, runoff can be decreased by maintaining an adequate vegetative cover on top of the soil that can intercept the heavy downpours. The impact of heavy rains is then reduced and the rapid flow of water to the soil is slowed down by the foliage of the trees.

In contrast to runoff, percolation and leaching, though harmful under certain conditions, are more difficult to control. They are natural processes necessary for the development of soil profiles. The differences in percolation and leaching bring about differentiation in soil horizons. For example, they are necessary for the transport of Ca down the pedon to form the calcic B horizons, characteristic for the *pedocals*, such as the mollisols in the American Midwest. Another example, showing the environmental necessity for percolation and leaching, is the formation of spodosols. Percolation and leaching of soluble Fe and humic matter are essential for the formation of spodic B horizons, identifying the spodosols. However, although percolation and leaching are sometimes undesirable, no satisfactory methods have been reported yet for their control. Perhaps, an environmentally sound method is to maintain a good cover crop on the soil surface that will reduce the effect of percolation and leaching, although it will not eliminate it. The loss of nutrient elements then can be reduced by the application of fertilizers.

On the other hand, it is possible to enhance the percolation process. Encouragement of percolation is defined by Brady and Weil (1996) as artificial *land drainage*. It is especially needed in soils where percolation is inhibited, hence must be increased for adequate crop production and for maintaining a low water table level under houses, roads, and highways. This is essential for preventing seepage and flooding that may damage roads and basements. Examples of poor drainage conditions posing hazards for flooding and damage to houses and properties can be seen in New Orleans, Louisiana. Land drainage is the process of removing superfluous water from the soil as rapidly as possible. Two general types of drainage systems are usually available: the *open* and *closed* drainage system. In the open drainage system, the drains are open to the atmosphere. A ditch drainage is an example of an open-drain system. In a closed-drain system, the drains are buried in the soil. Tile drains are examples of a closed-drain system.

5.4 DISSOLVED SOLID SUBSTANCES

Water is the main solvent in soil for a variety of compounds and contains, therefore, an assortment of dissolved materials. Many of the dissolved materials are in ionic form. If a compound with a general formula B_nA_x is in contact with soil water, it dissociates into its ionic components B^{n+} and A^{x-}, in which B is a metal cation with charge n, and A is the anion with charge x. As is the case with protons (H^+), which exist as H_3O^+, a metal ion cannot exist by itself. In soil water the metal reacts with water molecules, which form a hydration shell. The number of water molecules attracted to the metal ion depends on the coordination number of the cation. The hydrated cation carries the original number of positive charges. Hence, the symbol for such a hydrated ion is $B(H_2O)_x^{n+}$. This concept will be illustrated below.

5.4.1 Aluminum

Aluminum is an important constituent of soils. It is the second most abundant element in rocks and minerals (Hunt, 1972) and is released in soils upon weathering of these rocks and minerals. It is chemically very active, hence the element will seldom occur free in nature and will usually be present as oxides or as a constituent of alumino-silicate minerals and sesquioxide clays.

The most important form of aluminum oxide is *gibbsite*, $Al(OH)_3$, *a sesquioxide clay*, and bauxite, $Al_2O_3.2H_2O$. As discussed in Chapter 2, gibbsite is an important clay mineral in oxisols and ultisols. More exotic forms of aluminum oxides are *corundum*, Al_2O_3, *ruby*, and *sapphire*. Ruby is in fact a red-colored variety of corundum, whereas sapphire is a blue-colored species of corundum. All three varieties belong to the hematite group of minerals, characterized by the formula X_2O_3 (Hurlbut and Klein, 1977), in which X represents Al or Fe. If X is Fe, the mineral is hematite, Fe_2O_3. Corundum is usually present as an accessory mineral in metamorphic and silica-deficient igneous rocks, such as syenite. The mineral occurs in the United States in the eastern part of the Appalachian Mountains of North Carolina and Georgia. Rubies are found in limestone affected by metamorphism in Burma, Thailand, and Cambodia, whereas most sapphires come from alluvial deposits of Sri Lanka, Thailand and Cambodia. Kashmir, India, central Queensland, Australia, and Montana, USA are additional regions where sapphires have been discovered. Corundum is usually used as an abrasive or polishing material, whereas rubies and sapphires are valuable as gemstones. Stones of gem quality also have applications in the watch industry and as bearings in scientific instruments.

The concentration of soluble Al in soil water is very small and amounts to only 1.0 mg/L (Manahan, 1975). When an Al compound dissolves in soil water, the Al^{3+} ion is quickly surrounded by 6 molecules of H_2O in octahedral coordination, forming $Al(H_2O)_6^{3+}$. This ion is called an *aluminum hexahydronium* ion, and is subject to hydrolysis, especially as soil pH increases. Protons are dissociated from the coordinated water by the hydrolysis reaction, yielding a series of dissociation products:

$$Al(H_2O)_6^{3+} \leftrightarrow Al(H_2O)_5(OH)^{2+} + H^+$$
$$\leftrightarrow Al(H_2O)_4(OH)_2^+ + 2H^+$$
$$\leftrightarrow Al(H_2O)_3(OH)_3^0 + 3H^+$$
$$\leftrightarrow Al(H_2O)_2(OH)_4^- + 4H^+$$
$$\leftrightarrow Al(H_2O)(OH)_5^{2-} + 5H^+$$

For simplicity, the aluminum hydroxide monomers above are usually written without the coordinated water:

$$Al^{3+} + H_2O \leftrightarrow Al(OH)^{2+} + H^+$$
$$Al^{3+} + 2H_2O \leftrightarrow Al(OH)_2^+ + 2H^+$$
$$Al^{3+} + 3H_2O \leftrightarrow Al(OH)_3^\circ + 3H^+$$
$$Al^{3+} + 4H_2O \leftrightarrow Al(OH)_4^- + 4H^+$$
$$Al^{3+} + 5H_2O \leftrightarrow Al(OH)_5^{2-} + 5H^+$$

These reactions show that hydrated metal ions with a charge of 3 or more are proton donors and may affect soil acidity. The charges of the aluminum products change from positive through zero to negative. As the degree of hydrolysis increases from 1 to $5H_2O$, the solubility of the aluminum hydrates decreases.

These aluminum hydroxide monomers tend to form polymers according to the reaction in Figure 5.1. These polymers are also capable of donating protons, and hence they behave as acids. The reaction is represented in Figure 5.2.

Free Al or hydrated Al ions are usually present in large amounts in acid soils. As can be seen from the reactions above, the presence of free Al is in fact the reason for the development of soil acidity. Mono-

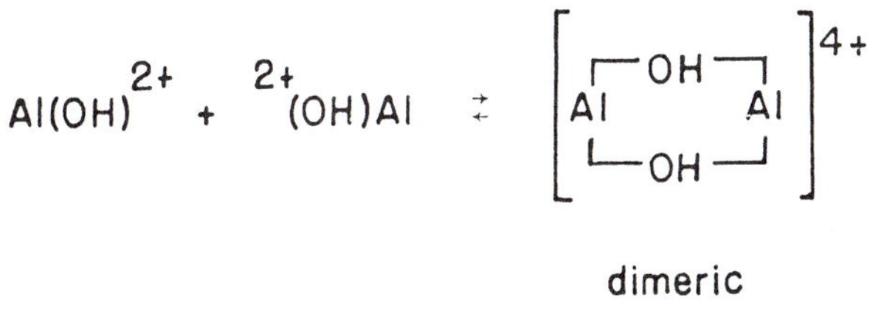

$$Al(OH)^{2+} + (OH)Al^{2+} \rightleftharpoons \left[Al \begin{array}{c} OH \\ \\ OH \end{array} Al \right]^{4+}$$

dimeric

Figure 5.1 Formation of an Al hydroxide dimer from two Al hydroxide monomers.

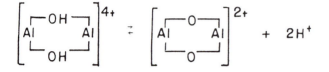

Figure 5.2   Dissociation of H^+ from an Al hydroxide dimer.

meric aluminum ions, Al^{3+}, adsorbed on the negative charged surfaces of clays can be released by exchange with other ions. They hydrolyze in solution forming a series of Al-hydroxy compounds as discussed earlier while releasing protons at the same time, which increases soil acidity. The hydrolysis products, readsorbed by the clay minerals, cause increased hydrolysis and production of more H^+ ions. Acid soils are, therefore, high in soluble aluminum accompanied by a complex series of its hydrolysis products. The degree of change in pH is controlled by the aluminum potential. In line of this concept, Kennedy (1992) presents a hypothesis that aluminum compounds in acid soils are in fact pH buffers. Free Al^{3+}, released by exchange of adsorbed Al, yields H^+ upon hydrolysis. When the latter is not readsorbed by clays but instead is neutralized by liming, the aluminum hydroxy ions precipitate as insoluble aluminum hydroxides. These processes are illustrated by the following reactions:

$$3M^+ + \text{Al-clay} \rightleftarrows M_3\text{- clay } + Al^{3+}$$

$$Al^{3+} + H_2O \rightarrow Al(OH)^{2+} + H^+$$

Liming then yields the following reactions:

$$H^+ + OH^- \rightarrow H_2O$$

$$Al(OH)^{2+} + 2OH^- + Al(OH)_3\downarrow$$

Since $Al(OH)^{2+}$ is eliminated, no more protons can be produced, and additional Al^{3+} ions will be released from the exchange sites to offset an increase in soil pH as a result of liming. Therefore, exchangeable Al^{3+} ions can be considered as a source of *reserve acidity* in the same sense as the concept of reserve acidity based on exchangeable H^+ ions. Since the hydrolysis of aluminum hydroxy compounds occurs at pK = 5.1, the buffering effect is felt at a soil pH $\leq$ 5.1. Kennedy (1992) believes that similar reactions also take place with Fe^{3+}. However, the buffering effect is felt at lower soil pH values, since hydrolysis reactions of iron hydroxy compounds occur at pK = 2.2.

Large amounts of these Al^{3+} ions may cause Al toxicity in plants. They also react with phosphates to form insoluble Al-phosphate compounds, which may cause P deficiency in plants. The reaction between Al and phosphate that results in the formation of an insoluble Al phosphate compound is called *phosphate fixation*.

5.4.2 Iron

Iron is another important element in soils. It is the third most abundant element in rocks and minerals. The central core of the earth is made up mostly of iron. All rocks, minerals, soils, and plants contain iron. In animals, it is present in the blood hemoglobin, which acts as a carrier of oxygen.

The most important iron mineral is *hematite*, Fe_2O_3, and *magnetite*, Fe_3O_4. The hydrated form of hematite is often called *limonite*, $2Fe_2O_3.3H_2O$. As discussed in Chapter 2, hematite is red and its presence gives to the soils a red color. Magnetite is black and crystalline in nature. It has strong magnetic properties. Another iron mineral is *pyrite*, FeS_2, which occurs in soils as yellowish crystals with a metallic luster similar to gold, hence the name *fool's gold* is often used for this mineral. *Ilmenite*, $FeTiO_3$, is also an important iron mineral. The name is derived from the Ilmen Mountains, Russia, where ilmenite has been discovered in large quantities. It is a common accessory mineral of igneous rocks, such as gabbros and diorites. It is found in the United States in several locations in the Adirondack Mountains and in humic acid deposits in Florida and southern Georgia. In Western Australia, ilmenite is currently mined from beach sands near Perth. This mineral is the major source of *titanium*, used in the plastic and paint industries as pigment, replacing older pigments such as lead, Pb. Metallic titanium is currently very valuable in the aerospace industry because of its high strength/weight ratio. It is the most desirable lightweight-but-strong metal for the production of aircrafts and supersonic spacecrafts.

When released in soil water, iron can exist in two different forms. The redox condition in soils determines the ionic form of iron in soil water. In anaerobic conditions, iron may be present as ferrous iron, Fe^{2+}. It is the dominant iron species in ground water, where its concentration may range from 1.0 to 10 mg/L. In aerobic conditions, iron is generally present as ferric iron, Fe^{3+}. In aerobic conditions, ferrous iron is unstable and will be oxidized into Fe^{3+}:

$$Fe^{2+} \rightleftarrows Fe^{3+} + e^- \tag{5.3}$$

Ferric iron behaves similarly as Al^{3+} ions and will be quickly surrounded by 6 molecules of H_2O in octahedral coordination, yielding $Fe(H_2O)_6^{3+}$. The latter is also subject to hydrolysis, which produces Fe hydroxide monomers and protons. Upon polymerization, the monomers yield Fe hydroxide polymers, which disperse in water and cause turbidity. With increased formation of ferric hydrates, the color of soil water becomes brown. This is perhaps the reason why too much iron in water is considered harmful for agricultural, engineering, and domestic purposes. In engineering, the use of water that contains iron may cause corrosion and the formation of unsightly deposits of Fe hydroxides. In the kaolin industry, kaolin deposits stained with iron have to be processed to remove the Fe so that the product is white in color. Dissolved iron at levels > 0.31 mg/L produces an unpleasant taste in drinking water. It will also stain laundry, bath tubs, sinks, and porcelain. In agriculture, iron is needed only in trace amounts by plants, and too much iron can cause Fe toxicity. When phosphate is present, Fe may react with phosphate to form insoluble Fe phosphate, a process called *phosphate fixation*. This may cause P deficiency in plants.

5.4.3 Silicon

Silicon is perhaps next to oxygen in abundance in the earth's crust. It occurs in the soil as six distinct crystalline minerals: quartz, tridymite, crystobalite, coesite, strishovite, and opal. As discussed in Chapter 2, they are classified today in the SiO_2 group of *tectosilicates*. The minerals quartz, tridymite, crystobalite, coesite, and strishovite are all characterized by the formula SiO_2, but differ from each other in their framework of geometric arrangements of the silica tetrahedron. Opal, on the other hand, is characterized by a formula of $SiO_2.nH_2O$. Flint is also a silica mineral, and a number of amorphous or noncrystalline and paracrystalline species can also be found.

Quartz is a common mineral of acid igneous rocks, such as granite and rhyolite. The mineral is used for a wide variety of purposes. It is used as sand in mortar and concrete and as an abrasive in the production of sandpaper. In powdered form, it is an important ingredient in the production of porcelain and crystal glass. Quartz also finds application in optical instruments because of its capacity to be ground into lenses and prisms. Because of its optical activity, its ability to rotate the plane of polarization of light, quartz is valuable for instruments used in the generation of *monochromatic* light of different wavelengths. Tiny quartz plates, serving as oscillators, are used in digital watches and clocks. Quartz oscillators are useful today in the measurement of instant high pressures resulting from atomic explosions and the firing of firearms. Quartz varieties that are of gem quality are very valuable ornaments. These include amethyst, rose quartz, agate, and onyx. Today, yellow, brown, blue and violet quartz are produced artificially by hydrothermal methods for gemstones and lenses. Tridymite is also a mineral of many acidic igneous rocks, whereas crystobalite is usually volcanic in origin. Volcanic ash often contains the mineral crystobalite. Opal, on the other hand, is considered biogenic in origin. The original materials are petrified plant materials or diatomites. For instance, *wood opal* is in fact fossilized wood with opal as the petrifying substance. The major colors of opal are white, yellow, or milky-white, in which a variety of blue, yellow, red, or black reflections can be seen. Opal with red reflections, called *fire opal*, is a very valuable gemstone. *Black opal* is another example of a precious gem.

Of the minerals discussed above, quartz is the most common SiO_2 mineral. The silicon in quartz is surrounded by 4 oxygen atoms in tetrahedral coordination. These silica tetrahedrons are linked together by sharing oxygen atoms. The Si-O-Si bonds, called the *siloxane bonds*, are the strongest bonds in nature and are of primary importance in crystalline minerals (Tan, 1993). Quartz minerals are, therefore, very resistant to weathering and very difficult to dissolve. Most of the soluble silica originates from amorphous silica and from primary minerals, such as anorthite ($CaAl_2Si_2O_8$) and albite ($NaAlSi_3O_8$), both plagioclase minerals, which are the least stable minerals to weathering. Soluble silica can also be introduced to the soil through artificial compounds. Silicates are used in detergents and as anticorrosive agents in antifreeze. In silicon chemistry, the term silica is used for Si compounds, whereas the term silicon is reserved for the element Si. The element Si itself is inactive at low temperatures and is very resistant to chemical attack. Si crystals are used today for the production of computer chips, semiconductors, and transistor devices in solar batteries. Elemental Si also finds application as a deoxidant in the manufacture of steel, copper, and highly acid-resistant alloys, such as *duriron*.

The concentration of dissolved silica in natural water is usually very small, ranging from 1 to 30 mg SiO_2/L, but concentrations of 100 mg SiO_2/L are also common (Garrels and Christ, 1965; Manahan, 1975; Moore and Moore, 1976). Seawater is usually low in silica content because Si is used by marine organisms in the formation of their skeletons and shells. Present in amounts below 140 mg SiO_2/L (25°C), soluble silica is found mainly as monosilicic acid, which is a true solution. Its formula can be written as H_4SiO_4, $Si(OH)_4$, or $SiO_2(OH)_2$ (Figure 5.3). This type of silicic acid is also known as *ortho-silicic acid*. Monosilicic acid forms *silica gel* by dehydration until the

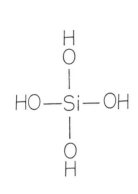

Figure 5.3 Schematic structure of monosilicic or orthosilicic acid, H_4SiO_4.

proper percentage of moisture is reached. The dehydration process, as-sumed to take place stepwise through formation of metasilicic acid, H_2SiO_3, *disilicic acid*, H_6SiO_7, *trisilicic acid*, $H_4Si_3O_8$, etc., is a type of polymerization.

The silica minerals are insoluble at low pH, and their solubility does not increase if pH is increased from 2 to 9. Only above pH 9.0 will silica minerals dissolve. In the range of pH 2 to 9, the solubility of silica remains constant at 140 mg/L (Tan, 1993; Krauskopf, 1956), and this solubility is generally determined by the law of polymerization. If the concentration of dissolved silica in the soil solution exceeds 140 mg/L, polymerization of silica usually occurs (Figure 5.4), and a mixture of polymers and monomers of $Si(OH)_4$ is found in the soil solution. Two or more monosilicic acids can react linearly or can form cyclic combinations with endless variations. Although not clearly understood, the formula of a polynuclear silicate ion is frequently written as $Si_4O_6(OH)_6^{2-}$, which differs somewhat from the one indicated in Figure 5.4.

Silica in the form of monosilicic acid can be adsorbed by sesqui-oxides. This reaction forms the basis for the formation of silicate clays.

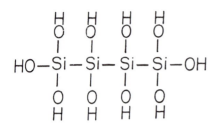

Figure 5.4 Polysilicic acid or polymeric silicic acid with linear structure. Cyclic structures and cross-linked polymeric structures are also possible.

It can be illustrated as follows:

$$H_4SiO_4 \rightleftarrows H^+ + H_3SiO_4^-$$

$$H_3SiO_4^- + Al(OH)_3 \rightleftarrows Al(OH)_2OSi(OH)_3 + OH^-$$

The monosilicic acid must first be converted into an anion by dissociating a proton. Only when it is negatively charged, can it react with positively charged sesquioxides.

In addition to the preceding reaction, monosilicic acid is able to form complexes with metal ions. Such a complexation reaction, as illustrated in Figure 5.5, is of importance in the weathering of primary minerals. The resulting metal-silicate complex is present in ionic form, hence remains in solution. This is perhaps one of the mechanisms in podzolization, by which Al and Fe may be transported as inorganic complexes to form Bs horizons, which are B horizons rich in Al and Fe compounds (s = sesquioxides). Such complex reactions and transport phenomena have received little research attention. The most commonly accepted version of podzolization is that Al and Fe move down the profile in the form of Al- and/or Fe-humic acid chelates (De Coninck, 1980; Tan, 1986).

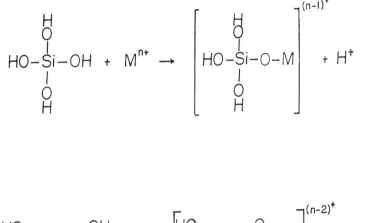

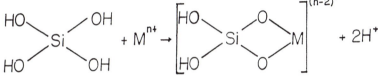

Figure 5.5 Top: Complex formation between monosilicic acid and a cation, M^{n+}, in which n^+ = the number of positive charges. Bottom: Chelation reaction between monosilicic acid and a cation, M^{n+}.

Monosilicic acid is also known to form complexes with soil organic compounds, especially with humic and fulvic acids. The complexation process can be represented by the reactions in Figure 5.6. These reactions play a significant role in the weathering of rocks and minerals. The high affinity of humic acids for Si and metal ions bring about the degradation of many rocks and minerals (Tan, 1986). The subsequent release of Si in the form of complexes or chelates has an important bearing on movement of Si in the pedon, hence in soil formation. It is a process of removal of silica by leaching which plays a significant role in formation of oxisols. This process is called *desilic-*

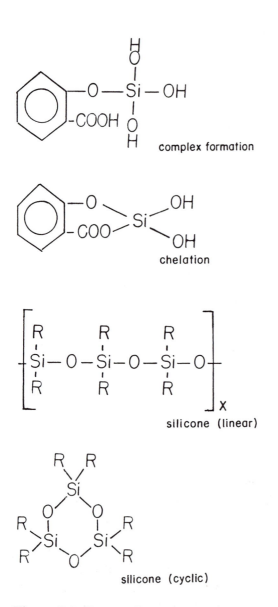

complex formation

chelation

silicone (linear)

silicone (cyclic)

Figure 5.6 Top two: Complex formation and chelation between monosilicic acid and a humic acid molecule. Bottom two: Formation of linear and cyclic silicones.

ification, and in the past it was known as *laterization*. Today this type of complex formation between silicic acid and organic compounds finds application in industry in the production of *silicones*. The general formula of silicone is R_2SiO_2, in which R is an organic radical. This could be a simple organic compound, such as a methyl group, CH_3, an ethyl group, C_2H_5, or a complex humic molecule. These silicones can be linear or cyclic in structure (Figure 5.6). Silicones, composed of cross-linked silica polymers, are also possible. Depending on their molecular complexity, silicones have the appearance of oily, greasy, or rubberlike substances. Consequently, they are used as lubricants, hydraulic fluids, and electrical insulators. Antifreeze for automobiles contains silicone for lubricating the water pump. In medical science, silicone finds application as filling material and implants in human bodies. The controversial use of silicone for breast implants has received a lot of attention in the media.

5.4.4 Nitrogen

Nitrogen, discovered as an element in 1772 by Rutherford, is an inert gas that does not support life and that can not burn. However, it is an essential constituent of proteins and amino acids in plants and animals. Natural gas may contain some nitrogen. Its formula is N_2, and this diatomic molecule is known to have the most stable structure in nature. This high structural stability is the reason for its property of being inert, and its lack of chemical reactivity. Inorganic sources of nitrogen include *saltpeter* (KNO_3) with deposits in India, and *Chile saltpeter* ($NaNO_3$) with deposits in Chile.

Because it is inert, gaseous nitrogen is used in chemical reactions that need to take place in an inert atmosphere. It is used in electric light bulbs to inhibit rapid oxidation of the filament. Modern light bulbs often contain argon gas, which is even more inert than nitrogen. Nitrogen finds extensive application in the food industry to reduce rapid deterioration of canned products. Canned foods, sliced meats, and hydrogenated vegetable oils are frequently sealed with nitrogen gas to prevent spoiling.

Nitrogen exists in soils in organic and inorganic forms and is present in several oxidation states, e.g., from -3 to +6. The total soil N content varies considerably — from 0.05% in desert soils and in humid region soils to 0.3% or higher, as in the soils of semi-humid regions, e.g., mollisols. In soil water, the concentration of nitrogen is considerably lower, constituting only a very small fraction of the amount present in soils. Under normal conditions, 2.4 mL nitrogen will dissolve in 100 mL water. Most of the nitrogen in soils (98%) is in organic form (Bremner, 1965). It is present in plant residue, barnyard manure, and industrial and domestic waste. Some of these organic nitrogen compounds, such as amino acids, are soluble in soil water. However, most of the nitrogen in soil water is in inorganic form, e.g., NH_4^+, NO_3^-, and NO_2^-. These ions are released in soil water by decomposition of soil organic matter. Inorganic nitrogen can also be added to soils by the application of fertilizers.

Ammonia. - One of the most common inorganic nitrogen species in soils is ammonia, NH_3, a product of ammonification. Ammonia, a gas, dissolves in soil water and becomes ammonium, NH_4^+. The reaction can be illustrated as follows:

$$NH_3 + H_2O \rightarrow NH_4^+ + OH^-$$

The amount of ammonium in solution depends on the partial pressure of the ammonia gas. The higher the partial pressure, the larger the amounts of ammonium in solution; the lower the partial pressure of the ammonia gas, the lower the ammonium concentration in solution. In the presence of negatively charged clays, most of the NH_4^+ ions will be adsorbed on the surfaces of the clay complexes, and become then exchangeable NH_4^+. Some the free NH_4^+ ions are subject to nitrification, and in well-aerated soils they are then converted into nitrate ions, NO_3^-.

Gaseous ammonia is colorless and has a strong odor. It is used as a heart stimulant and people who have fainted are often revived by letting them inhale ammonia gas. However, some people may be allergic to the gas and some may die by inhaling it. Ammonia gas can be liquified by compression and/or by cooling. This property is used in the manufacture of liquid N_2 fertilizers, such as *anhydrous ammonia*, or

NH_3. This fertilizer is soluble in water, since 1.2 L of gaseous ammonia can dissolve in 1 L of water (0°C). Other nitrogen fertilizers are also produced from ammonia, e.g., sulfate of ammonia. In addition, ammonia gas is used in industry as a refrigerant and in the production of ice because its *heat of vaporization* (328 calories/g) is the highest among many liquids, except water. It also finds application in domestic chemicals in the form of *household ammonia* for cleansing and washing.

Ammonium is generally stable only under anaerobic conditions, whereas nitrate prevails in aerobic conditions. Ammonium ions are chemically more reactive than nitrate ions. They react quickly with negatively charged clay minerals and form complexes with metals and organic compounds. The cationic nature of NH_4^+ is the reason for its adsorption and retention by soil clays. This process prevents NH_4^+ from being leached to deeper layers of the pedon. When NH_4^+ is adsorbed in intermicellar spaces of expanding types of clays, such as smectite, entrapment of NH_4^+ occurs, a process called *ammonium fixation*.

Ammonium is believed to dissociate according to the following reaction (Manahan, 1975; Moore and Moore, 1976):

$$NH_4^+ \rightleftarrows H^+ + NH_3 \tag{5.4}$$

The pK_a value of the dissociation reaction is 9.26, which means that at pH 9.26 this dissociation produces equal concentrations of NH_3 and NH_4^+. In soils with a pH value above 9.26, reaction (5.4) proceeds to the right. However, most soils and waters in lakes and streams have pH values below 9.26; hence, the reaction goes to the left. In other words, the dissociation does not take place, and NH_4^+ is stable.

Nitrates and Nitrites. – Other species of inorganic nitrogen compounds are nitrite, NO_2^-, and nitrate, NO_3^-, which are the products of nitrification of NH_4^+. Nitrate is used in industry for the manufacture of drugs, plastics, rayon, and dyes and for curing meat products. Ham and bacon are cured with $NaNO_3$. Nitrate is also an important ingredient for the production of explosives, e.g., *TNT* (trinitrotoluene) and gunpowder. An explosive, composed of ammonium nitrate and aluminum powder, is called *ammonal*. The acid form of nitrate is called *nitric acid*, HNO_3, whereas the acid of nitrite is called *nitrous acid*, HNO_2. When nitric acid reacts with human skin, it may make the skin yellow. The reaction is called a *xanthoprotein reaction* and causes the development of yellow stains on the fingers of people using nitric acids in the laboratory.

As indicated earlier, nitrite is usually quickly converted into nitrate, so that the most important inorganic species of N in soil water are the ammonium and nitrate forms. In contrast to ammonium, nitrate ions are anionic in nature, and consequently are not attracted by negatively charged clay minerals. Therefore, NO_3^- tends to be leached into the ground water, rivers, and lakes. The presence of nitrate in amounts above 30 mg/L is considered hazardous to human health by EPA standards and is particularly harmful to babies. When consumed, NO_3^- may not only be carcinogenic, it will also be reduced into NO_2^- in the anaerobic environment of the digestive system. The nitrite reacts with hemoglobin and interferes with its function as an oxygen carrier, as is shown by a bluish tint in a baby's skin. A similar reduction process can also take place in the digestive system of ruminants.

In soil water, nitrate is stable in the absence of organic compounds. Several of these organic compounds may act as biological catalysts, which induce the reduction of nitrates. This process finds application in sewage treatments, as discussed before in the denitrification process.

5.4.5 Phosphorus

Phosphorus is a nonmetallic element and exists in two allotropic varieties, called *white phosphorus* and *red phosphorus*. White phosphorus is a white, soft, waxlike solid, but it will turn yellow when it is exposed to light. Red phosphorus has a red color and vaporizes when heated. White phosphorus is very poisonous, and inhaling the fumes of white phosphorus may cause necrosis of the bones in the jaw and nose, and may also cause death. It has been used in the past for making matches, but since the fumes are very toxic, match heads today contain *tetraphosphorus trisulfide*, P_4S_3, a less toxic substance. Those matches were the once-famous *strike-anywhere* matches. Phosphorus is still used today as rat poison, in making explosives, fireworks, chemicals, fertilizers, and as spoil retardants in food substances.

Phosphorus in soils exists as organic and inorganic compounds. Humus, manure, and other types of nonhumified organic matter are the major sources of organic phosphorus in soils. Some of the compounds in the soil organic fraction considered potential sources of phosphorus are phospholipids, nucleic acids, and inositol phosphates. Bones, teeth, and muscle tissue in animals are rich in phosphorus. Inorganic phosphorus is derived mostly from the apatite minerals, which are accessory minerals in all types of rocks. The most important minerals in the apatite group are fluorapatite, $Ca_5(PO_4)_3F$, chlorapatite, $Ca_5(PO_4)_3Cl$, and hydroxyapatite, $Ca_5(PO_4)_3(OH)$. Apatite is an accessory constituent in igneous, sedimentary, and metamorphic rocks. Large deposits have been detected in the Kola Peninsula near Kirovsk, Russia, on the southern coast of Norway, and in Kiruna, Sweden, and Tahawas, New York. Large amounts of apatite have also been found in Ontario and Quebec, Canada. Present as large deposits, apatite is valuable in the manufacture of phosphate fertilizers. By treatment with sulfuric acid, the apatite is converted into *superphosphate*. A by-product of this process is *gypsum*, $CaSO_4$, a liming material. Phosphoric acid can also be used for treatment of apatite, and the product is then called *double-* or *triple-superphosphate*.

In general, the inorganic P content in soils is higher than the organic P content (Tisdale and Nelson, 1975, 1993). The total phosphorus content in surface soils of the United States may range from 0.04% to 0.3% P_2O_5 or higher. High phosphorus content is found in soils of the northwestern region, whereas low phosphorus content is usually detected in soils of the southeastern region of the United States.

In solution, phosphorus is present only in the inorganic form, as either the primary, $H_2PO_4^-$, or the secondary, HPO_4^{2-} orthophosphate ion. The concentration of these ions in the soil solution depends on the pH. In acid soils, $H_2PO_4^-$ will be more dominant than HPO_4^{2-}. At pH

6–7, both forms are equally represented in the soil solution, whereas at pH >7.0, HPO_4^{2-} will be dominant together with some PO_4^{3-} ions. Scientifically, the stability of these ions can be explained by the pK_a values of orthophosphoric acid, H_3PO_4, which exhibits three steps of dissociation reactions, each characterized by a specific pK_a:

$$H_3PO_4 \rightleftarrows H^+ + H_2PO_4^- \qquad pK_{a1} = 2.17$$

$$H_2PO_4^- \rightleftarrows H^+ + HPO_4^{2-} \qquad pK_{a2} = 7.31$$

$$HPO_4^{2-} \rightleftarrows H^+ + PO_4^{3-} \qquad pK_{a3} = 12.36$$

As defined previously, the pK_a value is the pH at which the compound dissociates producing equal concentrations of anions and original material. The soil pH or water pH is usually above pH (pK_{a1}) 2.17, but below pH (pK_{a3}) 12.36. Therefore, the first and second step of dissociation will occur in soils and natural waters. The final step of dissociation, characterized by pK_{a3} 12.36, will occur only if the soil pH is above 12.36. These chemical considerations support the idea above indicating that under normal conditions $H_2PO_4^-$ and HPO_4^{2-} are the dominant species of phosphorus in the soil solution.

The concentration of these ions is very small and is estimated to be 1 mg/L or less. Maximum availability of these phosphate ions for plant growth occurs within the pH range of 5.5-6.5 (Tisdale and Nelson, 1975, 1993). Large amounts of phosphorus ions are detrimental to the environment. The use of household detergents and the overuse of phosphate fertilizers are the main reasons for accumulation of phosphate in soil water. In lake water, excessive amounts of phosphorus result in excessive growth of unwanted aquatic plants. This phenomenon is called *eutrophication*.

Phosphate reacts readily with metals present in soils. Acid soils contain large amounts of Al, Fe, and Mn, which form complexes or insoluble metal-phosphate compounds. The reaction was previously called *phosphate fixation*. The ultimate end product of the reaction between Al and phosphate is *variscite* ($AlPO_4.2H_2O$), and that of Fe and phosphate is *strengite* ($FePO_4.2H_2O$). A series of intergrade compounds between variscite and strengite is usually present in soils, which is called the *variscite-strengite isomorphous series* (Lindsay et al., 1959). Phosphate fixation can also occur at high pH values. Many aridisols with high pH contain large amounts of Ca, which can form insoluble $Ca_3(PO_4)_2$, and/or other Ca phosphate compounds in the apatite group.

5.4.6 Sodium and Potassium

Sodium, Na, and potassium, K, are alkali metals. They are very reactive and react vigorously with water, releasing hydrogen gas while producing at the same time the strong bases NaOH and KOH, respectively. The heat of reaction may cause the evolved hydrogen to ignite, hence sodium and potassium elements should never be touched with bare hands. The amount of moisture on the skin is sufficient to cause the exothermic reaction. The Na and K elements are usually stored under kerosene in sealed containers. Sodium is used in industry as a reducing agent in the production of drugs, perfumes, and dyes. It is used for making *sodium light*s that produce yellow light which penetrates through dense fog. These lights are frequently used for automobile headlights and streetlights. Potassium is generally more reactive than sodium. However, it is not used much in industry, because it can be easily replaced by sodium. Sodium and potassium compounds are widely distributed in nature. The sodium content of normal soils is on the average 0.63%, whereas that of potassium is approximately 0.83%. The mineral sources for Na and K are Na aluminum silicates, e.g., albite, and K aluminum silicates, e.g., orthoclase, respectively. They can also occur in nature as chloride, sulfate and borate minerals. The dissolution of Na and K from the aluminum silicates is attributed to hydrolysis reactions, which are weathering processes. The chemical reactions can be represented as follows:

$$NaAlSi_3O_8 + H_2O \rightarrow HAlSi_3O_8 + NaOH \qquad (5.5)$$
albite   soluble

$$KAlSi_3O_8 + H_2O \rightarrow HAlSi_3O_8 + KOH \qquad (5.6)$$
orthoclase   soluble

The NaOH and KOH dissolve in soil water and release their sodium and potassium in the form of Na^+ and K^+ ions, respectively. The concentrations of these ions in soil water are relatively low if compared with their contents in soils. On the average, the K concentration in the soil solution is 5 mg/L, whereas that of Na is 10 mg/L or larger. Both ions are stable in soil water and are very difficult to precipitate. Only by complex formation can K ions be precipitated. In soils, these ions exist mostly as exchangeable cations. Sodium reacts rapidly with Cl^- to form NaCl, which can accumulate in arid regions or where drainage is inhibited. Accumulation of sodium salts in aridisols gives rise to crust formation on the surface soil and/or the development B_n horizons (see Chapter 1). In other soils, such an accumulation of NaCl contributes toward development of salinity, resulting in a decrease in the quality of soil and water.

5.4.7   Calcium

Calcium, Ca, is an element that belongs to the *alkaline earth metal* group. The primary sources of Ca are calcite, aragonite, dolomite and gypsum. Calcite, $CaCO_3$, is the major constituent of limestone, calcareous marls, and calcareous sandstone (see Chapter 2). It comes in many varieties, and Mexican onyx is a common variety of calcite. The mineral calcite possesses an optical property called *birefringence* or *double breaking* or *double refraction*, which means that a light beam entering the crystal is broken up into two beams. Hence, calcite crystals find useful application as lenses in petrographic or polarizing microscopes. However, the most important use of calcite is in the production of cements. Aragonite is another $CaCO_3$ mineral. Under normal atmospheric conditions, it is less stable than calcite. The pearly layer of seashells is composed of aragonite. Pearl itself is built up of concentric layers of aragonite formed around a nucleus of a foreign particle, which can be a small grain of sand. The name aragonite is derived from the town Aragon, Spain, where the mineral was first discovered. Dolomite is a calcium-magnesium carbonate mineral, $CaMg(CO_3)_2$, belonging to the dolomite group. In addition to being a source for Ca, it is also a common source of Mg. The name is derived from the name of the French chemist *Dolomieu*. This mineral is a potential source of metallic Mg. Gypsum is a calcium sulfate mineral with a composition of $CaSO_4.2H_2O$. It is a common mineral of sedimentary rocks and has been found in large amounts in the United States, Canada, England, France, and Russia. Gypsum is usually used for the production of *plaster of Paris* and gypsum boards to make walls in the housing industry. In agricultural operation it is a liming material.

The element calcium itself finds application as a dehydrating agent for organic chemicals. It is also used as an adsorbent of gases in metallurgy and in the steel industry for making alloys with silicon. Calcium is often applied as a hardening material in cables and for the production of storage battery grids and bearings.

Calcium is a very important cation in soils, soil water, and waters in lakes and streams. The average Ca content in soils is estimated to be 1.4%. Depending on climatic conditions and parent materials, the Ca content may vary considerably from soil to soil. Soils in desert climates may be high in Ca, often containing $CaCO_3$ in the B horizon, which is then called a Bk horizon (see Chapter 1). This Bk horizon is thick and close to the surface in aridisols, but is relatively thinner and located deeper in the pedon of mollisols. In humid regions, drastic leaching has removed most of the Ca minerals from the soils. Therefore, the soils in humid regions, e.g., alfisols and ultisols, do not exhibit Bk horizons and are low in Ca. They are often acidic in reaction.

Calcium carbonate is insoluble in water, but will dissolve in water containing CO_2, a process called previously *carbonation* (Chapter 4). The partial pressure of CO_2 in soil air controls the amount of CO_2 gas dissolved in soil water, and therefore determines the rate of dissolution of $CaCO_3$. The main source of CO_2 is the respiration process of plant roots and microorganisms, which accounts for high CO_2 partial pressures in soil air. In soil water calcium occurs as a Ca^{2+} ion or as ion pairs with HCO_3^-, SO_4^{2-}, or Cl^-. It contributes to the hardness of water, which may prevent soap from producing foam and may cause a white crust to form on glass and on pots and pans. In agricultural operations, it is common practice to lime acid soils to a pH of 6.0. Four major benefits of liming are (1) decreasing soil acidity, (2) replenishing Ca, (3) making nutrients more available to plants, and (4) decreasing micronutrient toxicity. The lime reaction, showing the decrease in soil acidity, can be illustrated as follows:

$$CaCO_3 + 2H\text{-clay} \rightarrow Ca\text{-clay} + CO_2 + H_2O \qquad (5.7)$$

Aluminum toxicity is reduced by liming because the Al^{3+} is converted into $Al(OH)_3$, which is a solid compound and chemically inactive. The reaction can be represented as follows:

$$3Ca(OH)_2 + 2Al\text{-clay} \rightarrow 3Ca\text{-clay} + 2Al(OH)_3 \qquad (5.8)$$

5.4.8 Magnesium

The minerals containing Mg are dolomite, Mg silicates, Mg phosphates, Mg sulfides and Mg molybdates. As indicated previously, dolomite, $CaMg(CO_3)_2$, the major constituent of dolomitic limestone, is the most common source of Mg in soils. It is a mineral usually found in sedimentary rocks. Large deposits of dolomite are detected in the dolomite region of southern Tyrol, Switzerland. In the United States, dolomite is found in sedimentary rocks of the midwestern states. Dolomite is used as building stones and ornaments and in the production of certain types of cements. The element Mg itself does not occur as a free element in nature, although it is widely distributed on earth as minerals and/or other types of compounds. Seawater contains abundant amounts of Mg in the form of $MgCl_2$ and $MgSO_4$. The element Mg, a white-silvery metal, is chemically very active and will react rapidly with air, water, and most nonmetal compounds. It is used for the production of lightweight alloys, flares, and flashlight powders. These powders produce an intense white light upon burning.

The average Mg content in soils is approximately 0.5%, whereas its concentration in soil water is estimated to be 10 mg/L. In soil water, Mg exists as the cation Mg^{2+}, or as ion pairs with HCO_3^-, SO_4^{2-}, and

Cl^-. The dissolution of dolomitic limestone is also affected by the CO_2 content in soil water and can be illustrated by the following reaction:

$$CaMg(CO_3)_2 + 2H_2CO_3 \rightarrow Ca^{2+} + Mg^{2+} + 4HCO_3^- \qquad (5.9)$$
dolomite

5.4.9 Sulfur

Sulfur has been known for a long time since it can occur free in nature. Yellow crystals of pure sulfur are deposited on the walls of active volcano craters by the gases coming out of the craters, loaded with sulfur particles. Sulfur gives these gases their pungent smell.

Sulfur is present in soils in organic and inorganic form. Organically, sulfur is an important constituent of proteins and amino acids. Inorganic sulfur comes from a number of different sources. It comes from the dissolution of gypsum, $CaSO_4$, and from the microbial oxidation of pyrite, FeS_2, as discussed previously. Gypsum is the most common source of S in soils. It is widely distributed in the world. Gypsum has been discovered in Canada, England, France, and Russia. Large amounts of gypsum in the United States occur in Arizona and New Mexico in aeolian sand deposits. As indicated previously, it is used primarily in the production of plaster, wallboard, fertilizers, and all kinds of casts. In medicine, gypsum is used for setting casts around broken bones. Sulfur can also be added to soils by the use of powdered S in agricultural operations, by industrial pollution causing acid rain, and by the use of chemicals containing S, such as insecticides and detergents.

Sulfur exists in soils and soil water as an SO_4^{2-} ion in combination with the cations Ca, Mg, K, Na or NH_4^+. Present in the form of elemental sulfur, S, it will be oxidized in aerobic condition and converted quickly into SO_4^{2-}. Under anaerobic condition, SO_4^{2-} may be reduced by microorganisms into SO_3^{2-} or H_2S. Hydrogen sulfide is formed especially in swamps and other areas with stagnant water. Waterlogged soils in coastal regions and paddy soils provide a suitable environment for formation of H_2S. The presence of hydrogen sulfide in paddy soils in Japan causes the development of so-called *Akiochi disease*. These paddy soils are reportedly low in Fe (Tisdale and Nelson, 1975, 1993). In soils rich in Fe, H_2S is usually precipitated as Fe_2S, which imparts to soils a black color, as is frequently noted in soils of the coastal regions and tidal marshes. In paddy soils, this reaction results presumably in Fe deficiency, causing the disease to appear. However, others believe that allelophatic organic acids, such as butyric acids, are suspect for the development of akioch or akiochi disease in rice plants (Stevenson, 1994).

Because most of the sulfur salts are soluble, sulfate is expected to be lost rapidly by leaching. The anionic nature of the sulfate ion prevents its attraction by clay colloids. However, soils containing hydrous oxide clays or sesquioxides have been reported to adsorb considerable amounts of sulfate (Tisdale and Nelson, 1975, 1993). The positively charged surfaces of these sesquioxide clays may cause electrostatic attraction of the negatively charged sulfate ions.

5.4.10 Manganese

Manganese is present in small quantities in many crystalline rocks. It is released into the soil by rock weathering and is redeposited in various forms of Mn oxides, e.g., *pyrolusite*, MnO_2, *braunite*, Mn_2O_3, and *manganite*, $MnO(OH)$ or $Mn_2O_3.H_2O$. Pyrolusite is a mineral that belongs to the *rutile* group, which is characterized by the formula XO_2. If the X is Ti, the mineral is rutile, but when X is Mn, the mineral is pyrolusite. Uranium oxide also belongs to this group. Pyrolusite occurs as an accessory mineral in crystalline rocks. Commercial deposits of pyrolusite are found in Russia, Gabon, South Africa, India, Australia, and Cuba. Small deposits have been discovered in the United States in Virginia, Tennessee, Georgia, Arkansas, and California and in the region of Lake Superior. This mineral is the most important source of Mn. The element Mn is used in the production of hard steel, chemicals, and disinfectants ($KMnO_4$). It is an important element for the production of batteries and dry cells. Often it is also used in the glass industry as a decolorizer of glass, or as a coloring material in glass, pottery, and bricks.

Manganese can exist in three oxidation states: Mn^{2+}, Mn^{3+}, and

Mn^{4+}. The divalent manganese ion is the main form of manganese in soil and soil water. Because it is a cation, Mn^{2+} is usually adsorbed on the negatively charged clay surfaces. The trivalent form usually exists as Mn_2O_3, which can be found in substantial amounts in acid soils. The tetravalent form, MnO_2, is perhaps the most stable and inert form of manganese.

The concentration of Mn^{2+} in soil water is very low, seldom exceeding 0.05 mg/L. Large amounts of soluble Mn in water cause Mn toxicity in plants, since it is needed only as a micronutrient. If it is used for household purposes, water containing Mn may stain clothes and bathroom fixtures. The stability of Mn^{2+} in solution depends on pH. At high pH values, Mn^{2+} precipitates as $Mn(OH)_2$ or is converted into MnO_2, or Mn_3O_4. Therefore, liming acid soils may frequently result in Mn deficiency in crops (Tisdale and Nelson, 1975, 1993). However, in the presence of high amounts of CO_2, manganese may exist as $MnHCO_3^+$. The concentration is, however, very low, and Mn is more likely to precipitate as hydroxides and oxides than to form $MnHCO_3^+$, especially when CO_2 partial pressures fluctuate suddenly.

5.4.11 Soil Colloids

A colloid is a state of matter consisting of very fine particles that approach but never reach molecular size. The upper size limit of colloids is 0.2 µm, and the lower size limit is approximately 50 Å or 5 nm, the size of a molecule. Colloids can be organic or inorganic in nature. On the basis of their interaction with water, colloids can be divided into hydrophobic and hydrophilic colloids. *Hydrophobic* colloids will not interact with water, in other words they repel water, whereas *hydrophilic* colloids will interact with water. Several organic colloids exhibit both hydrophobic and hydrophilic properties in one molecule. Such colloids are called *amphiphilic* and include phospholipids and many amino acids. Detergents are examples of synthetic amphiphilics. Plant and soils contain large amounts of colloidal particles. Most plant colloids are hydrophilic, whereas soil inorganic colloids are hydrophobic. The inorganic fraction of soils consists of sand, silt and clay. Clay comprises all inorganic solids with an effective diameter of <0.002 mm (<2 µm) and exhibits colloidal behavior. Organic colloids include carbohydrates, amino acids, protein, lipids, lignin, humic and fulvic acids.

Many of these inorganic and organic colloids, such as clay, carbohydrates, amino acids, protein, bacteria, and algae, are present in soil water. However, the amount dispersed in soil water is smaller than the amount present in the soil solid phase. When clay is added to water, it forms a suspension. The clay is then dispersed and exists as single, individual particles. Since clay usually carries negative charges, it attracts cations. This maintains electroneutrality in the soil system. The cations are held on the clay surfaces by electrostatic bonds, which are weak bonds. Consequently, the adsorbed cations can be replaced or exchanged by other cations from the soil solution. The process of exchange is called *cation exchange*. The ions involved in the exchange reaction are referred to as *exchangeable cations*, and the clay particles that adsorb these ions are frequently called the *exchange complexes*. Under natural conditions, H^+, Ca^{2+}, Mg^{2+}, K^+, and Na^+ are the most common cations in soils, and they can replace each other. When Ca, Mg, K, and Na are exchanged by H^+, the exchange complex is then saturated with H^+. This produces an acid reaction in soils. Because of the high surface area of clays and because of their electrochemical properties, colloids may adsorb a variety of chemical compounds dissolved in soil water, e.g., products from microbial reactions, organic waste, gases, heavy metals and other pollutants. By adsorption to clay surfaces, many toxic metals can be immobilized or precipitated. The clay fraction, especially in the solid phase of soils, also provides for a buffer and intercepts pollutants. For example, heavy metals carried by soil water trickling through the pedon are removed because of the cation exchange capacity of clays. Consequently, clay acts as a purifying agent in nature, improving in this way the quality of soil water. However, this property of clay depends on its *cation exchange capacity* (CEC), which may vary with the different types of clay. Exchange reactions and their importance in plant growth and the environment are discussed in more detail in Chapter 7.

The organic colloids of importance are carbohydrates, amino acids, protein, lipids, nucleic acids, lignin, and humic substances. The carbohydrates may range from simple sugars to complex poly-saccharides. The simple sugars are readily soluble in water and are sweet in taste. They are the primary choice as a food and energy supply of microorganisms. Polysaccharides are insoluble in water, amorphous in nature, and frequently tasteless. They are high in molecular weights. Cellulose, the most important constituent in plant residues, is insoluble in water. It is often associated with hemi-cellulose and lignin, is semicrystalline in nature, and has a molecular weight between 200,000 and 2 million. Plant starches and animal glycogen are important examples of polysaccharides. In water, they exhibit *imbibition*, or become dispersed, but they are not strictly soluble. The amount of carbohydrates has been estimated to be 5 to 25% of soil humus. Their concentrations in the soil solution are expected to be a small fraction of those in the soils. Today the presence of extracellular polysaccharides (EPS) in soils has attracted considerable attention, because of their importance in forming organo-mineral associations in soils. As discussed earlier, EPS is especially important in contributing to soil aggregation. Extracellular poly-saccharide is synthesized by microorganisms when carbon is available on a per cell basis (Chenu, 1995). The amounts produced by algae or bacteria are very small, and are estimated to range from 0.02 to six times the dry weight of the cell. No doubt, the amounts in the soil solution will be a very small fraction of those synthesized.

As indicated previously in Chapter 3, amino acids are relatively abundant in soils. The amount of free amino acids in soils is about 2 mg/kg soil and may be sevenfold higher in the soil rhizosphere. The term free amino acid refers to amino acids that are not linked together into peptides or other compounds. Amino acids in soils can also be present in the cell walls of microorganisms or as structural constituents of humic and fulvic acids. These are bound amino acids. With certain exceptions, free amino acids are generally soluble in soil water, but are insoluble in nonpolar organic solvents, such as ether, chloroform, and acetone. Since amino acids contain both carboxyl and amino groups, they can react with acids and bases. Such compounds are said to be *amphoteric*. The amphoteric character is the result of amino acid dissolving in water in the form of a *zwitterion,* as discussed previously. In alkaline soils this zwitterion is transformed into an anion, whereas in acidic soils the zwitterion becomes a cation. From the foregoing, it follows that in acid soils, the amino acid is positively charged and will react with clays and organic anions. On the other hand, in basic soils, the amino acid is more likely to be negatively charged, and hence will undergo reactions with metal ions and organic cations. The presence of negative charges contributes toward also increasing the cation exchange capacity of basic soils. These reactions with clay and metal ions contribute toward accumulation of amino acids in soils. Oxisols and ultisols are known to be very strongly acid soils, hence their amino acids are most likely positively charged, and may be a factor in the high anion exchange capacity of these soils.

The lipid concentration varies considerably from soil to soil and generally it is very small, ranging from 0.4 to 4.7 % in the soil organic matter fraction (Waksman, 1936). Undoubtedly, its concentration in soil water must be a fraction of the amount present in soil organic matter. They are hydrophobic in nature, and a high lipid content will make the soil *water repellant*. Induced hydrophobicity is reported as a problem in turf and golf greens.

Not much data are present on nucleic acids in soil and soil water. This is one of the organic compounds that decomposes rapidly once it is released in the soil from plant cells by decomposition. It is perhaps also incorporated in the structure of humic acids during formation of humic matter. These complexed or chelated nucleic acids are believed to account for the presence of hydrolyzable unknown nitrogen (HUN) in soils.

Lignins and related aromatic substances in soils are very resistant to microbial decomposition, though certain fungi, called *lignolitic fungi*, have been reported to attack lignin. The bulk of lignin comes from the secondary cell walls of plants, and the lignin content increases with plant age and stem content. It is a very important constituent of woody tissue, and it contains the major portion of the methoxyl content of the wood. A large amount of lignin is also detected in the vascular bundles of plants. Because tropical grasses are characterized by larger amounts of vascular bundles than temperate region grasses, soils under tropical grasses are expected to contain higher lignin content. Lignin is insoluble in water, in most organic solvents, and in strong sulfuric acid. It is a major organic compound in the formation of humic matter.

Humic matter is considered to be the major fraction of soil humus. It is by definition made up of organic compounds that are amorphous, colloidal, polydispersed with yellow to brown-black color. These compounds are hydrophilic and acidic in nature and have high molecular weights, ranging from several hundreds to thousands of atomic units. Based on solubility in bases and acids, humic matter can be distinguished into several humic fractions (see Chapter 3). Of these fractions, humic and fulvic acids are considered the two major fractions. Fulvic acid is believed to be soluble in water, whereas humic acid is usually insoluble in water and in acids. The concentration of fulvic acid is estimated to vary from 20 µg/L in ground water to > 30 mg/L in surface water, including water in rivers and swamps (Aiken, 1985). The water in the swamps and rivers in southern Georgia, is frequently black in color, caused by humic matter in solution. The humic matter, isolated from this so-called *black water,* is composed of 72% fulvic acid and 28% humic acid. (Lobartini et al., 1991). The present author assumes that black water in other parts of the world, e.g., Rio Negro in Brazil, also contains considerable amounts of fulvic acid. As discussed in Section 3.6, this humic matter has important implications for the quality of soils and the environment. Due to its enormous chelating capacity, humic matter is capable of detoxifying lakes and swamps from metal pollution. It is also suggested for use in the control of the eutrophication process in rivers and lakes.

5.5 DISSOLVED GAS

5.5.1 Henry's Law

The two most important gases in soil air that affect soil properties and plant growth are CO_2 and O_2. As indicated in Chapter 4, these gases may dissolve in soil water. The amount dissolved is usually very small and is controlled by the partial pressure of the gas in contact with soil water. *Henry's law* dictates that the solubility of a gas in a liquid is proportional to the partial pressure of that gas in contact with the liquid, as expressed by the equation:

$$C_{aq} = k\,P_t \qquad\qquad (5.10)$$

In this equation, C = concentration of gas dissolved in moles/L, k = Henry's law constant, and P = partial pressure of gas at temperature = t, in atm. The k values for the most important gases in soil air are very small:

	k (moles/L/atm)
CO_2	3.38×10^{-2}
O_2	1.28×10^{-3}
N_2	6.48×10^{-4}

Since the medium, water, in which the gas dissolves also exhibits a vapor pressure, both the water vapor pressure and partial pressure of the gas then determine the amount of gas dissolved. The vapor pressure (or partial pressure) of water at 25°C is 0.0313 atm. Using these parameters, the amount of gas dissolved can be easily determined. An example of the calculation of the amount of oxygen dissolved is given below. Soil air in equilibrium with atmospheric air has a pressure of 1 atm, and since atmospheric air contains 21% O_2, at 25°C the following is valid:

P_{oxygen} = (atmospheric pressure - water vapor pressure) x 0.21
        = (1 - 0.0313) x 0.21 = 0.2034 atm

Using Henry's law, the concentration of oxygen dissolved in soil water is

$C = k\,P_t$ t = 25°C
$C = (1.28 \times 10^{-3}) \times 0.2034$
   $= 0.2604 \times 10^{-3}$ moles/L

Since 1 mole of O_2 = 2 x 16 = 32 grams, the concentration of oxygen dissolved in grams is

$C = (0.2604 \text{ x } 10^{-3}) \text{ x } 32$
$\quad = 8.33 \text{ x } 10^{-3} \text{ g/L} \text{ or } 8.33 \text{ mg/L}$

The amount calculated above is the maximum concentration of dissolved O_2 at a water temperature of 25°C. The solubility usually decreases as the temperature increases. The amount of CO_2 dissolved in water can be calculated in a similar fashion and amounts to 0.984 x 10^{-5} moles/L. This is the amount of CO_2 in water of high quality, which is in equilibrium with atmospheric air.

5.5.2   Dissolved CO_2 Gas

Carbon dioxide is produced when carbon is burned with O_2. An example is the burning of fuel, such as methane, CH_4:

$$CH_4 + 2O_2 \rightarrow CO_2 + 2H_2O \qquad (5.11)$$

In nature, CO_2 is produced by respiration of roots and the aerobic decomposition of organic matter, as discussed earlier. It is a colorless and odorless gas. It is 1.5 times heavier than atmospheric air. Carbon dioxide is considered one of the polluting gases that causes the greenhouse effect and the destruction of the ozone shield. Large amounts of CO_2 can be produced by large-scale burning of the tropical rain forests and by the burning of fossil fuels in industry and by automobiles. Supersonic aviation also generates large amounts of CO_2. Another source of considerable amounts of CO_2 is active volcanism. The latter has received little attention. All these activities and processes contribute to the generation of excessive amounts of CO_2, causing global warming and the destruction of the ozone shield. The amount of CO_2 produced by natural respiration and decomposition processes is usually of no harm, and is in fact a necessary process in the *carbon cycle*. Excess CO_2 in the Earth's atmosphere is usually absorbed by green plants and recycled through photosynthesis into carbohydrates. The oceans with their salt water and high pH are also sinks for CO_2. They are not only capable of absorbing huge amounts of CO_2 and converting it into carbonates and bicarbonates, but the large populations of green marine plants, such as seaweed and kelp, absorb CO_2 and function similarly as terrestrial green plants. However, this formidable capacity of the green plants and ocean water to prevent such pollution can be destroyed by excessive and accelerated production of CO_2.

Carbon dioxide forms *carbonic acid*, H_2CO_3, with water, which is a weak acid that gives a weak sour taste to water. Because of this, it is a valuable gas for the manufacture of carbonated water and other beverages, such as Coca Cola, Pepsi Cola, Dr. Pepper, and other soda pops. Carbonic acid is very unstable and has never been isolated. Its stability depends on the partial pressure of the CO_2 gas above. When the pressure in a can of soda water is released, *effervescence* takes place, and the soda water tastes *stale* due to the disappearance of carbonic acid. Carbon dioxide is also an important ingredient for certain types of fire extinguishers. Fire extinguishers containing CO_2 are preferred for fighting fires from explosive materials, such as gasoline and aviation fuel. Carbon dioxide is one of the materials that does not support the combustion. Air, containing 2.5% CO_2, is able to extinguish such fires. It is also used in the production of *baking soda*, $NaHCO_3$, and *washing soda*, $Na_2CO_3.10H_2O$. Today baking soda is used in toothpaste and as a water softener. When frozen, carbon dioxide forms a solid, called *dry ice*. It is often used as a refrigerant in coolers and for cooling of beverages.

As discussed in Chapter 4, atmospheric air contains 0.03% CO_2. However, because of respiration and aerobic decomposition of organic matter in soils, the CO_2 content in soil is usually above 0.03%. Therefore, the amount of CO_2 dissolved will be above 1.0×10^{-5} moles/L.

Dissolved carbon dioxide, or CO_2, affects many biological and chemical reactions in soil water. It is used by aquatic plants, such as algae, in photosynthesis. It is a chemically active compound and interacts with water to form carbonic acid; hence, it affects the pH of soil and soil water. The reaction is usually represented by the following equations:

$$CO_2 + H_2O \rightarrow H_2CO_3 \tag{5.12}$$

$$H_2CO_3 \rightarrow H^+ + HCO_3^- \tag{5.13}$$

Carbonic acid is a weak acid and will dissociate some of its protons in soil water, as indicated by reaction (5.13). Such a reaction normally increases the acidity of soil water. However, the HCO_3^- is unstable and may dissociate, producing OH^- ions as follows:

$$HCO_3^- \rightarrow OH^- + CO_2^{'} \tag{5.14}$$

The OH^- ions neutralize the H^+ formed by reaction (5.13). Because carbonic acid is capable of producing H^+ and OH^- ions at the same time, it is considered a buffer compound. Therefore, it is capable of stabilizing the acidic conditions of soil and water to a certain extent. This stabilizing effect is especially enhanced when it is present as a bicarbonate of calcium.

Water containing CO_2 is a powerful solvent for limestone and other calcareous materials. Calcium carbonate, which is otherwise insoluble in pure H_2O, becomes soluble in carbonated water. Such a reaction, earlier called *carbonation,* can be illustrated by the reaction below:

$$CO_2 + H_2O + CaCO_3 \rightarrow Ca^{2+} + 2HCO_3^- \tag{5.15}$$

This reaction is the main weathering process by which limestone is decomposed in soil formation, and is also the reaction that makes liming material reactive.

5.5.3   Dissolved Oxygen

Another important gas dissolved in soil water is oxygen. In nature, oxygen is the most abundant and widely distributed element. Atmospheric air theoretically contains 23% oxygen. In the form of compounds, oxygen is present in rocks and minerals and makes up a large part of the bodies of plants and animals. It is a colorless and odorless gas, which is slightly more dense than air. Liquid oxygen, however, is pale blue, and will boil at -183°C at atmospheric pressure. The element consists of two atoms, hence the formula O_2. In the liquid and solid state, oxygen has *paramagnetic properties*, meaning that it can be attracted by a magnetic field. Chemically, oxygen is the most active element. When oxygen combines with another element, the product formed is called an *oxide*.

Oxygen is needed for the slow oxidation of compounds that generates the energy required by all living organisms for normal functions and maintenance of body temperature. The intake of O_2 by roots and microorganisms for respiration is an example of such a production of energy through burning of tissue carbohydrates. Oxygen is in fact considered the key substance in both soil air and soil water for the existence of all forms of life in soils. Oxygen deficiency in soil air causes many undesirable effects, as has been discussed in Chapter 4. Oxygen deficiency is detrimental to organisms living in soil water. In lakes and streams, lack of oxygen is fatal to fish and aquatic plants.

The amount of oxygen dissolved in soil water depends on the partial pressure of O_2 and water vapor, as discussed previously. It is very small; the calculated value of dissolved O_2 is 8.33 mg/L at 1 atm and 25°C. This amount of dissolved oxygen can be depleted very rapidly unless some mechanism for aeration is present. In natural streams and lakes, oxygen content is replenished by aeration through turbulent flows and waterfalls. Artificially, aeration of water is achieved by pumping air in water or creating fountain sprays, such as in aquariums, hydroponic systems, and sewage treatment plants. Several processes are responsible for the depletion of dissolved oxygen. Oxygen can be consumed rapidly in the oxidation process of organic matter as discussed previously. Soil water may contain a lot of organic pollutants, which are washed down from manure and industrial and domestic wastes during heavy rains. Part of the dissolved organic compounds is sometimes referred to as *dissolved organic carbon* or *DOC*. Microbial decomposition of these organics consumes most of the dissolved oxygen. Moreover, a lack of O_2, rather than the presence of pollutants, may cause aquatic organisms to suffocate in polluted streams, lakes, and reservoirs. Pollutants are frequently not the primary source for fish kill, but their presence complicates the effect of other adverse conditions. This is evidenced by the fact that fish and other aquatic organisms thrive in turbulent streams and lakes that are loaded with clay suspensions. Other microorganism-mediated processes depleting the dissolved oxygen content are the *biochemical oxidation* of:

Iron compounds

$$4Fe^{2+} + O_2 + 4H^+ \rightarrow 4Fe^{3+} + 2H_2O \tag{5.16}$$

Sulfur compounds

$$2SO_3^{2-} + O_2 \rightarrow 2SO_4^{2-} \tag{5.17}$$

Ammonium

$$2NH_4^+ + 3O_2 \rightarrow 2NO_2^- + 2H_2O + 4H^+ + energy \tag{5.18}$$

$$2NO_2^- + O_2 \rightarrow 2NO_3^- + energy \tag{5.19}$$

The oxidation of ammonium into nitrate, called *nitrification,* has been discussed previously.

5.5.4 Oxygen Demand

The rate at which oxygen is consumed by the oxidation processes, discussed above, is called *oxygen demand* (OD). Three types of ODs are distinguished:

Biological oxygen demand (BOD). This is defined as the amount of oxygen consumed by the microbial decomposition of organic matter during a 5-day incubation period. The test was developed in England, where it was considered that dissolved organic matter not decomposed within 5 days would be transported into the sea.

Chemical oxygen demand (COD). This is defined as the amount of oxygen consumed in the chemical oxidation of organic matter by $K_2Cr_2O_7$ in the presence of H_2SO_4. The analysis for determination of C_{org} by the wet oxidation method with $K_2Cr_2O_7$ is today called the *Walkley-Black method*.

Total oxygen demand (TOD). This is the amount of oxygen consumed in the catalytic oxidation of carbon. The amount of CO_2 produced in the reaction is measured.

Of these three types of ODs, the BOD is the best known, although its determination is more difficult than the determination of COD, whereas the 5-day incubation period is subject to much argument. Nevertheless, many scientists prefer the use of BOD for determining the quality of stream and lake water and the amount of pollution in soils and the environment. Organic wastes are frequently introduced into streams and lakes through runoff and/or leaching. The BOD demand of organic waste decreases the dissolved O_2 content during the decomposition process. The lower the BOD values, the better is the water quality. A BOD value of 1 ppm means that 1 ppm of oxygen was consumed in the decomposition process during a 5-day incubation period. This indicates that only a small amount of organic pollutant was present; hence, the water being analyzed is of high quality. On the other hand, high BOD values (5-20 ppm) in an analysis suggest that the water contains high amounts of organic contaminants, or is water of low quality (Stevenson, 1986). Since the maximum amount of dissolved oxygen is 8.33 mg/L (at 25°C, 1 atm), a value of 8 ppm as the lowest limit of a high BOD value is perhaps closer to reality than 5 ppm (Tan, 1993). It is expected that the soil can take care of the amount of dissolved organic substances on its own as long as the level of pollutants can be oxidized by 8.33 mg O_2/L. As soon as this amount of dissolved O_2 is depleted, organic contaminants start to accumulate, decreasing the quality of soil water. Animal waste, runoff from barnyards, and effluent from food processing plants are notorious for their high BOD values. The BOD may run as high as 10,000 ppm in runoff from feed lots (Stevenson, 1986). Similar and even higher BOD values have been noted by the author in waste (pulp) of palm oil factories. The value of BOD generally increases with an increase in algae, manure, and plant residue content in soil water. On the other hand, a controversy seems to exist on the BOD of digested sewage sludge, which is reported low by some, because of microbial oxidation during the digestion process (Stevenson, 1986), but high by others (Miller and Gardiner, 1998). Runoff containing organics with BODs ≥ 20 ppm entering streams and lakes is generally considered very harmful, because water contaminated with substances possessing high BOD values will soon run out of oxygen. However, when recycled as a soil amendment, the organic waste with extremely high BOD values does not necessarily result in degradation of the environment. It is a valuable source of N and other plant nutrients that will be released upon decomposition. The pulp waste from palm oil factories has been noted to increase the growth of oil palm seedlings.

CHAPTER 6

SOIL PHYSICS
IN THE ENVIRONMENT

6.1 PHYSICAL PROPERTIES

The physical properties of soils are usually discussed as fundamental properties in distinguishing soils with respect to their behavior in agricultural and engineering operations. They are used in soil taxonomy for placement of soils in different categories, but very little is present in the various soils textbooks on the role of physical properties in the environment. The physical characteristics of soils are the products of environmental factors. They are affected by environmental conditions but at the same time they also affect many of the environmental factors in the soil ecosystem. For example, soil texture is an important physical property that is formed by products derived from the disintegration of rocks and minerals as conditioned by environmental factors, such as precipitation and temperature. Rocks and minerals will produce large amounts of clays only under intense weathering conditions, as a result of high humidity and high temperatures. In arid regions, where water is a limiting factor, normally less clay is formed and the texture is generally coarse. The types of clays formed are also synthesized in nature under the influence of environmental factors. Kaolinite, for example, is formed when conditions favor the loss of Si by leaching, whereas smectite can be produced only when environmental conditions are suitable for accumulation of Si.

The texture formed, coarse or fine, affects plasticity, stickiness, and dispersion of soils, factors of importance in erosion and denudation. These are natural forces that form the landscape on the surface of the earth. Together with soil structure and several other physical properties, soil texture affects pore spaces, aeration, water flow, percolation and leaching, hence drainage conditions, topics that have been addressed in Chapters 4 and 5. Rate of water intake, water-holding capacity and water-supplying power are all influenced by soil texture and soil structure. Bulk density, another soil physical property, is a good measure of the density and porosity of soils. However, the concept of bulk density takes into account both the mass of the solid particles and the volume of pore spaces. Consequently, soil texture and soil structure are indirectly connected to bulk density. In fact many of the physical properties are interrelated, and in particular in their functions it is often difficult to say that they are the result of the action of a single property. For example, percolation is affected not only by soil texture; soil structure and soil density have an equal important role in this flow of water. Several different physical properties are usually complementing each other in their effect on the flow of soil water.

This chapter is an attempt to address the major physical properties of soils in relation to the environment. Names such as *environmental soil physics* and *physical edaphology* have been suggested for such a topic. Though the basics of soil physical properties are presented in the following pages, they are included only as background material, hence only the necessary information will be provided. Therefore, the discussion will always be related to some environmental aspect, conforming to the title of the book. If not needed to enhance comprehension, material that can be found readily in many other soil science textbooks is not duplicated here. For a more detailed discussion on the physical properties of soils related to soil survey and soil taxonomy, reference is made to Miller and Gardiner (1998), Brady and Weil (1996), Foth (1990), Hillel (1980), and Soil Survey Staff (1962).

6.2 SOIL TEXTURE

6.2.1 Soil Separates

Physically, the soil is a mixture of mineral matter, organic matter, water, and air. The mineral matter is composed of inorganic particles varying in size from stone and gravel to powder. These inorganic particles, separated according to size, are referred to as *soil separates*. The USDA Soil Survey division (Soil Survey Staff, 1951) recognizes three major groups of soil separates: sand (2.0 - 0.050 mm in diam.), silt (0.050 - 0.002 mm in diam.), and clay (<0.002 mm in diam.). The three groups can be subdivided into finer size fractions (see Tan, 1995). The soil separates make up the texture of the soil. They are soil constituents and not a physical property like soil texture.

Sand. - Sand particles are derived from the weathering of rocks and minerals and are usually composed of primary minerals, such as quartz, feldspars, micas, and ferromagnesians. They are chemically inert, have a low water-holding capacity, but promote good drainage and a loose friable condition to the soil. Upon weathering, the sand particles produce clays.

Silt. - Silt particles are the intermediate weathering products of sand, and are in fact micro-sand particles. If they are coated with some clay, these silt particles exhibit some chemical activity, e.g., plasticity, cohesion, cation exchange, and water-holding capacity.

Clay. - Clay is considered a secondary mineral, since it is produced by weathering of the primary minerals. It is the smallest particle in soil and exhibits colloidal properties. Clay is the chemically active inorganic fraction in soils and is negatively charged. Because of this charge, clay has a high water-holding and cation exchange capacity. Clay is plastic and sticky when wet due to its plate-like structure. Many of the clays are crystalline in structure, whereas others are amorphous. Some of the clays, like the iron oxide clays, play an

important role in soil aggregation, and in addition impart red to
yellow colors to soils.

6.2.2 Soil Texture

The relative proportions of the three major soil separates are
referred to as the *soil texture*. Twelve types of soil textures are
recognized on the basis of % sand , % silt, and % clay for division of
soils into twelve textural classes. Percentage organic matter is not
included in soil textural classes, since texture is based only on the
inorganic fraction following the concept:

% sand + % silt + % clay = 100%

The twelve types of soil textures are usually grouped in a textural
triangle (Figure 6.1) to make their determination possible. The
amount of sand, silt, and clay is estimated by a chemical analysis,
called *particle size distribution analysis*. In the past, this analysis was
called *mechanical analysis* (Tan, 1995). By plotting the percentages
of sand, silt, and clay in the triangle, the textural class can be found.
The terms soil texture and soil classes are frequently used interchangeably. However, it has to be realized that scientifically soil
texture is a soil characteristic, whereas a soil class is a group of soils
determined by the soil texture. Hence, the andosol has a sandy loam
texture, but the soil can then be called a sandy loam.
For easy comprehension these soil classes can be grouped into four
major categories:

Sands. - This category includes all the soils with sand content above
75%. The group is subdivided into the sands and loamy sands, as can
be noticed in the left-hand bottom corner of the textural triangle. The
soils will usually be single grained, and the individual grains can be
seen and felt. Squeezed when wet, a cast will form that will crumble
easily when touched. A loamy sand is more coherent at wet or dry
condition than a sand.

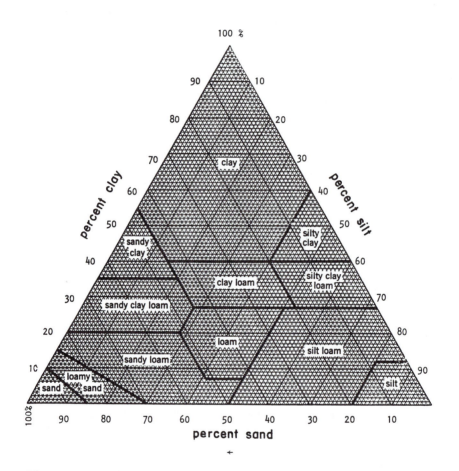

Figure 6.1   USDA textural triangle.

Loams. - This category includes the sandy loams, loams, and silt loams. When squeezed in moist condition a loam soil will form a cast or ribbon that will bear careful handling. A wet loam will be slightly plastic. A sandy loam will be gritty. The presence of sand makes the wet cast less plastic and less stable. A silt loam will form a ribbon when squeezed in moist condition. When dry, it appears cloddy, but the clods can be readily broken, and when pulverized they feel soft and floury.

Clay Loams. - Clay loams are soils, which include the sandy loams, clay loams, and silty clay loams. When squeezed in moist condition, the soil will form a ribbon, which will break, barely sustaining its own weight. The soil is plastic when wet. A sandy clay loam is more gritty, whereas a silty clay loam feels velvety when moist.

Clays. - Clays are all the soils with clay contents ≥35% clay. The soil classes are then sandy clays, clays, and silty clays. When wet, clays form a thick, flexible ribbon. When dry, the soil forms hard lumps and clods. A sandy clay feels gritty, but often this grittiness is not evident due to the high clay content. A silty clay feels plastic and velvety at the same time. Care should be taken since unwetted clay aggregates are often confused with sand grains. This is especially true in the case of oxisols, where the clay particles are aggregated forming stable structures due to cementation and coating by amorphous iron oxides. Squeezed in dry condition, the oxisols feel gritty, and the heavy clay texture appears only after a sufficient amount of water is added, which apparently dissolves the iron coatings.

As can be noticed above, the names for the variations in soil textural classes are formed by adding the prefixes sandy or silty to the major soil classes loams, clay loams, and clays.

6.2.3 Environmental Significance of Soil Texture

When clay is present in dominant proportions relative to silt and sand concentrations, the soil is described as having a *fine* or *heavy* texture. The term fine refers to the clay content, which provides the finest particles in soils. On the other hand, when sand has the dominant concentration in clay and silt, the soil is said to have a *coarse* or *light* texture. Sand particles are the coarsest inorganic soil constituents, and soils with coarse textures are loose and friable, and easy to work with, hence the terms coarse and light texture, respectively. Soils containing relatively equal proportions of sand, silt, and clay are medium in texture. Fine-textured soils are generally plastic and sticky when wet, and hard and massive when dry. These conditions restrict agricultural operations. The soils are very difficult (hard, heavy) to plow, hence the term heavy texture. Clayey soils tend to disperse easily, and if they remain dispersed, the soils will be puddled. When a puddled soil becomes dry, caking or soil compaction may occur, since the clay particles will settle (precipitate) in a close dense packing. Therefore, compaction reduces pore spaces, which inhibits aeration necessary for root and microbial growth. Infiltration rate of water is also expected to be harmfully affected, resulting in increased run-off and erosion. A flocculating concentration of electrolytes must be present to prevent dispersion of soils. In agricultural practice lime is added to the soil to flocculate the clay, which makes aggregation possible. The critical concentration of elec- trolytes at which flocculation of clay particles takes place is called the *flocculation value* (Levy, 1996), and will be discussed at the end of this section on soil conditioners. In nature, a flocculating concentration of electrolytes is present in the form of Al, Fe, and Mn concentrations. The Al, Fe, and Mn content is especially high in acid soils. This is one reason for the presence of stable structures in oxisols in tropical regions. Oxisols are known to be high in clay content, yet they lack the plasticity and stickiness of clayey soils. These soils can be cultivated even during heavy downpours because of their strong structure. Their friable consistence regardless of their high clay content is assumed to be caused by iron, coating the structural units into a strong and stable structure. In contrast, coarse-textured soils are not sticky or plastic when wet. They are usually poorly aggregated because of lack of clay. However, they are characterized by a rapid percolation rate, even more rapid than in fine-textured soils. This is in part due to the concentration of pore spaces. The arrangement of solid particles determines the amount of pore spaces. Sandy surface soils have a *total* pore space ranging from 35 to 50% (by volume), whereas finer textured soils may have pore spaces ranging from 40 to 60%. Thus, fine-textured soils have in fact a greater percentage of total pore spaces than coarse-textured soils, but do not exhibit the high rates of percolation. As explained earlier, two types of pores are present in soils: macro- and micropores, though no sharp line of demarcation is present in nature between these two types of pores. Macro- (large) pores function in air and water movement, whereas micro- (small) pores retain or hold soil moisture. Sandy soils have a preponderance of macropores. Hence, they have a fairly rapid movement of air and water in spite of the small amount of total pore space. Fine-textured soils have large amounts of micropores. These soils have, therefore, a greater water holding capacity than sandy soils, but the movement of air and water is comparatively more restricted because of the lack of macropores.

The type of soil texture will determine the growth of agricultural crops or natural vegetation alike. Especially, the yields of crops and forest trees are affected by soil texture. This is an environmental aspect that has been known for quite some time but that is not fully realized by most soil scientists. Root crops are not producing to their full capacity when grown in heavy-textured soils. For example, Irish potatoes are seldom grown in clayey soils because of their poor yield. Therefore, sweet potatoes are grown in the sandy loams of south Georgia. Peanuts give also better yields when grown in light-textured than in heavy-textured soils. This is one of the reasons why the peanut-growing area is located in the southern part of Georgia, where the soils are sandy loam and loamy sand in texture. Even forest trees, growing in nature or grown for pulp and timber production, do better in light- than in heavy-textured soils. The limiting factor for adequate plant growth in sands and loamy sands is not the texture, but the water content. Because of the rapid rate of percolation, sandy soils tend to become dry very soon. Where water can be applied through rains or irrigation, the growth and production of crops and timber are noted to be as good as those in sandy loams.

Resistance to root and tuber growth by *soil impedance* is a very common occurrence in clayey soils. Some uses the name *mechanical impedance* (Taylor and Ashcroft, 1972; Wild, 1993), whereas others use the term *soil strength* or *bearing capacity* for the resistance of soils to penetration (Spangler and Handy, 1982). It is closely related to soil density, and its analysis is also conducted by using penetrometers, instruments for measuring soil resistance to penetration or soil densities. Impedance values of 1.55 to 1.85 g/cm^3 are reported by Wild (1993) to inhibit root growth, but Spangler and Handy (1982) indicate that by jabbing the thumb into the soil, values of soil strength are obtained in the range of 2 to 0.5 kg/cm^2. A soil penetration strength of 1 kg/cm^2, when the thumb can be jabbed to the first knuckle into the soil, is a relatively low impedance level, easy for roots to overcome. Growing roots are known to be capable of heaving objects as heavy as 10,000 kg or more. The problem of soil impedance is of less significance in coarse-textured soils because of the abundance of macropores. Roots grow more easily through soil pores with diameters larger than the sizes of the roots. As indicated previously in Section 3.3.3, the size of feeder roots is in the range of 100 to 400 μm, whereas root hairs may be 10 to 50 μm in diameter. Therefore, the soil must contain an abundant amount of macropores for the growth of feeder roots. However, these pores are usually not present in clayey soils. In the process of growing, the roots must also displace soil particles. They will be capable of pushing aside the soil particles only if sufficient energy can be expended by the elongating cells of the root tips. Imagine, the energy that has to be used by peanuts and potato tubers to overcome the compression exerted by a clayey soil matrix. Production of peanuts and potatoes will be inhibited because most of the energy has to be spent by the plant to overturn the resistance of the soil matrix. Taproots of forest trees are frequently stunted when growing in heavy-textured soils. Soil impedance is usually decreased in crop production by cultivation methods. It is common practice in modern agriculture to use heavy equipment, which often results in compaction of surface and subsoils. Soil compaction will increase soil impedance to root growth. Subsoiling is often suggested to reduce soil compaction brought about by the heavy equipment. In the natural environment, soil impedance is generally noticed in heavy-textured soils where the soil cakes and becomes hard in dry conditions. It can be reduced by enhancing the biological activity in soils. Decomposition of organic matter produces humus that will increase aggregation in soils, making the soils softer and friable. In addition, the *cultivation effect* of the macro- and microfaunae in soils has a similar effect as plowing in agriculture. The soil is constantly mixed and kept in a friable condition by these organisms. As discussed in Chapter 3, the pores created by earthworms facilitate rapid air and water movement but can also function as channels for root growth.

6.3 SOIL STRUCTURE

Soil structure is defined in many textbooks as the arrangement of soil particles into aggregates. A more correct definition is that soil structure is an aggregate formed by the arrangement of soil particles. The arrangement of the soil particles is a process that results in formation of aggregates or structural units. When the aggregates are formed by cultivation, they are often referred to as *soil clods*, but when they are formed by natural processes, they are called *soil peds*. Well-granulated soils are reported to be 70 to 80% aggregated, and aggregation may run as low as 20% in soils with sandy textures (Foth, 1990; Brady, 1990). The remainder of the soil particles is present more or less as individual particles in the soil.

6.3.1 Types of Soil Structure

Single-Grained Structure. - Soils composed of only individual particles are said to have a single-grained structure. Sands are usually single-grained in structure. There is little or no cementation, except for mere touching at the points of contacts between the sand particles. A soil with a single-grained structure may feel loose or compacted by changing the degree and number of points of interparticle attraction. The latter is generally transient and disappears on wetting and drying. These soils are in fact structureless.

Massive Structure. - Another structureless condition is the massive structure, where all the particles are cemented together without producing any identifiable units. Massive structures generally occur in soils that contain sufficient amounts of clays to form a matrix totally enclosing the coarser grains, as is the case of the coarse particles in a concrete wall. For the purpose of soil taxonomy and agricultural operations, seven types of soil structures are recognized, e.g., crumb, granular, platy, subangular blocky, blocky, prismatic ,and columnar structures (Soil Survey Staff, 1962). In engineering and geology, different types of soil structures are recognized, and massive and single-grained structures are accepted as major types of soil structures (Spangler and Handy, 1984; Hunt, 1972).

Crumb and Granular Structures. - These structures are irregularly shaped aggregates, the size of a wheat grain or a bread crumb. When the units are porous, the name crumb is used, in contrast to granular, which appears more solid and less porous. They are characteristic structures for the soil's A horizon with high organic matter content, hence a horizon affected by high biological activity. Mechanical granulation of soils often yields granular structures, but seldom crumb structures.

Subangular and Angular Blocky Structures. -   These types of structures are typical for the B horizons of soils. Apparently they are affected by natural compression or natural compaction, since the units are more solid and look like irregularly shaped blocks. The structure is called *subangular* when the corners of the small blocks are rounded, hence not pointed or sharp. Subangular blocky structures in fact look like granular structures, but are somewhat coarser in size and occur in the B horizon. They may range in size from 0.5 cm to 1 cm. Aggregates 10 cm in size are called by some authors subangular blocky structures. The structure is called *angular* blocky, because the corners of the aggregate units are sharp and pointed

Prismatic and Columnar Structures. -   These aggregate units are longer in the vertical than in the horizontal direction. If sharp corners are present, the name *prismatic* applies, but when the units are round like the pillars of an antebellum home, they are called *columnar*. These types of structures are characteristic in the B horizons of aridisols in particular, where accumulation of salt is considered to affect their formation. Humid region soils may occasionally exhibit these types of structure in the B horizons when 2:1 expanding types of clays are present.

Platy Structures. - This is a type of structure that can occur in the A as well as in the B horizon. The aggregates are usually flat in shape like a sheet of paper or like a plate. They are assumed to be formed by precipitation or deposition of the particles from water, or by compression due to ice formation.

6.3.2   Environmental Significance of Soil Structure

The formation of soil aggregates or soil structures is often called *aggregation* or *granulation*, a process which is closely influenced by environmental and especially biological factors. All soil particles, be it inorganic or organic, are included in the aggregation process. The climate is known to play an important role in aggregation and deaggregation processes. Freezing is noted to exert compression that enhances cementation of soil particles, whereas thawing causes swelling of the aggregates, which may weaken the cementation action. Roots are recognized to contribute toward structure formation by binding together the soil particles. Plants with a dense root system are of particular importance in binding the coarse particles (>2 mm). Grass is reported to be the most efficient type of vegetation in binding the soil particles together, because of the completeness of the root system in filling the soil (Foth, 1990). Humus and secretions from roots and microorganisms are cementing agents in the aggregation process. Root and microbial secretions have long been suspected of being capable of flocculating and cementing soil aggregates. Polysaccharides and extracellular polysaccharides have been discussed earlier as increasing aggregation due to their cementing capacity and by reacting with the inorganic particles. Extracellular polysaccharides are produced by a variety of microorganisms, such as bacteria, fungi, and algae. As discussed previously, they occur as a mucous coating constituting the outermost layers around the cells, providing in this way protection to the cells, and at the same time forming the interfaces between the cells and the surrounding soil or soil water. The interaction of extracellular polysaccharides with the inorganic soil particles yields organo-mineral complexes that contribute toward formation of stable soil aggregates. In contrast to plant roots, humus, plant and microbial secretions, polysaccharides, and extracellular polysaccharides are particularly effective in cementing the smaller-sized inorganic particles (0.2-0.002 mm). The aggregates formed are reported to be more stable than those formed by the binding effect of roots and fungal hyphae (Baver, 1963; Chenu, 1995).

The types of structure present in soils are conditioned by environmental factors, e.g., climate, biological activity, activity of soil colloids, and the presence of exchangeable cations.

Biological Activity. -   The effectiveness of humus and microbial secretions in soil aggregation has been discussed above. Living roots of plants, especially grass, were indicated as quite efficient in binding soil particles together. However, little is known on the direct influence of microorganisms in producing stable soil aggregates. The very few reports available in the literature assume that bacteria, fungi, and actinomycetes are capable of inducing aggregation (Hubbell and Staten, 1951). Depending on the type of microorganisms, three kinds of aggregates can be distinguished: (1) *bacterial type*, which is composed of small (< 0.56 mm), compact, and angular units that are fragile and easily crushed, (2) *fungal type*, formed by the binding action of the hyphae. The aggregates are larger in size (2 - 4 mm), more resistant to crushing, and spongy in nature, (3) *actinomycetes type*, which is nonangular, very compact, and fairly resistant to crushing. In the 1950s, it was believed that the granules were cemented together by metal bridging as illustrated by the reaction:

clay - metal - OOC - R - COO - metal - clay

The metal ion can be Ca, Al, Fe, Mn or any other metal cation. Today, polysaccharides and extracellular polysaccharides are recognized as the major organic substance contributing to the aggregation reaction above. Van der Waals and hydrogen bonding are also possible mechanisms for binding the soil particles into aggregates.

Inorganic Cementing Agents. - The role of Al, Fe, and other cations in cementing soil particles together has been known for some time and discussed in the preceding pages. Iron was explained earlier as a major contributor in cementing and coating aggregates in oxisols, making these soils very friable regardless of their high clay content. However, Deshpande, Greenland, and Quirk (1964) believe that iron oxide does not act as a cementing agent and is present in soils as a discrete crystal. In their opinion, Al oxide forms interlayers with clay minerals, and in this way acts a cementing agent. Though this may be true, the authors forgot that iron oxide can also be present in soils in amorphous form, and it is this form of Fe that is very active. Silica is another cementing agent recognized in Chapter 5. The metal cations, Al and Fe, chelated by silica can function as bridges to bind clay particles together, similarly as in the metal-bridging reaction illustrated above for the binding of clay by organic compounds.

Climate. - Because of the effect of biological activity, the surface soil in humid regions often exhibits granular or crumb structures, whereas the subsoil, with less biological activity, is mostly subangular blocky in structure. In the event that the subsoil is heavier in texture, its structure can be angular blocky. In arid region soils, the surface soil may also be characterized by granular structures, but due to accumulation of salts, prismatic or columnar structures are prevalent in the subsoil. Water must be present in sufficient amount for structure to be formed. Too much water leads to puddling of the soil and collapse of soil structures. Excessive drying by prolonged dry periods may lead to caking of surface soils and formation of massive structures. The soil structure is noted to modify the effect of soil texture, which is very important in heavy-textured soils containing expanding lattice type of clays, such as the vertisols. Vertisols are known for their poor physical properties, due to smectite in their clay fractions. However, because of the presence of well-developed granular structures formed by biological activity in the surface soil, these soils can still support good forest stands and crop production, as can be noticed in India, South Africa, and in the *blacklands* of Texas, Alabama, and Georgia.

Soil structure affects the pore space content, hence air and water flow in soils, soil strength, and soil compaction. Its effect is complemented by the effects of soil texture and bulk density. Root penetration is increased by the development of granular and crumb structures. Soil impedance is drastically reduced by the presence of crumb and granular structures. Crumb structures are porous, hence provide for a lot of intra-aggregate and inter-aggregate pores. On the other hand, granular structures are more solid, hence lack the intra-aggregate pores. However, because of their irregular shapes an *edge-to-face* settling occurs so that a considerable number of inter-aggregate pores are formed.

Soil Conditioners. - The stabilizing effect of organic matter on soil structure has attracted the attention of industry to produce artificial compounds that will produce a similar, but more long-lasting cementing effect in soil aggregation. These chemical compounds are known under the names *soil conditioners* or *soil stabilizers*. Because of the stabilizing effect on soil structure, natural and artificial soil stabilizers are of utmost importance in soil conservation practices and erosion control. Two different groups of soil conditioners are recognized: (1) gypsum and phosphogypsum, and (2) organic polymers (Levy, 1996). Gypsum and phosphogypsum are applied to provide an electrolyte concentration in the soil solution above the critical concentration that forces flocculation of clay particles. This critical concentration was earlier called the *flocculation value*. Surface application of gypsum or phosphogypsum is noted to be as effective as plowing under in controlling *surface sealing* of the soil, hence maintaining high infiltration rate and reducing runoff. Surface sealing is a phenomenon caused by destruction of the structure of soils exposed to the impact of raindrops. Two mechanisms are assumed to be responsible in seal formation: (1) destruction of soil aggregates as pointed out above, and (2) dispersion of the released clay particles and their consequent clogging of soil pores. The two mechanisms are interrelated and the first usually enhances the second (Agassi et al., 1981). The second group of soil conditioners is *organic polymers*, synthesized by chemical companies. A well-known soil conditioner used in the past is *krilium*, produced by Monsanto Co. It is a trade name for a mixture of polymers, containing vinyl acetate, malic acid, and polyacrylonitrile. The synthetic conditioners of today are hydrolyzed polyacrylonitrile (HPAN), a copolymer of vinyl acetate and maleic acid (VAMA), polyacrylamide (PAM), and polysaccharide (Taylor and Ashcroft, 1972; Levy, 1996). The synthetic polymers function in the aggregation process through interaction or chelation with clays, and by stabilizing soil structure to prevent dispersion of clays, hence control surface sealing of soils. The latter then results in increasing the infiltration rate, reducing runoff, and thereby preventing loss of soil by erosion. Applying the polymers in solution (dispersed) form rather than in powdered form is suggested. Adding small amounts of the polymers in water for use in overhead sprinklers, so that a concentration of approximately < 50 g/m^3 is attained, is suggested by several soil physicists.

6.4 SOIL DENSITY AND SOIL POROSITY

The density of a soil refers to the amount or mass of soil per unit volume of soil. It is dependent in part on the mineral and organic matter content, and in part on the amount of pore spaces or soil porosity. Therefore, the concept of soil density takes into account both mass of the solid particles and the volume of pore spaces. This mass of a given volume of soil can be expressed in terms of bulk density and particle density.

6.4.1 Bulk Density

Bulk density is defined as the mass (weight) per unit volume of undisturbed soil, called *bulk volume*, and is usually formulated as follows:

Bulk density = (Ovendry weight of soil)/Volume of soil

Bulk volume includes the volume of solid particles and the enclosed pore spaces, as occurring in a soil ped or a soil clod. The unit of bulk density is normally expressed in grams per cc (cm^3 or mL) of soil. In the US system, this unit can be converted into lb (pounds) per cubic foot of soil by multiplying by 62.4. Today, the USDA recommends using Mg/m^3 as the unit of bulk density, which in fact equals g/cc.

Bulk density is a good measure of the physical condition of a soil and can be used to estimate soil porosity. Low bulk density, 1- 1.5 g/cc, is an indication of the presence of large amounts of pore spaces, and in general terms, is a favorable physical condition for plant growth. Very low bulk density values, 0.5 g/cc or lower, suggest the presence of extremely high organic matter contents, and such low values are found only in organic soils (peat soils). Environmentally, the bulk density must not be too low or too high for adequate support of the growth of plants. Very low bulk density values are detrimental to the growth of trees, which tend to lodge when blown by heavy winds, because of failure of their roots to anchor firmly in the soil. High bulk density values, 1.8 - 2.0 g/cc, indicate the presence of low amounts of pore spaces, hence poor physical conditions and can be attributed to compaction of the soil. Generally, sandy soils exhibit higher bulk densities than finer textured soils. Sands and sandy loams vary in bulk density from 1.2 to 1.8 g/cc, whereas silt loams, clay loams, and clays may have bulk densities ranging from 1.0 to 1.6 g/cc. Coarse-textured soils are comparatively lower in total pore space than fine-textured soils. Bulk density values also increase with increasing depth in the pedon, because of more compaction, lower organic matter content, and less aggregation in subsoils. Dense or compacted soils may have bulk density values of 2.0 g/cc or greater. Compaction destroys pore spaces by forcing solid material into the pores. Plowing and tillage operations are usually designed to increase pore spaces and consequently decrease bulk density. However, the use of heavy equipment in today's agricultural operations produces compaction. Especially, the tractor tracks are notorious for increasing bulk density. This increase in bulk density will increase soil resistance to root penetration (soil impedance), and root growth to deeper layers of the soils will be inhibited. This effect of bulk density on root growth appears to be dependent also on soil texture, since the maximum values for bulk densities that restrict root growth vary from a high of 1.75 g/cc for sands to a low of 1.45 g/cc in clays (Veihmeyer and Hendrickson, 1948).

6.4.2 Particle Density

Particle density is defined as the mass (weight) of a unit volume of solid particles. In this concept, the pore spaces in the bulk volume of soil are destroyed, so that only the weight and volume of the solid particles are measured. The formula can be written as:

Particle density = (Ovendry weight of solid particles)/volume of solid particles

It is usually expressed in grams per cc. Particle density values can be used to replace specific gravity of soils, and particle density is considered by many a better unit than specific gravity. Although the density of individual soil particles varies considerably, the particle density of mineral soils is fairly constant, and varies between 2.60 and 2.75 g/cc. The average of 2.67 g/cc is commonly referred to as the *specific gravity* of soils. The density of water is 1.0 g/cc at 25°C. A soil with a particle density of 2.60 to 2.75 g/cc then means that the soil is 2.60 - 2.75 times heavier than water.

Quartz, feldspar, mica, and colloidal silicates make up the major portion of mineral soils in the United States, and their densities fall within the range above. When unusual amounts of heavy minerals, such as magnetite, zircon, tourmaline and hornblende, are present, the particle density of the soil may be greater than 2.75 g/cc. Size and arrangement of soil particles do not affect the value of particle density. However, organic matter, which weighs much less than an equal volume of mineral solids, will decrease particle density of soils. Surface soils with high organic matter contents may exhibit particle density values of 2.4 g/cc. Organic soils (peat and mucks) have extremely low particle density values.

6.4.3 Pore Spaces

Pore space is that portion of the soil volume not occupied by solid particles, but occupied by air and water. The arrangement of solid particles determines the amount and size of pore spaces. The total amount of pore spaces in soils is referred to as the *soil porosity*. The percentage pore spaces or soil porosity can be calculated from the values of bulk and particle densities as follows:

% Pore space = [(Particle density – Bulk density)/Particle density] x 100%

The arrangement of solid particles determines the amount of pore space, and so does soil texture. As discussed earlier, the surface of sandy soils has a total pore space usually ranging from 35 to 50%, whereas finer textured surface soils may have pore spaces ranging from 40 to 60%. Thus, finer textured (silty and clayey) soils have a greater percentage of total pore space (larger porosity) than coarse-textured soils (sands). Compacted subsoils may have total pore spaces as low as 25 to 30%. These differences are caused by the relative proportion of macro- and micropores that together make up the total pore space. Sands usually have large amounts of macropores but very small amounts of micropores. In contrast, clays contain large amounts of micropores, but very few macropores are present. The amount of micropores generally far exceeds that of the macropores; hence when considering the porosity, the total pore space in clayey soils is larger than that in sandy soils. Thus, sandy soils, containing larger amounts of macropores, have fairly rapid movement of air and water in spite of the lower amount of total pore spaces. On the other hand, fine-textured soils, with a preponderance of micropores, have a slower movement of water and air, but have higher water-holding capacities.

6.4.4 Environmental Significance of Soil Density and Soil Porosity

The bulk density and pore spaces are interrelated. Development of low bulk density values also means the development of large amounts of pore spaces. In nature, low bulk density values are usually found in soils with high organic matter contents. High biological activities are necessary for formation and large accumulations of organic matter. Together with the effect of soil organisms, the high humus content will encourage aggregation, increasing in this way soil porosity, and thereby decreasing bulk density values. The cultivation effect of the soil macro- and microfauna produces an intricate system of macropores, which is a major factor for lowering the bulk density of the soil. Continuous cropping is noted to decrease the amount of organic matter in soils, and lowering the organic matter content is expected to decrease soil aggregation. Especially the macropores are destroyed by the destruction of soil structure. Tillage by plowing is designed to increase pore spaces in soils, but is in fact decreasing organic matter content. The latter seems to defeat the purpose of plowing, and Brady and Weil (1996) reported that the amount of macropores could be reduced by one-half by most cropping systems using the plow. Recently, conservation tillage and no-till farming have been introduced to alleviate these problems and at the same time to control soil erosion. The results were received with mixed reactions. Though many claimed that conservation tillage or no-till methods have increased organic matter content in soils, others indicated that the latter have not always resulted in an increase in total pore spaces. Brady and Weil (1996) indicated that soils under conservation tillage contained less pore space than those tilled by conventional methods.

6.5 SOIL CONSISTENCE

Soil consistence is considered a soil property caused by the combined effect of other soil properties which are dependent upon the forces of attraction between soil particles as influenced by soil moisture. The Soil Survey Manual (Soil Survey Staff, 1962) defines soil consistence as a soil characteristic that is expressed by the degree and kind of cohesion and adhesion, or by the resistance to deformation or rupture.

6.5.1 Types of Soil Consistence

Depending on the moisture content, five major types of soil consistence are distinguished: *hard, friable, plastic, sticky*, and *viscous*. The water content increases from the hard to friable, to plastic, sticky and viscous consistence. Each of these types of consistency forms can be divided again into several subtypes. For example, the sticky consistence can be distinguished into nonsticky, slightly sticky, sticky, and very sticky.

Hard Consistence. - A dry soil has a hard consistence and shows great resistance to deformation. Interparticle attraction and forces of adhesion are believed to be dominant. Water content is at a minimum.

Friable Consistence. - Friability characterizes the ease of crumbling of a soil. Soils with a friable consistence contain enough water so that they are easy to crumble. The individual aggregates feel soft and mellow, and cohesion is at a minimum. The moisture content of a friable soil is also the range in which soils are in optimum condition for tillage.

Plastic Consistence. - Plasticity characterizes the ease of pliability of soils or the ease of soils being molded. Most soil will exhibit plasticity when wetted above the field capacity. This property is due to the plate-like nature of the clay minerals. When sufficient amounts of moisture are present, these plates easily slide over each other. The moisture serves as a lubricating and at the same time as a binding agent. Usually, expanding lattice clays with a high hydration capacity exhibit the greatest plasticity. Thus, smectite or montmorillonite is substantially more plastic than kaolinite. Plasticity encourages a ready change in soil structure.

Sticky Consistence. - Sticky consistence develops when the soils are wetted to the point that they become sticky. The water content is larger than that at plasticity. It seems that the sticky consistence is also closely related to types of clay minerals.

Viscous Consistence. - Viscous consistence is reached when the water content reaches the point at which dispersion of soil particles starts. The soil becomes puddled and starts to flow. At this consistence structural changes are also apparent.

6.5.2 Atterberg's Plasticity Index

From the above, it is perhaps clear that soil consistence varies over a wide range of soil moisture contents. The water content at which a particular type of consistence is produced is called the *Atterberg's plasticity limit*. Atterberg recognizes a series of plasticity limits and five of them are presented as follows: (1) liquid limit, (2) sticky limit, (3) plastic limit, (4) roll out limit, and (5) transition limit (Wisaksono and Tan, 1964; Grimshaw, 1971).

Liquid Limit. - This is the point at which the soil almost flows like a liquid. Literally translated from the German language, this limit is the *limit of fluidity*, the term used in Grimshaw's (1971) book.

Sticky Limit. - This is the point at which the soil becomes sticky. The soil contains less water than at the liquid limit.

Plastic Limit. - This is the point at which the soil starts to exhibit pliability. Grimshaw (1971) called this the adhesion limit.

Roll Out Limit. - This is the point at which the soil ceases to exhibit plasticity and will be fragmented by rolling out. Because of the latter, this limit is also called the *rolling out limit*, which is in fact the exact translation from Atterberg's book.

Atterberg's Number or Index. -   The difference in water content between the liquid limit and the roll out limit reflects the degree of plasticity of the soil, which is called the *Atterberg's number* or *index*. On the basis of these plasticity indices, Atterberg proposed several plasticity classes:

	Atterberg's index	
Class I	0 - 5	Nonplastic
Class II	6 - 10	Slightly plastic
Class III	11 - 17	Plastic
Class IV	18 - 30	Very plastic

Thus, if 30% of water is required to reach the liquid limit, and 20% is needed to enable the soil to be rolled out, the plasticity index equals 30 - 20 = 10, which falls in class II. The soil is then slightly plastic. However, some doubt exists about the usefulness of Atterberg's indices. A number of authors argue that the indices do not fully describe plasticity, but define only the range of water contents over which plasticity can appear at various degrees. Grimshaw (1971) believes that Atterberg's index corresponds closely to the binding power of clays, rather than to the true plasticity. The binding power of clays is defined by Grimshaw as a property of retaining particles in their mass in a state of suspension. The binding strength of kaolinite ($250 \ g/cm^2$) is ten times less than that of smectite ($2500 \ g/cm^2$).

6.5.3 Environmental Significance of Consistence

Soil consistence is a physical property that is closely related to the other properties of soils. Its function cannot be separated from the effects of water content and soil texture. In the aforementioned sections, it is pointed out that with increasing moisture content, the soil's consistence will change from hard to friable, sticky, plastic, and viscous. At a viscous consistence, soils tend to be easily puddled by mechanical operations in crop production or by the trampling effect of animals in nature. Both will increase the chances of accelerated soil erosion. A friable consistence will appear at the appropriate moisture content, after excess moisture, supplied by rain or irrigation, has drained by gravity. Structural changes will be at a minimum at this consistence. Organic matter content affects the development of a friable consistence. Generally, soils with high humus contents are more friable than soils with low humus contents. On the other hand, heavy-textured soils generally tend to be less friable than light-textured soils. However, in the presence of high amounts of organic matter, even clayey soils will exhibit friable consistencies at the right water content. However, oxisols are heavy-textured soils, and are low in organic matter, yet they are very friable. The latter is caused by their stable structure, due to cementation and coating of the aggregates by amorphous iron oxides, which subdues the effect of the soils' high clay content.

Plasticity has attracted considerable attention in industry because of its importance in the production of pottery, ceramics, bricks, and even dentures. The degree of pliability, displayed by the various clay minerals as discussed in Chapter 2, has an important bearing in forming the various shapes of ceramic products. Kaolin, with lower degrees of plasticity, is often mixed with montmorillonite or smectite, which are characterized by higher degrees of plasticity, for enhancing pliability. However, it should be realized that it is not only the pliability that accounts for the quality of the product. The swell-shrink capacity of the clay minerals also determines the quality of ceramic products, and it is kaolin's low swell-shrink capacity together with plasticity that produces high-quality porcelain. Bricks, pots, and especially dentures will also be inferior if they contain large cracks due to the use of materials with high swell-shrink capacities, though in their production, materials are needed that are plastic enough for producing the various shapes and forms.

6.6 SOIL TEMPERATURE

Soil temperature is the least known physical property, though it has a pronounced influence on almost all biochemical and biological processes in soils. Seed germination and seedling and root growth will be affected by soil temperature. Almost all reactions in soils go faster at higher than at lower temperatures. Soil temperature determines the rate of respiration, transpiration, evaporation, and decomposition processes. Weathering and soil formation are additional processes that are affected by soil temperature. Under the influence of very low temperatures, such as in northern Alaska and Siberia, the soils are continuously in a frozen state, called *permafrost*. Under this condition, microbial activity is at a minimum and decomposition proceeds anaerobically, if any at all, favoring accumulation of tundra peat. More available water is present in soils at higher than at lower soil temperatures. Evaporation from soils, plants, rivers, and lakes is all dependent on surface temperatures of soil, earth, or water.

6.6.1 Thermal Properties of Soils

Some of the thermal properties of soils are soil temperature, heat capacity, and thermal conductivity. These properties vary considerably with soil and environmental factors. The soil factors affecting the thermal properties are the water content, compaction, and mineralogical composition. Compaction increases the area of contact, hence improves conductivity. Transmission of heat in porous soil is very poor, because air in the pore spaces is a poor conductor. An important environmental factor is, for example, the incoming radiation from the sun. Dry sand particles transmit heat more slowly than wet sand particles, because of water present at the areas of contact between the sand particles. Water is an excellent heat conductor.

Soil Temperatures. - Measuring soil temperatures appears to be not as simple as might be expected, because of the many factors to be considered. The best method of analysis, using thermometers or thermocouples, and the soil depth at which the temperature is to be measured remain to be solved. Temperatures are suggested to be measured at 5, 10, 20, 50, 100, and 300 cm depth of the soil by some (Taylor and Ashcroft, 1972), whereas others assume that the *standard temperature* is measured at a depth of 50 cm (Miller and Gardiner, 1998). However, the use of an *average mean annual temperature* by the latter, measured at a depth of 6 m does not add to making the problems more understandable. Measurements of soil temperature at a considerable depth of the soil are in this author's opinion meaningless from an environmental standpoint. Biological activity, if any at all, is minimal at great depths. Except for an occasional strand of a root, root systems of most plants seldom extend to 1 m or to 6 m depth in the soil. The root zone is usually confined to 10 - 20 cm depth, which is also the zone of maximum microbial activity. Hence, the purpose of measuring soil temperatures at great depth is not clear.

Most of the sun's radiation is absorbed by the soil surface. Only a small fraction is reflected by the soil and vegetation. This part which is reflected is referred to as the *albedo*. Values for albedo ranges from 5 to 20% from forested surfaces, and from 5 to 35% from bare soils (Taylor and Ashcroft, 1972). That part absorbed by the soil is used for warming up the soil's surface. Generally the increase in soil temperature is greater in dry soils than in wet soils. This is because water has a stabilizing effect on soil temperature, since loss of heat by evaporation of water is high. The larger the water content of the surface soil, the more difficult it is to increase its temperature. The extent of the temperature increase also depends on its heat capacity and thermal conductivity.

Heat Capacity of Soil. - The heat capacity of soils varies considerably depending on the conditions imposed when the heat is absorbed, and is usually larger for dry than for wet soils. In wet soils, the energy introduced is also used for volume expansion of soil water, hence a smaller rate of a temperature increase is obtained. In other words, more energy is needed to increase the temperature of wet soils, hence the heat capacity of wet soil is higher than that of dry soil. According to Baver (1956), the specific heat capacity for most soil solids is in the range of 0.2 cal/g °K, and that of soil moisture is 1.0 cal/g °K.

Thermal Conductivity. - Thermal conductivity is the capacity of soils to transmit heat through their body. Soils are known to be slow _heat conductors_, and because of this the soil surface is expected to be at a different temperature than the subsoil. Not much data is available on this aspect. However, the soil surface, exposed to the sun, will be warmer by day than the subsoil. Due to the _low heat conductivity_, the heat absorbed from the sun's radiation will not be transmitted rapidly to the subsoil, hence the surface soil is warmer than the subsoil. At night, the surface soil will be cooler than the subsoil, because radiation of heat from the surface soil causes the surface soil to cool off more rapidly than the subsoil. However, daily fluctuations in temperature are reported seldom to affect the soil deeper than 30 to 40 cm (Miller and Gardiner, 1998). These authors maintain that the difference in mean summer and winter temperatures at 1 m depth in the pedon is $\leq 5°C$. This is a considerable difference, since the _law of Van 't Hoff_ indicates that the rate of reaction increases two to threefold with an increase of 10°C in temperature.

6.6.2 Significance of Environmental Factors for Soil Temperature

In temperate region soils, soil temperature is lowest at night and in the morning and increases by day to reach a maximum by 3:00 P.M. In the tropics, these diurnal variations in temperature are somewhat less, but still noticeable. The soils in the tropics are also cooler in the morning and warmer by day, with maximum values occurring around noon to 1:00 P.M. Such a fluctuation in soil temperature due to heating by day and cooling at night is reduced considerably by trees and other kinds of vegetation cover. Leaves of plants themselves absorb large amounts of radiation energy from the sun through photosynthesis, which is released as heat during the process of respiration. The heat released would increase the temperature of the leaves if evapotranspiration did not take place in the leaves. This heat is now used for the conversion of water into water vapor, resulting in cooling the leaves to normal temperatures.

The canopy cover intercepts part of the radiation of the sun to the soil. This is part of the reason why it feels cool under the shade of trees. The air within and below the canopy also acts as a very poor conductor for heat transmission, and because of this the soil underneath remains relatively cool. Mulch, composed of organic matter, acts in the same way as a canopy cover. The surface of the mulch, which is heated by the sun, is usually at a higher temperature than the soil underneath. However, plastic mulch, composed of plastic sheets laid over the soil, will warm up the soil. It was noted that black plastic sheets will warm up the soil more than white or clear plastic sheets. The process of heating is due to a process similar to that in the *greenhouse effect*.

CHAPTER 7

ELECTROCHEMICAL PROPERTIES OF SOLID CONSTITUENTS

7.1 ELECTRICAL CHARGES

As discussed in Chapters 2 and 3, the solid constituents of soils consist of a variety of inorganic and organic compounds. Most of the coarse materials, such as sand, silt, and undecomposed organic matter are chemically inert. They are important constituents for building up the soil and affect many soil properties. Since they are coarse in size, they have low specific surface area and do not exhibit colloidal properties. They may participate in a number of soil reactions and exhibit some adsorption capacities, but they are not really chemically active. However, sand and silt may weather to form clay, whereas the undecomposed organic matter will be broken down by decomposition processes into humus. Clay and humus are the smallest constituents of soils and exhibit colloidal properties. They have large specific surface areas and display a surface chemistry different from the coarse materials. The surface chemistry of clay and humus is attributed to the presence of electrical charges in their molecules. Because of these charges they are chemically very active and are considered the *seat of soil activity* (Brady, 1990), causing many soil chemical reactions, such as cation exchange, anion exchange, and complex and chelation reactions.

7.2 SOIL CLAYS

Soil clays ordinarily carry electronegative charges, which are the result of one or more of several different reactions. Two major sources for the origin of negative charges are *isomorphous substitution* and *dissociation of exposed hydroxyl groups*.

7.2.1 Isomorphous Substitution and Permanent Negative Charges

Isomorphous substitution is the substitution of atoms in the crystal structure for other atoms without affecting the crystal structure. It can take place in both the silica tetrahedrons and the aluminum octahedrons of the clay mineral. For example, in the absence of isomorphous substitution, kaolinite is electrically balanced. Assuming that the unit cell formula of kaolinite is $Al_2O_3.2SiO_2.2H_2O$, the following simple calculation shows that the positive charges carried by the cations are completely neutralized by the negative charges carried by the oxygen atoms:

Al^{3+}	Si^{4+}	H^+	O^{2-}
2x	2x	4x	9x
18^+			18^-

A replacement of one octahedral Al by Mg by the isomorphic process yields one negative charge unbalanced in the crystal. Since Mg is divalent, it can only contribute two positive charges for the neutralization of the negative charges in the crystal, as shown by the balance sheet below:

Al^{3+}	Mg^{2+}	Si^{4+}	H^+	O^{2-}
1x	1x	2x	4x	9x

$$17^+ \qquad\qquad 18^-$$

A similar substitution can also occur in a Si tetrahedron, where Si can be replaced by Al. This reaction also leaves one negative charge not neutralized. Isomorphous substitution will occur only between atoms of almost equal sizes, and also when the difference in valences does not exceed one unit (Tan, 1998). These types of negative charges are called *permanent negative charges* and are independent of pH. The replacement process is a very gradual process occurring during the synthesis of the secondary minerals and once the replacement has taken place, it is not subject to further modification, though in nature exceptions always occur. This is the reason why the electrical charge developed is called *permanent* or sometimes *constant charge*. This type of permanent charge is of special importance in 2:1 types of clays, e.g., smectites. A number of authors object to using the name permanent charge, since they believe that it cannot be used for soil systems containing high amounts of organic matter, intergrade minerals, and allophane. The Soil Science Society of America subcommittee on soil chemistry terminology even drops the use of the term permanent charge, and redefines it as the net negative (or positive) charge of clay particles inherent in the crystal lattice of the particle, and that is not affected by pH changes or by ion exchange reactions (Mehlich, 1981). However, other authors fail to see the improvement in the definition and feel that the new definition is too long and too wordy. In addition, to name it *inherent charge* is also very confusing and ambiguous. Some suggest naming it *permanent structural charge*, σ_p (Sposito, 1989; Tan, 1998).

7.2.2 Isomorphous Substitution and Permanent Positive Charges

Isomorphous substitution not only produces negative charges, but can also yield electropositive charges when the replacing ion has a larger positive charge than the ion being replaced. However, such a substitution process cannot occur in the aluminosilicates due to their composition. Al and Si ions are ions that possess the highest charges in soils. Only Mn^{4+} has a higher charge than Al, but Mn^{4+} has never been noticed to replace Al ions in 1:1 or 2:1 clay minerals. The development of positive charges by isomorphous substitution is commonly known to occur in vermiculite or in the 2:2 lattice type of clay minerals. These clay minerals are by composition magnesium silicates, hence the magnesium that is present as Mg^{2+} can be replaced by ions of larger charges, such as Al^{3+} or Fe^{3+} ions. For example, vermiculite, with the following composition, $Mg_3Si_4O_{10}(OH)_2$, is completely neutral. The positive charges are balanced by the negative charges:

Mg^{2+}	Si^{4+}	H^+		O^{2-}
3x	4x	2x		12x
	24^+			24^-

However, if one Mg ion is replaced by an Al ion, the vermiculite mineral is positively charged:

Mg^{2+}	Al^{3+}	Si^{4+}	H^+		O^{2-}
2x	1x	4x	2x		12x
		25^+			24^-

This type of isomorphous substitution is common in chlorite minerals.

The silicon in tetrahedral positions may be replaced by Al, whereas Fe and Al may replace Mg in octahedral positions. The resulting negative charges are then neutralized by the positive charges. This is the reason why chlorite minerals are known to be very low in negative charge or carry no charges at all.

7.2.3 Dissociation of Exposed Hydroxyl Groups

Exposed hydroxyl groups are OH groups present on the surface of the Al octahedron sheets. They are prevalent in 1:1 types of clays, sesquioxides, and amorphous clays. These OH groups are in contact with the soil solution and tend to dissociate, releasing thereby their protons as illustrated in the reaction below.

$$- Al - OH \quad \rightleftarrows \quad - Al - O^- + H^+ \tag{7.1}$$

neutral   negatively charged
octahedron   octahedron

This dissociation of the H^+ leaves one negative charge in the octahedron not neutralized. Such a dissociation reaction is dependent upon pH. The dissociation reaction occurs at high pH and decreases at low pH. Therefore, the magnitude of the negative charge also increases and decreases accordingly with the change in pH. Because of this, this type of negative charge is called *pH-dependent charge* or *variable charge*. It is of importance in the minerals with OH groups present on their surfaces, as indicated above. For example, kaolinite exhibits variable charges. Permanent charges are also present in kaolinite, but to a lesser amount. The minerals with variable charges are frequently called *variable charge minerals*, and the soils containing these minerals are called *soils with variable charges*. Mehlich (1981) prefers the use of the term *pH-dependent charge* over that of *variable charge*. However, Sposito (1989) retains the name variable charge and adds a quantitative interpretation to it by formulating the variable charge in terms of *proton charge*, σ_H:

$$\sigma_H = m_H - m_{OH}$$

where m_H = mol/L of H^+ ions, and m_{OH} = mol/L of OH^- ions complexed by surface functional groups.

7.2.4 Protonation of Exposed Hydroxyl Groups

Protonation of exposed OH groups is the addition of H^+ ions to the exposed OH groups of the clay minerals. The H^+ is adsorbed with a relatively weak bond. Because of this addition of H^+, the OH group is oversaturated with protons and the clay surface becomes positively charged, as illustrated by the reaction below.

$$- Al - OH + H^+ \quad \rightleftarrows \quad - Al - OHH^+ \qquad\qquad (7.2)$$
 neutral   positively charged
 octahedron   octahedron

Protonation of exposed OH groups occurs only at low pH, because acid conditions are required for the supply of the extra protons. This kind of positive charge is called a *variable* or *pH-dependent positive charge*, in contrast to the permanent positive charge discussed above. This kind of pH-dependent positive charge is important in kaolinite, sesquioxides and amorphous minerals. Smectite does not have exposed OH groups, hence exhibits only very small amounts of variable charges.

7.2.5 Zero Point of Charge

The *zero point of charge (ZPC)* is the *pH* at which the mineral has no charge, or has equal amounts of negative and positive charges. It has a similar meaning as *isoelectric point*. As discussed above, at high pH values the mineral carries a negative charge, which decreases with a decrease in pH. When the pH is continuously decreased, a point will be reached at which the negative charge equals zero. The pH at which this occurs is called ZPC, or *zero point of charge*, as indicated in the definition above. Some authors call it PZC or *point zero charge*, which perhaps boils down to semantics. Other authors recognize three types of ZPCs (Sposito, 1989):

1. ZPC, which is the general ZPC as defined above.

2. ZPPC, which is the *zero point proton charge*. Consequently, $\sigma_H = 0$.

3. ZPNC, which is defined as the *zero point net charge*, where the total charges must equal zero.

The ZPC is a specific characteristic of the clay mineral, and is assigned the symbols pH_0. Its value differs from one to another mineral (Table 7.1). At soil $pH > pH_0$, the clay is negatively charged. At soil $pH < pH_0$, the clay is positively charged. When the net charge is zero at ZPC, clay particles in soil water will not repel each other and tend to aggregate, forming larger particles. These aggregates precipitate and form the soil structure. In contrast, at soil $pH > ZPC$,

Table 7.1 ZPC Values of Selected Minerals

Mineral	ZPC
Hematite, $\alpha\text{-}Fe_2O_3$	2.1
Corundum, $\alpha\text{-}Al_2O_3$	2.2
Goethite, $\alpha\text{-}FeOOH$	3.2
Gibbsite	4.8
Lepidocrocite	5.4 - 7.3
Maghemite	6.7
Ferrihydrite	8.1
Amorphous $Al(OH)_3$	8.3
Amorphous $Fe(OH)_3$	8.5

Sources: Van Schuylenborgh and Sänger
(1949); Van Schuylenborgh and Arens (1950).

the negatively charged clay particles repel each other, resulting in dispersion. Under such a condition, the clay suspension is considered stable. If clays remain dispersed, the soil is puddled and sensitive to water erosion. Puddled soils are also sticky when wet, and become hard and dense upon drying. A compacted soil is undesirable for plant growth. Root growth requires a porous soil for adequate aeration. Such a soil can be formed if aggregation of clay can be enhanced naturally or artificially by maintaining a critical electrolyte concentration in the soil solution, called the *flocculation value* (see Section 6.3.2). In natural conditions, aggregation is enhanced by the presence of high amounts of Al and Fe, which usually occurs in acid soils. Although aggregation is enhanced, the high amounts of Al and Fe may create toxicity to plants. Artificially, aggregation is increased by liming practices or by the application of soil conditioners. Calcium and Mg are known to flocculate clay, while at the same time they reduce the toxicity of Al and Fe. The polymeric soil conditioners are no doubt also effective in the aggregation of clays, but their effect on Al toxicitiy is not known and needs to be investigated.

7.2.6 Electric Double Layers in the Environment

Because of the presence of electrical charges, colloidal materials in suspension can attract ions with opposite charges. Negative charged clay surfaces will attract cations, whereas positively charged clay surfaces attract anions. These reactions also occur with organic colloids. Therefore, cations are held on or near the clay surface, and some are free to exchange with other cations. The latter are called *exchangeable cations* and will be discussed in the next section, on cation exchange. Consequently, the negative charge clay surfaces are screened by an equivalent swarm of counterions that is electropositive. The negative charge on the clay surface and the swarm of positive counterions are called the *electric double layer*. The counterions are attracted to the clay surface, but at the same time, they are free to distribute themselves by diffusion throughout the solution phase. Attraction and diffusion will come to an equilibrium, and the resulting distribution zone of counterions varies according to the existing theories on electric double layers. At present, four theories are available, e.g., Helmholtz theory, Gouy-Chapman diffuse double-layer theory, Stern double-layer, and Yates triple-layer theory (Tan, 1998). According to these theories, two particles in suspension approaching each other will repel each other because the outer zones of their double layers are equally positive in charges. Flocculation by interparticle attraction can only take place when the double layers are suppressed to very thin layers by increased concentration of counterions. The thin double layers then decrease the interparticle distance and make a close approach possible. If the interparticle distance decreases to ≤20 Å, it is always assumed that *van der Waals attraction* becomes bigger than the repulsive forces and this results in flocculation of the clay.

The author now suggests that the presence of electric double layers on colloidal surfaces is only possible in extremely dilute or thin soil suspensions, containing only very small amounts of particles. This condition allows the particles to remain in suspension as true individual particles, separated from each other by considerable distances. In the case of a clay suspension in nature, this is almost impossible. In natural conditions, even minor puddling of soils causes relatively large amounts of clay to disperse. The clay particles, each surrounded by their electric double layer, are at close distances to each other. The double layers are in fact not repelling the particles, but two double layers confronting each other are more likely to fuse together to become just one layer. This fused double layer is shared by the two adjacent particle surfaces in question. Consequently, in natural conditions, two particles at close distance share a counterion layer. Like a root that is unable to distinguish between Na and K in nutrient uptake, the negative charge surface of one clay is also unable to distinguish whether the counterions belong to its own or to the neighbor's surface. Neither can the counterions. Located between two adjacent particles, the counterions are unable to distinguish to which surface they belong. Moreover, the concept of cation exchange dictates that Na^+ from one surface can exchange for Na^+ from the other surface. This process of sharing can be called *counterion bridging,* similar to *metal bridging*, which is very important in the interaction between negatively charged humic matter and clay. If metal bridging is accepted as a process for the interaction between negatively charged humic matter and clay particles, it is very conceivable that this counterion bridging will also take place between two clay particles. The suspension is still considered stable, if the concentration of the counterions is too low to shield completely the negative charges on the opposing clay surfaces. Hence, it is not the counterion charge that produces the repulsion, but it is in fact the *unshielded* negative charges of the opposing clay surfaces that bring about repulsion. Though not really similar, this can perhaps be compared to a hand, shielded by a thin glove, touching a hot plate. Since the glove is too thin, shockwaves after shockwaves of heat are striking the hand. Such a process explains the dispersion effect of Na^+ ions when present in low concentrations. As illustrated in Figure 7.1 (on the left), the nega-

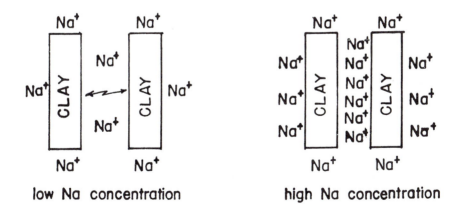

low Na concentration   high Na concentration

Figure 7.1 Counterion bridge between two clay particles. Left: Penetration of the thin Na^+ shield by the negative charges of clay surfaces, resulting in repulsion. Right: Salting out the suspension results in large amounts of Na^+ in the double layers, shielding completely the negative charges.

tive charge penetrates the thin shield of Na^+ ions, resulting in repulsion of the particles, or in other words in dispersion of the clay. In the presence of large amounts of Na^+ ions (Figure 7.1, on the right), the negative charges of the opposing clay surfaces are completely shielded, and the Na^+ ions function as metal bridges to bind the two particles together, resulting in flocculation of the clays. Such a process is called in laymen's language *salting out* a suspension.

7.2.7   Cation Exchange Capacity

The *cation exchange capacity* is the capacity of clays to adsorb and exchange cations. The negative charges of clays are usually attracting a swarm of cations (positively charged ions). This is nature's way of maintaining electroneutrality in the soil. A system is then created in the soil composed of particles with negatively charged layers neutralized by layers of positively charged ions. Such a system is called in soil chemistry an *electric double layer*. The cations, held electrostatically on the clay surface, can be replaced by other soil cations. The adsorbed cations are, therefore, called *exchangeable cations*, and the process of exchange is referred to as *cation exchange*. The soil particles responsible for adsorption and exchange of cations are called the *exchange complexes*. Exchange reactions are instantaneous. To maintain electroneutrality, the reaction is stoichiometric, meaning that the exchange occurs in equivalent amounts, as illustrated by the reaction below:

$$Soil\text{-}Ca \quad + \quad 2H^+ \quad \rightleftarrows \quad Soil\text{-}2H \quad + \quad Ca^{2+} \tag{7.3}$$
adsorbed Ca^{2+} adsorbed H^+ free Ca ion

As shown in reaction (7.3), it needs two monovalent ions, e.g., two H^+ to replace one Ca^{2+}, which is divalent.

The following reactions are believed to also be cation exchange reactions (Sposito, 1989):

$$\text{KAlSi}_3\text{O}_{8(s)} + \text{Na}^+_{(aq)} \rightarrow \text{NaAlSi}_3\text{O}_8 + \text{K}^+_{(aq)}$$
orthoclase $\qquad\qquad$ albite

$$\text{CaCO}_{3(s)} + \text{Mg}^{2+}_{(aq)} \rightarrow \text{MgCO}_{3(s)} + \text{Ca}^{2+}_{(aq)}$$
calcite $\qquad\qquad\quad$ magnesite

Unfortunately, these reactions have never been noticed to occur in nature as exchange reactions, and such a concept sends an improper message on the principles of cation exchange. Conversion of a primary mineral into another primary mineral has not been known to occur by cation exchange. Exchangeable cations are by definition cations held by electrostatic forces on negatively charged colloidal surfaces. Neither orthoclase nor calcite nor any other primary mineral is colloidal in nature. The K in orthoclase, though considered a nonframework cation, is held very strongly by ionic or covalent bonds by silica tetrahedrons in the crystal structure. This type of tecto-silicate is next to quartz in its resistance to weathering. Nevertheless, it may weather to form clays. Perhaps, the conversion of a primary mineral into another primary mineral as illustrated by the reactions above can be called *isomorphous substitution*. However, isomorphous substitution is believed to occur during weathering processes, such as in the synthesis of clay minerals. The reaction proceeds in one direction and is more likely to be a terminal reaction. On the other hand, an exchange process goes either way and reaches equilibrium without producing something permanent. If primary minerals exhibit cation exchange reactions, the minerals must also have cation exchange capacity (CEC) values. In Soil Science and Soil Mineralogy, orthoclase, albite, calcite, magnesite, and other primary minerals are not known to display CEC values of any significance.

The cation exchange capacity (CEC), expressed in milliequivalents (me) per 100 g (=cmol(+)/kg), differs from soil to soil depending on (1) clay content, (2) types of clays, and (3) organic matter content. The higher the clay content, the larger will be the CEC of the soil. Hence, sandy soils exhibit low CECs. Similar reasoning can also be given for organic matter content. The types of clays may also affect the CEC of soils because different types of clays exhibit different values of CECs, as can be seen in Table 7.2.

Table 7.2 CEC Values of Major Soil Colloids

Soil colloid	CEC (me/100g)
Humus (humic acids)	> 150
Vermiculite	150
Smectite	100
Illite	30
Kaolinite	10
Sesquioxides	4

7.2.8 Environmental Importance of Cation Exchange

Cation exchange is of great importance in (1) soil fertility, (2) fertilizer application, (3) nutrient uptake, and (4) environmental quality.

Soil Fertility. - Under natural conditions, H^+, Ca^{2+}, Mg^{2+}, K^+, and Na^+ are the most common cations in soils. They can replace each other, depending on the conditions. When conditions are favorable for the presence of high amounts of protons, H^+ replaces the other cations in the exchange complex. Such an exchange with H^+ ions produces acid soils, which occur especially in humid regions. The bases released in the soil solution are leached by the percolating waters. Liming such soils is then necessary to replace the adsorbed H^+ by Ca^{2+} and thereby increase soil fertility.

Fertilizer Application. - Fertilizers are added to soils to increase soil fertility and improve plant growth. The major nutrients in fertilizers are N, P, K, Fe, Cu, Zn, and Mn. Calcium and Mg are also major plant nutrients, but they are found in lime materials. The N in fertilizers can be in cationic (NH_4^+) or anionic form (NO_3^-), whereas P is always anionic ($H_2PO_4^-$, HPO_4^{2-}). The other nutrients are in the form of cations. These cations are released upon the dissolution of the fertilizer in soil water. After release, they are either adsorbed or they replace the original cation on the exchange complex. In this way the fertilizer elements needed for plant growth are stored in the soil, and are less subject to leaching. As such, the adsorption complex gives to the soil the storage and buffering capacity of cations.

Nutrient Uptake. - The exchangeable cations serve as storage for large quantities of available nutrients for plant growth. Plant roots obtain the adsorbed cations by cation exchange. The exchanged material used by roots is H^+, which is produced as a by-product of root respiration, as discussed previously. Such an exchange reaction is illustrated in Figure 7.2.

Environmental Quality. - A lot of material may be added to soils by agricultural, industrial, and domestic operations. Some of the com-

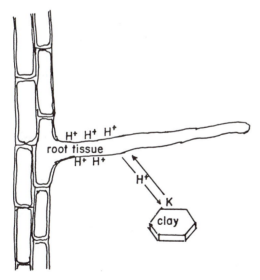

Figure 7.2 Cation exchange reaction between H^+ ions in the root rhizosphere and K adsorbed on the clay surface.

pounds are hazardous to humans and animals. They may leach with the percolating waters and pollute the groundwater. However, the presence of clays with their cation exchange capacity provides a buffering capacity. Because of the electrical charges and high surface areas, clays adsorb a variety of chemical compounds entering the soils, e.g., products of microbial reactions, organic waste, gas, heavy metals, and other pollutants. By adsorption to clay surfaces, many toxic metals are immobilized and/or precipitated into less harmful solid compounds. The clay will intercept a variety of pollutants trickling down the soil with the percolating water. These pollutants will be removed by clay because of cation exchange before they reach the groundwater. Consequently, clay acts as a purifying agent in nature and increases the quality of soil and groundwater.

7.3 ANION EXCHANGE CAPACITY

The *anion exchange capacity* is the capacity of clay to adsorb and exchange anions. The clay must be positively charged to be able to adsorb negatively charged ions. Positive charges in clays are of importance only in acid conditions, when the soil pH is below the ZPC of the clay. It can also develop because of broken bonds on the broken edges of the clay mineral. In general, the positive charge, hence the anion exchange capacity of clays, is considerably smaller than the cation exchange capacity. The major anions in soils subject to anion exchange reactions are SiO_4^{4-}, $H_2PO_4^-$ SO_4^{2-}, NO_3^-, and Cl^-. Adsorption of phosphate ions will usually result in phosphate *fixation* and *retention* (Tan, 1993).

When adsorption of anions is caused by electrostatic forces, it is called *nonspecific adsorption*. Two types of nonspecific adsorption can be recognized: *negative* and *positive adsorption*. When the adsorption is caused by nonelectrostatic forces, such as ligand exchange or complexation, it is called *specific adsorption, chemisorption,* or *ion penetration* (Tan, 1998). In specific adsorption, the anion enters the crystal or the structure, and the reaction takes place within the coordination sphere.

Negative Adsorption. - The repulsion of anions from negatively charged surfaces is called *negative adsorption*. The anions are expelled from the electric double layer, which is composed mostly of cations. The exclusion of anions from the double layer causes the anions to concentrate in the *bulk solution*. Negative adsorption of anions is affected by (1) soil pH, (2) negative charge density of soil colloids, and (3) the valence of cations (Parfitt, 1980). The effect of soil pH is important in soils with variable charges. Decreasing the soil pH will decrease the negative charges, which consequently decreases the rate of negative adsorption. The greater the charge density of the colloids, the greater will be the negative adsorption of anions. Because of its larger negative charge, smectite exhibits a larger negative adsorption than kaolinite. The valence charges of anions also have some influence on negative adsorption. Anions with higher valence charges tend to be repelled more than anions with lower valence charges.

Positive Adsorption. - Positive adsorption of anions is the adsorption and concentration of anions on the positively charged surfaces or edges of clay minerals. Here, negative adsorption of cations or repulsion of cations occurs. The mechanisms of cation exchange also apply to adsorption of anions, but the anion exchange capacity (AEC) of soils is usually substantially smaller than the CEC. As with cations, lyotropic series of anions are also present in anion adsorption, as can be noticed from the following decreasing order of anion adsorption:

$$SiO_4^{4-} > PO_4^{3-} > > SO_4^{2-} > NO_3^{-} = Cl^{-}$$

The lyotropic series above indicates that SiO_4^{4-} and PO_4^{3-} ions are strongly adsorbed, whereas SO_4^{2-} and NO_3^{-} ions are adsorbed in considerably lower amounts, and Cl^{-} may even not be adsorbed at all. The nitrate and chloride ions are generally subject to negative adsorption. These truly electrostatically adsorbed anions are exchangeable with other anions in just the same manner as cations are exchanged by other cations:

$$- \text{Al- OHH}^+\text{Cl}^- + \text{NO}_3^- \rightleftharpoons - \text{Al- OHH}^+\text{NO}_3^- + \text{Cl}^-$$

Such an exchange of anions is the reason for the development of an anion exchange capacity (AEC) in soils. Phosphate ions can also be adsorbed in a similar fashion by positive charges on the edge surfaces of clay minerals:

$$- \text{Al- OHH}^+ + \text{H}_2\text{PO}_4^- \rightleftharpoons - \text{Al - OHH}^+\text{-H}_2\text{PO}_4^-$$
protonated
octahedral OH

The phosphate ion held in this way is also subject to exchange reactions. The adsorption above is prevalent in acid soils.

Specific Adsorption. - This type of adsorption takes place through an interaction process, and is important with oxides, hydrous oxides, amorphous and other types of variable charge clays. Because of its unique mechanism, the exchange of the adsorbed anions is called *ligand exchange*, *chemisorption*, or *anion penetration*, as indicated above. The process of reactions involves ionic and especially covalent bonding. The anions involved are mostly phosphate ions, and to a lesser extent also silicate, fluoride, and sulphate ions. The reaction can be illustrated as follows:

$$- \text{Al - OH} + \text{H}_2\text{PO}_4^- \rightarrow - \text{Al - H}_2\text{PO}_4 + \text{OH}^-$$
octahedral OH

As can be noticed, the reaction is a displacement reaction, resulting in the release of a structurally bonded OH^- ion. The increase in OH^- ion concentration in the soil solution may cause the soil pH to increase, a phenomenon called *exchange alkalinity*. Other anions, such as F^- ions, are also capable of displacing structural ions from the octahedrons of silicate and hydrous oxide clays.

In view of the above, it is deemed of utmost importance to make a distinction between the true anion exchange, as a result of positive adsorption, and displacement reactions of structurally bonded ions. In the latter reaction, the anion (phosphate) is held by covalent bonding, which is a stronger bond than electrostatic bonds. Exchange of a covalent bonded anion is more difficult than exchanging electrostatically held anions. Though the exchange is still possible, often only part of the structurally bonded phosphate can be recovered.

7.3.1 Environmental Significance of Anion Exchange

From the environmental aspect, negative adsorption of anions is a harmful process. Instead of being attracted by the colloid surfaces, the anions are repelled. Such a repulsion causes the anions to be pushed away from the interface solution into the bulk solution, where they are free to move with the percolating water. This increases the chances of leaching of anions that are important sources as nutrients for growing plants. Leaching also increases the chances for anions to be transported and contaminate our groundwater system. When anions, such as phosphates and nitrates, are transported and accumulated in this way in streams, lakes, and swamps, they may cause an excessive growth of unwanted aquatic weeds, an environmental degradation process called *eutrophication* that chokes out all fish and other animal life. Because such a degradation process is brought about by human activities, Brady and Weil (1996) suggest calling it *accelerated* or *cultural eutrophication* in contrast to *natural eutrophication*. These authors believe that some lakes and wetlands are already eutrophic by nature and the overabundance of dead organic residues will eventually fill the lakes and wetlands. The latter will then gradually develop into bogs, peats, and muck, and eventually form histosols. In nature, eutrophication is a very slow process and it takes centuries for the development of histosols. Both natural and cultural eutrophication apparently have a similar effect on the environment by choking out all animal and fish life. The difference is that cultural eutrophication is a rapid process in contrast to natural eutrophication. In particular, growth of algae is stimulated by the high nutrient concentrations, and the water often becomes extremely thick and slimy from the overgrowth of algae, a process often referred to as *algal bloom*. The growth of higher plants may also become stimulated by the higher nutrient contents. The dead masses of algae and higher plants make conditions worse, because decomposition of dead organic matter under anaerobic conditions not only consumes all the dissolved oxygen content but also releases a number of toxic compounds as previously discussed. Most scientists consider phosphate to be the major cause of eutrophication, and currently phosphate has been banned by EPA in household detergents. However, no agreement can be found yet on the minimum limit of phosphate in soil water at which degradation of environmental quality starts to become a problem. In Canada, water containing a maximum of 0.15 mg/L of P is considered acceptable; however, reports from the United States indicate that a concentration of P as low as 0.01 mg/L is sufficient to increase algae growth (Miller and Gardiner, 1998).

Today, cultural eutrophication is applied in fish farming in Thailand and the United States, where fish ponds are deliberately fertilized to produce plankton and algae for fish food. This will be discussed in more detail in Section 9.6.2.

7.4   SOIL ORGANIC MATTER

A number of soil organic compounds also carry electrical charges, which cause the many biochemical reactions in soils. In contrast to the clay fraction, many of the soil organic substances are amphoteric, whereas their electrical charges are mostly *pH-dependent* or *variable charges*, as will be explained in more detail in the following sections. Numerically, their charges are often very high, frequently exceeding those of even vermiculite, which is characterized by the highest negative charge among the clay minerals (see Table 7.2). The pH-dependent negative charge of humic and fulvic acids, in terms of CEC, may vary from a low of a few hundreds me/100g at low pHs to a high of 1500 me/100g at high soil pHs (Tan, 1998; Stevenson, 1982), values that are astronomically high. The contribution of soil organic matter to the soil's negative charge has not been fully realized and even today many prominent soil scientists consider the soil clays as the most important soil constituents responsible for the negative charges in soils. However, with the increased knowledge of soil organic matter and especially soil humic acids, the electrochemical properties of humic matter have started to attract more attention. Although many organic substances are known to possess electrical charges, for instance organic acids, only two major organic compounds, amino acids and humic matter, will be discussed as examples. The electrochemical properties of amino acids and humic acids are unique and are responsible for many of the interactions, complexation, and chelation reactions affecting numerous aspects of importance in soil problems and in issues on environmental quality.

7.4.1 Amino Acids

Amino acids are amphoteric compounds because they contain both carboxyl and amino groups. The carboxyl groups behave as acids, whereas the amino groups are basic. Therefore, they can be negatively and positively charged depending on the conditions. The pH at which an amino acid exhibits equal amounts of positive and negative charges is called the *isoelectric point* (pH_0). At the isoelectric point, the amino acid is electrically neutral. The neutral ion of amino acid is called a *zwitterion* (zwitter is German for double or pair), and has the following formula:

$$H_3C - \underset{\underset{R}{|}}{\overset{\overset{NH_3^+}{|}}{C}} - COO^-$$

Note that the zwitterion carries both positive and negative charges, which neutralize each other. When amino acids are present in soils with a pH > pH_0, the condition is considered "alkaline" for the amino acids, and the excess OH^- ions present in the soil solution react with the NH_3^+ group of the zwitterion. Such a reaction yields a negative charge as can be noticed from the equation:

$$R - \overset{\overset{\displaystyle COO^-}{|}}{\underset{\underset{\displaystyle CH_3}{|}}{C}} - NH_3^+ + OH^- \rightarrow R - \overset{\overset{\displaystyle COO^-}{|}}{\underset{\underset{\displaystyle CH_3}{|}}{C}} - NH_2 + H_2O \qquad (7.4)$$

zwitterion   anionic form

On the other hand, in acid soils (soil pH < pH$_o$), the system contains large amounts of H$^+$ ions. The protons will then react with the carboxyl group of the zwitterion and the amino acid is positively charged

$$CH_3 - \overset{\overset{\displaystyle NH_3^+}{|}}{\underset{\underset{\displaystyle R}{|}}{C}} - COO^- + H^+ \rightarrow CH_3 - \overset{\overset{\displaystyle NH_3^+}{|}}{\underset{\underset{\displaystyle R}{|}}{C}} - COOH \qquad (7.5)$$

zwitterion   cationic form

In the anionic form, amino acid reacts with cations, such as metal ions, whereas as a cation it will react with anions or undergo interactions with clays.

Environmental Significance of Amino Acids. - Oxisols and ultisols in tropical and subtropical regions are known to be strongly and very strongly acid soils; hence their amino acids are most likely positively charged and will undergo interactions with negatively charged organics and clays. As a complex compound, amino acids are more resistant to microbial attack. This is one reason why they can accumulate in soils and the environment. On the other hand, mollisols and aridisols in the temperate regions are neutral to slightly alkaline in reactions, and contain high amounts of Ca and/or Na. Their clay fractions are usually composed of high amounts of negatively charged 2:1 clays. In these soils, characterized by high pH values, the amino acids are expected to be negatively charged, and, therefore, will not be able to react with the also negatively charged clays. However, the negatively charged amino acids will react with Ca and/or Na ions. Under certain conditions the Ca adsorbed by the amino acids or by the clays can function as a bridge connecting amino acid to the clay surfaces. Such an interaction reaction is called *bridging* or *coadsorption* (Tan, 1998). This is then the way in which amino acids can be protected against decomposition and hence accumulate in soils.

7.4.2 Humic Matter

Humic matter, composed of humic and fulvic acids, also consists of amphoteric compounds. Although they can be positively and negatively charged, the negative charges are usually of more importance in humic matter than the positive charges. As discussed in Chapter 3, the negative charge in humic matter is expressed in terms of *total acidity*, which is defined as the sum of the carboxyl and phenolic-OH groups. In general, these two functional groups control the electrochemical behavior of humic matter. Dissociation of protons from the carboxyl groups starts at pH 3.0, and the humic molecule becomes electronegatively charged (Figure 7.3). At pH < 3.0 the humic mol-

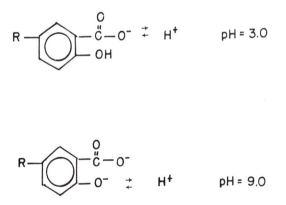

Figure 7.3 Development of negative charges in a humic acid molecule by dissociation of protons from carboxyl (COOH) groups at pH 3.0 and from phenolic-OH groups at pH 9.0.

ecule exhibits a small negative charge, but this charge increases with increasing pH. At pH = 9.0, the phenolic-OH group also starts to dissociate its proton, and the humic molecule attains a high negative charge. Since the development of this negative charge is pH dependent, this charge is called *variable charge* or *pH-dependent charge*.

Several reactions or interactions can take place because of the presence of these charges. At low pH values, the humic molecule is capable of attracting cations, which leads to cation exchange reactions (Figure 7.4). The magnitude of the cation exchange capacity (CEC) of humic matter is reflected from the total acidity value. As discussed in Chapter 3, the total acidity, hence the CEC, can range from 500 to 1500 me/100g, which is the highest CEC value of all the colloids in

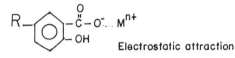

Electrostatic attraction

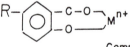

Complex reaction (chelation)

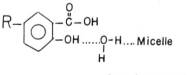

Co-adsorption (H$_2$O bridge)

Figure 7.4   Electrostatic adsorption of metal ions by humic acid (top), complex reaction between humic acid and metal ions (middle), and water bridging or coadsorption (bottom). M^{n+} = cation with charge n$^+$; R = remainder of the humic molecule.

soils. When both the carboxyl and phenolic-OH groups are completely dissociated, humic matter is also able to undergo complex reactions or chelation reactions with metal ions or other soil constituents. These reactions have been discussed in other chapters of this book, as they play an important role in soil fertility, in plant nutrition, and in enhancing environmental quality. Both adsorption and complex reactions can also take place by a water or metal bridging process. This is the process by which two negatively charge soil constituents can attract each other. The interaction between humic acid and clay is made possible by metal or water bridging, also called *coadsorption*. Water or any of the metal ions, such as Ca, Al, Fe, or Mn, can serve as a bridge between the organic ligand (humic acid) and the clay micelle. Sodium bridging, formed by fusing of two opposing electric double layers, was explained to play an important role in interparticle repulsion and attraction.

Environmental Significance of the Electrochemical Properties of Humic Matter. - The interaction reactions, such as electrostatic adsorption, complex reaction, chelation, and water and metal bridging, affect the physical, chemical, and biological properties of soils. Stability of soil organic matter against decomposition and consequent accumulation in soils is made possible by formation of organo-clay complexes or chelates through metal and water bridging. The high organic matter contents in mollisols and andosols are believed to be attributed to such reactions. In mollisols, the organo-mineral complexes are formed by Ca-bridges, whereas in andosols Al and allophane bridges are assumed to be responsible for the formation of organo-mineral complexes (Tan, 1998a). Formation of stable soil structures is often associated with complex formations between clay particles and soil organic compounds. Not only humic matter but especially polysaccharides and extracellular polysaccharides play a very important role in enhancing aggregation of soil particles. The difference between humic acid and polysaccharide is that humic acid encourages structure formation through the metal bridging process, whereas polysaccharides compete with water molecules for adsorption sites on the clay surface. The force of such an exchange process creates so-called *hydrophobic bonding*. Because of the displacement of water, wetting and swelling of clay particles are reduced, thereby increasing adhesion and cementation.

In relation to the chemistry of the environment and soils, chelation increases the mobility of plant nutrients and other elements. It is beneficial because the process provides the carrier mechanism by which plant nutrients can be transported to the depleted areas in the root zone. Complexing of heavy metals by humic acids may also reduce the toxic effects of many metals. Bonded as chelates, the chemical activity of toxic metals is known to be drastically reduced. Depending on conditions, toxic pesticide residues can also be detoxified by the interaction with humic acids. Since most of them are organic in nature, the pesticide can be chelated by humic acid in different ways. When it is chelated peripherally exposing its active sites, the pesticide residue remains very active and can still do much damage. This is the method suggested for increasing the effectiveness of pesticides. It is said that the amount of pesticides required can be reduced substantially by mixing them with humic acid before application. However, in the soil, these residues can also be incorporated into the humic acid molecule during the chelation reaction. Since they become structural constituents of the humic acid, they lose their toxic properties, and their identity as pesticides (Tan, 1998a; Stevenson, 1994). This can perhaps be considered as a form of *biodegradation* of pesticides.

Currently, reports have increasingly been presented on the adsorption of microbial cells by clays and humic acids. Although not much is known yet, the interaction between microorganisms and soil clays or humic matter may affect changes in eletrochemical properties of the soils. It is apparent that complex formation between live organisms, such as bacteria, fungi, humic acids, and soil clays have changed completely the perception of soil biology or biochemistry in soil science. It is a process of great environmental importance because of its influence on many of the biochemical cycles vital for continuation of life. Live microbial cells can be adsorbed by organic and inorganic soil colloids, or may interact in much the same way as humic acid interacts with clay. This kind of interaction is expected to change the charge properties of soils. Most soil scientists are used to the idea that the charge characteristics of soils are mainly due to the clay and humic matter. However, it has been known for some time that the cells of live organisms possess significant amounts of charges (Paul and Clark, 1989; Burns, 1986). Since the cell charges are also negative, the interaction with humic acid or clay is assumed to take place by metal and water bridging. In analogy to the preservation of organic matter in soils as a result of complex formation with clays, the interaction between bacterial cells and clays or humic acids also ensures the survival and accumulation of specific groups of bacteria, their enzymes, and metabolic products. The adsorption of protein by clays and their subsequent protection from decomposition, known for many years, is one example. Clay-enzyme and humic-enzyme interactions are processes by which the activity of the enzymes can disappear or can be preserved. When the enzyme is chelated so that it is incorporated in the humic acid structure, similarly to the incorporation of pesticide residues, the enzyme and its activity are dissolved and the enzyme ceases to function. However, when in the interaction process, the enzyme is attached peripherally to the humic acid molecule, it remains stable and functionally active. These enzymes make organic and nutrient cycling possible. Much of the organic C and N entering the soil is polymeric; in other words, the molecules are very large and complex substances. If these large molecules cannot be broken down into smaller molecules, by the process called decomposition, they will not be recycled and cannot become available for uptake by plants and microorganisms. As discussed in Chapter 3, decomposition is an enzymatic process, and a variety of enzymes are required to break the bonds within the structure of the numerous chemical substances in the cell tissue. It would be an environmental disaster if the enzymes did not survive in soils. Without interaction with clay and humic acids, free enzymes are usually rapidly *denatured* by a host of physical and chemical soil reactions, or they may serve as substrates for *proteolytic* microorganisms.

7.5 SOIL REACTION

7.5.1 Concept of Soil Reaction

Soil reaction is a soil parameter which is also closely controlled by the electrochemical properties of soil colloids. The term is used to indicate the acidity or alkalinity of a soil. The degree of acidity or alkalinity is determined by the hydrogen ion, H^+, concentration in the soil solution. In acidic soils the H^+ concentration is greater than the OH^- concentration, whereas in alkaline soils the H^+ concentration is smaller than the OH^- concentration. In a soil with a neutral reaction, $H^+ = OH^-$. These conditions are usually expressed in pH values, ranging from 0 to 14. The p refers to negative logarithm and the H means the H^+ ion concentration in grams/L. Thus,

$$pH = - \log (H^+)$$

or

$$pH = \log 1/H^+$$

At an H^+ concentration of 0.001 g/L, the pH = -log 0.001 = -(-3)= 3. At an H^+ concentration of 0.000001 g/L, the pH = -log 0.000001 = -(-6)= 6. Notice that a solution with fewer H^+ ions has numerically larger (higher) pH values.

Acidity and alkalinity reflect both the H^+ and OH^- ion concentrations. The mass action law states that the product of the concentrations of H^+ and OH^- ions is always constant. This works out to be

$$(H^+) (OH^-) = 10^{-14}$$

By taking the negative logarithms of both sides, this equation changes into

$$pH + pOH = 14$$

Since the sum of pH and pOH is constant, the concentrations of H^+ and OH^- ions vary inversely. Therefore, only one needs to be determined to know the other. It is customary to determine the pH to indicate both acid and basic soil conditions. Though pH was defined on the basis of H^+ concentration, the above pH + pOH relation states that the pH value not only indicates the H^+ ion, but also reflects automatically the OH^- ion concentration.

Soil acidity or alkalinity is affected by the types of cations adsorbed on colloidal surfaces. The major soil cations adsorbed on colloidal surfaces are Al^{3+}, H^+, Na^+, K^+, Ca^{2+}, and Mg^{2+} ions. An overabundance of adsorbed Al^{3+} and H^+ ions decreases soil pH, whereas a saturation of the exchange sites with the bases Na, K, Ca, and Mg increases soil pH.

7.5.2 Types of Soil Reactions

Based on soil pH values, the following *types of soil reactions* are distinguished:

Soil Reaction	pH	Soil Reaction	pH
Slightly acid	7.0 - 6.0	Slightly alkaline	7.0 - 8.0
Moderately acid	6.0 - 5.0	Moderately alkaline	8.0 - 9.0
Strongly acid	5.0 - 4.0	Strongly alkaline	9.0 - 10.0
Very strongly acid	4.0 - 3.0	Very strongly alkaline	10.0 - 11.0

These terms can be used to identify the degree of acidity or alkalinity in soils. If one says that the soil has a very strongly acid reaction, it means that the soil pH is between 4.0 and 3.0. Similarly when we say that the soil has a slightly alkaline reaction, we are referring to soils in the pH range of 7.0 to 8.0. The use of the terms acidic pH or basic pH should be avoided, since the pH is expressed in numerical values, ranging from 0 (low) to 14 (high). Figures are neither acidic nor basic, hence, the pH is either low or high.

Most soils have a pH between 5 and 9. In a humid region, the surface soil usually exhibits a pH of 5 to 7 because most of the bases are exchanged and leached, as can be illustrated by the following reaction:

$$\text{Clay-Ca} + 2H_2O \rightleftarrows \text{Clay-2H} + Ca^{2+} + 2OH^- \qquad (7.6)$$

The percolating water, provided by the high amount of precipitation in humid regions, removes the soluble Ca^{2+} and OH^- and forces the reaction above to go to the right. The soil is left with clay saturated with H^+ ions. Therefore, due to the leaching of bases, a humid region soil will seldom exhibit a pH > 7.0. The pH can decrease below 3.0 under very special soil conditions, e.g., the presence of S compounds as discussed before. On the other hand, soils in arid regions are characterized by a pH of 7.0 to 9.0 in the surface soil. Here, the adsorbed bases are not leached away, and some of them may even form salts, e.g., $CaCO_3$, Na_2CO_3, and NaCl. These salts provide a reserve of cations that can maintain the saturation of the clay complex. The process can be illustrated as follows:

$$\text{Clay-2H} + CaCO_3 \quad \rightleftarrows \quad \text{Clay-Ca} + H_2O + CO_2{\uparrow} \qquad (7.7)$$

Hydrolysis of the carbonate salts contributes toward increasing the soil pH level over the value that would have been expected from a 100% base saturation. The hydrolysis reaction of $CaCO_3$ can be written as follows:

$$CaCO_3 + H_2O \quad \rightarrow \quad Ca^{2+} + 2OH^- + CO_2{\uparrow} \qquad (7.8)$$

The Ca^{2+} ions released in the hydrolysis reaction above can also be used in reaction (7.7) for exchanging two exchangeable H^+ ions. Therefore, because leaching is not important in arid regions, the soil remains saturated with bases, hence the soil pH seldom decreases below 7.0.

7.5.3 Sources and Types of Soil Acidity

A number of compounds contribute to the development of acidic and alkaline reactions in soils. Water is a source of a small amount of H^+ ions. Inorganic and organic acids, produced by the decomposition of soil organic matter, are common soil constituents that also contribute to soil acidity. Respiration of plant roots, hydrolysis of Al, nitrification, oxidation of S, fertilizers, and acid rain are additional sources of H^+ in soils. All these compounds and their reactions contributing to soil acidity have been discussed in previous chapters or will be discussed in the following chapters.

The H^+ ions may be present in soils as adsorbed ions on the surface of the colloidal complex or as free H^+ ions in the soil solution. The adsorbed H^+ ions make up the exchangeable H^+ ion concentration and create the so-called *potential* or *exchange acidity*. The free H^+ ions are the ions free in solution and create the *active acidity*. Taken together, the active and potential acidity make up the *total soil acidity*. The system composed of free and exchangeable (adsorbed) H^+ ions is illustrated in Figure 7.5. The concentration of H^+ ions free in solution

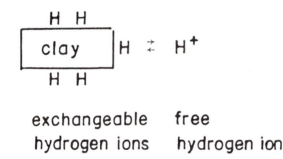

Figure 7.5 Dissociation of exchangeable H^+ ions from clay surfaces.

when measured is expressed in pH. The exchangeable H^+ ion concentration is not measured as pH, but can be determined by titration, and its value is useful for liming purposes. The free H^+ ion concentration of the soil solution at any particular time is relatively small compared to the reserve H^+ ion concentration. Most soil chemists believe that the potential acidity may be 1000 times greater than the active acidity in acid sandy soils, and it may be even 50,000 to 100,000 times greater in acid clayey soils. In humid region soils, the removal of bases from the colloidal complex by leaching, especially of Ca, occurs constantly through cation exchange. Their place on the exchange sites is taken by H^+ ions. These adsorbed H^+ ions will dissociate free H^+ ions, and the concentration of these free H^+ ions determines the soil pH.

The reaction of acid soils can be influenced only when enough lime is added to react with the *total acidity*. The greater the exchange capacity of the soil, the greater is the reserve acidity, and the more difficult it is to reduce the soil pH. The resistance to this change is called *buffer capacity*.

Brady and Weil (1996) reported a third type of soil acidity, which they called *residual acidity*. It is defined as soil acidity that remains after active and exchange acidity have been neutralized. They believe that this residual acidity is related to Al^{3+} and aluminum hydroxy ions bound in nonexchangeable forms by organic matter and silicate clays.

7.5.4   Effect of Soil Reaction on Plant Growth

The soil reaction has a direct and indirect nutritional effect on plant growth. The direct effect is manifested by the toxic effect of H^+ and OH^- ions, whereas the indirect effect is made by controlling availability of plant nutrients.

Most plants grow best in soils with a slightly acid reaction, although some variations may be noticed. For example, alfalfa, clover, and cedar require a soil pH closer to 7.0 for maximum growth, whereas pine trees need a soil pH of 5.0. On the other hand, azaleas and rhododendrons do better in a more acidic medium.

In the pH range of 6.0 to 7.0, nearly all plant nutrients are available in optimal amounts. Soils with a pH below 6.0 will more likely be deficient in some available nutrients for optimal plant growth. Calcium, Mg, and K are especially deficient in acid soils. In strongly to very strongly acid soils, the microelements Al, Fe, Cu, Zn, and Mn may exist in very high quantities, creating micronutrient toxicity. In strongly alkaline soils Ca, Mg, and K concentrations are very high, whereas the soluble microelement concentration is very low. For example, Al^{3+} ions, when present, tend to be precipitated in the form of insoluble $Al(OH)_3$. The concentrations of Fe, Mn, Cu, and Zn are decreased by a similar reaction and become so low that the very strongly basic soils tend to exhibit micronutrient deficiencies. In general, high soil pH and Ca content are closely related, but there are exceptions, especially when Na is present.

The type of phosphate ion and its concentration in soils are also controlled by the soil reaction. At very strongly acidic conditions, phosphate ion concentration is very low because most of the phosphate is precipitated in the form of insoluble $AlPO_3$ or $FePO_3$, called variscite and strengite, respectively. The major phosphate ion present in soils with acid reactions is $H_2PO_4^-$, and its concentration increases from very low at pH 3.0 to sufficiently high values at pH 6.0 to 7.0. At pH 6.0, 94% of the phosphate in solution occurs as $H_2PO_4^-$, but its amount drops to 60% at pH 7.0 (Stevenson, 1986). Phosphate ion concentration is also low in the soil solution at very strongly basic conditions because of precipitation as insoluble tricalcium phosphate, $Ca_3(PO_4)_2$. Tricalcium-phosphate is a constituent of animal bones, and *bonemeal* is a source of phosphate for plants, animals, and also humans. It is commonly applied as a fertilizer in soils exhibiting high phosphate fixation. In slightly to moderately alkaline soils the phosphate ion is present in the form of HPO_4^{2-} and PO_4^{3-}. Stevenson (1986) believes that at pH 5.0 - 8.0, the amounts of undissociated phosphate, H_3PO_4, and trivalent PO_4^{3-} ions are negligibly small. The discussion above about low and high phosphate concentration pertains to phosphate in solution. In general, the P content in solution ranges from 0.001 to 1.0 mg/L (Brady and Weil, 1996). However, the total P content in soils may provide a different picture. Mollisols and aridisols, soils with minimum leaching, are known to contain high amounts of P, and even the highly leached ultisols in the Atlantic and Gulf coasts of the United States may contain 100 - 400 µg/g of P. This is a relatively high amount of P, though considered very low by US soil scientists when compared to the mollisols and aridisols, where contents of 900-1000 µg/g are detected (Stevenson, 1986).

Soil acidity also affects microbial life and their activity in soils. In general, both bacterial and fungal life are present in all soils, but depending on soil acidity one tends to dominate over the other. It is common knowledge in soil microbiology that fungi become dominant in strongly acid soils, whereas bacteria will be prevalent in strongly basic soils. In soils exhibiting slightly acid reactions, both bacteria and fungi are equally present in sufficient amounts. An example of this effect of pH on soil organisms in the natural environment is the soil under coniferous (pine) vegetation. The litter and the surface soil below are usually interspersed with a dense system of fungal hyphae. The soils under pine trees tend to be acid.

Induced Micronutrient Toxicity and Deficiency. - By conventional standards, micronutrient toxicity and deficiency are harmful in plant growth or crop production since they reduce yield and quality of the crops or even kill the plants. However, in an unorthodox way, the author believes that a slight toxicity or deficiency is required for the production of certain ornamental plants. Like the induced stunted and crooked condition of _bonsai plants_, the often _sickly_ plant appearances due to micronutrient toxicities or deficiencies enhance their appeal and quality as ornaments. Brilliant colors, reddish or purplish colors, and the yellow chlorotic colors between the leaf veins increase the aesthetic value of the plants as ornaments, hence the plants become more appealing to many houseowners than plants exhibiting a healthy uniform green color. The author notices that many plants growing in the swamps and mangrove forests in Southeast Asia often display brilliant or appealing mottled colors. The soils are usually strongly to very strongly acid in reactions and have Fe and Mn contents present to toxic levels. The colors are not genetic in origin, since they change to uniform green when the plants are completely fertilized after transplanting in pots. The healthy form of the plants loses its appeal for sale in the market. Therefore, growing plants under slightly toxic or nutrient-deficient condition becomes very important in nurseries in Southeast Asia. The art is to induce the same colors or chlorosis as displayed by plants growing in a natural hostile environment, but without reducing plant growth or without killing the plants. Though not similar, the process is like the induced reddish or purplish colors on the cheek of a feverish child. Used as decoration, chlorotic plants provide for more variety, style and decor than the usual uniform green plants, hence demand higher prices.

CHAPTER 8

SOILS AND CROP PRODUCTION

Soil has been defined previously as the natural portion of the earth's surface that supports plant growth. It is the medium for crop and animal production. Soils, providing good media for an adequate supply of food and fiber, have been one of the reasons for the establishment of large human populations in many parts of the world. In Mesopotamia and Egypt, long considered the cradle of civilization, the fertile soils in the valleys and deltas of the Tigris, Euphrates, and Nile rivers were responsible for the creation of flourishing civilizations. Similarly, the Ganges and Indus rivers in India, Bangladesh, and Pakistan, and the Mekong, Yangtze, and Hwang Ho rivers in Southeast Asia and China created sites with rich soils where stable and organized human communities could grow food and thrive. The presence of water for irrigation and periodic flooding contributed to the continuous buildup of the nutrient supply of the soils for good soil fertility.

As the world's population has expanded and living standards have improved, the demand for food and fiber has increased considerably. Because recent advances in medical science have drastically improved world health, many countries have experienced and are still experiencing an unprecedented population growth. Predictions have been made that by the year 2025 the world population will be 8 to 9 billion, which is 25% higher than it is today, and may even reach 10 billion in another century (Brady and Weil, 1996; Wild, 1993). The population increase is reported to be less in the well-developed countries, but much greater in the nations of the third world, where the existing economic and food conditions already pose serious problems. The need for more food is apparent, and to feed the growing population will be a big challenge for these countries. This staggering number of people will consume perhaps more food than all other animals combined (Deevey, 1960). The increased demand for food places an increasing load on soil productivity, and several methods to increase soil productivity have been proposed. Brady (1990) suggested three routes to increase food production:

1. Clear and bring new lands into cultivation.
2. Intensify production on lands already under cultivation.
3. Increase the number of croppings per year.

Recent advancements in agricultural technology indicate that three more methods should be added today as possibilities:

4. Employ *biotechnology* to create the most high yielding crops.
5. Use *hydroponics, aeroponics,* or *soilless* agriculture to grow food and fiber without putting pressures on soil resources.
6. Develop *aquaculture*, which utilizes the resources of freshwater bodies, the seas, and oceans.

The first three routes will be discussed below in more detail. The problem of *soilless agriculture* and *aquaculture* will be examined in Chapter 9, and *biotechnology in soil science* will be the topic of Chapter 10.

8.1 CLEARING NEW LANDS

This method of increasing food production depends on the quantity and quality of soils. Quantity refers to the available acreages of new soils that can be cultivated. In Europe and the United States most of the soils are already under cultivation, and not much is left for further expansion (World Resources, 1987; FAO, 1987; President's Science Advisory Committee, 1967). However, only 25% of the soils are presently being used for crop production in Africa. In South America and Oceania (Australia and New Zealand), 20% and 31% of the soils, respectively, are under cultivation. Most of the unused areas providing a possibility for agricultural expansion are covered by rain forests or savannah-rain forests. In the six Amazonian countries in South America, the total forested area is estimated to be 825 million ha, and an average of 4 million hectares per year is expected to be cleared during the next 6 years. The cultivated area in Brazil alone must be increased by the year 2050 by 70-90 million hectares to support the food requirement of a population increase from an estimated 135 million at present to 250 million in the year 2050 (Fearnside, 1987).

The second point to be considered in clearing new areas for crop production is the quality of soils. Quality refers to the fertility of soils, or the capacity of soils to supply nutrients in adequate amounts and in the proper balance for crop production. The President's Science Advisory Committee report (1967) indicated that many of the soils in the world were low in fertility. Approximately 3.5 billion ha of these infertile soils are entisols. Of this group, 80% are too sandy and too shallow for cultivation. Another 2.9 billion ha are ultisols and oxisols, located mostly in the humid tropics of Africa, South America and Southeast Asia. These soils are very acidic and low in nutrients. They are highly weathered soils, but their clay mineralogy, composed of 1:1 lattice type of clays (kaolinitic clay) and sesquioxides, and their stable structure permit them to be cultivated even under very humid conditions or under the heavy monsoon rains of the tropics (Arnold and Jones, 1987). Soils with high fertility, such as the mollisols in the United States and Russia, are estimated to be 1.2 billion ha in extent, but 90% of these soils are already under cultivation.

When the virgin forest has been cleared and the soil is exposed, all kinds of changes on the soils are going to take place. The changes not only will affect the quality of soils, but will also affect the environment. Some of these changes are beneficial, but others can be harmful.

Air temperature, precipitation, natural drainage and wildlife habitat, for example, are going to be seriously affected regardless of efforts in damage control. Usually, the habitat for wildlife is lost forever after the destruction of the forest, and seldom will reforestation produce an environment similar to that found originally. Many of the plant and animal species have also been destroyed with the destruction of the forest and cannot be replaced.

The most striking effect is the decrease of soil organic matter content, which has a serious impact on nutrient cycling. Then, the degradation process of soils starts to increase, though it can be slowed down somewhat by the use of proper management techniques. The issues on nutrient cycling and soil degradation will be discussed in somewhat more detail in the following sections, and some techniques offsetting the loss in organic matter content and nutrient cycling will be given at the end of the chapter.

8.1.1 Nutrient Cycling

Nutrient cycling, a natural process that preserves and maintains soil fertility, is examined here in the context of clearing new lands for crop production in the tropical rain forest. The term *nutrient cycling* is used instead of *element cycling* or *mineral cycling*. Element cycling refers to the cycling of elements regardless of their importance as plant food. Mineral cycling is an incorrect term, since minerals are inorganic compounds, such as kaolinite and feldspars, which are not recycled in nature.

In a tropical rain forest, the vegetative cover produces high amounts of organic matter that are very rich in plant nutrients (Follet et al., 1987; Stewart et al., 1987). The organic matter in the form of leaf fall under the canopy of a tropical rain forest in Nigeria is estimated to be annually 7 Mg (tons)/ha. The rate of leaf fall in Nigeria is largest during the dry season in December and January (Ghuman and Lal, 1987). In the nutrient cycling process, this organic residue decomposes, releasing nutrients into the surface soil where they are taken up again by growing plants; in other words, recycled (Stevenson, 1986). This recycling process entails the maintenance of a thick forest cover, which preserves the fertility of the surface soil, the soil's A horizon. It is estimated that generally it takes 50 years for the soil under a forest in the tropics to accumulate a litter layer approximately 10 cm thick. While the bottom part is continuously decomposing in the recycling process, and is incorporated into the soil by biological activity, the top portion of this organic layer is constantly being replenished with new types of dead vegetative materials. The thickness of the organic layer is, therefore, maintained during the growth of a healthy forest stand. A number of factors affect nutrient cycling, e.g., climatic and biotic factors (Jordan, 1985). Important climatic and biotic factors are temperature, soil moisture, and rate of decomposition. Differences in these factors produce differences in nutrient cycling in tropical and temperate region forest. The year-round high temperature in the humid tropics insures a year-round growing season, resulting in high annual nutrient uptake by plants, and high annual return of nutrients to the forest floor. In the temperate region, seasonality in both temperature and precipitation is the reason why production of biomass and decomposition are not as high as in the humid tropics. Decomposition is very rapid in the tropics, and leaf litter is mineralized within approximately 8 months on the average, and little is converted to soil humus (Lavelle, 1987). Humus is usually formed by a decomposition process, called *humification*. The humus accumulated is very beneficial and is composed of a mixture of complex organic compounds, e.g., humic and fulvic acids. As colloids, humus exhibits properties similar to those of soil clays. Chemically, humus increases the cation exchange and water-holding capacity of soils, and also provides for a large buffer capacity which offsets pH changes. Physically, humus enhances formation of a stable soil structure, which helps to improve poor drainage and control erosion. As indicated above, mineralization is a stronger force than decomposition in the humid tropics, resulting in low humus content of the soils.

8.2   SOIL DEGRADATION

As soon as the vegetative cover is destroyed and the soil is exposed, the process of soil degradation starts to accelerate. Soil degradation is in fact a natural process and all soils will be affected by this process. In nature it is a very slow process, and it can perhaps be considered as a process of becoming of age, or growing old in terms of animal and plant life. However, because of human interference the degradation process of soils is suddenly accelerated, which is reflected by what we call *accelerated soil erosion*. Accelerated or natural soil erosion can take different forms but generally has the same result, that is carving the landscape of the earth by denudation. The only difference is that it will take natural erosion millions of years to carve the landscape, whereas accelerated erosion will do it, if not within several years, within a person's life span. Soil erosion is not just an agricultural issue but an environmental issue. As soon as the soil is exposed by deforestation, it is subject to attack by wind and rain. The wind will blow dust particles in the air, decreasing the quality of the air. Air loaded with dust particles is unhealthy and reduces visibility. The impact of raindrops destroys soil aggregates and disperses clay particles, which clog the soil pore spaces. Runoff, created during heavy or light rains, may carry sand, silt, and clay particles that will choke rivers and lakes. In turn, this process reduces the storage capacity of lakes and rivers and increases the chances for flooding. Soil degradation can be distinguished into (1) physical soil degradation, (2) chemical soil degradation, and (3) biological soil degradation.

8.2.1 Physical Degradation

Physical soil degradation is a degradation process that relates to changes in soil physical properties, such as changes in soil structure due to problems in aggregation, changes in bulk density and pore spaces due to compaction, and changes in infiltration due to formation of soil crust and seals. In the context of growing old, unfortunately the changes are seldom beneficial.

Soil Structure. - Structure formation and changes are both physical and chemical. As discussed in the aforementioned chapters, formation of soil structure is caused by aggregation of soil particles. Aggregation of clay particles in suspension is made possible by the presence of a flocculating concentration of ions in the soil solution, earlier called the flocculation value. The ion concentration in soils under a forest cover is expected to reach the flocculation value due to the large supply of ions released from the organic residue by the recycling process. In addition, the high biological activity, producing secretions mostly in the form of extracellular polysaccharides, is an important factor in assisting structure formation. With the disappearance of the forest cover, soil organic matter content decreases, whereas leaching of electrolytes or ions increases, reducing the flocculation value of the soil solution. The exposed soil is subject to easy attack by raindrops and dispersion by water. In agricultural operations, the damage of soil structure is usually controlled by liming the soils. However, this is a temporary effect. In humid regions, the Ca and Mg ions supplied by the liming materials will be subject to leaching, and in order to maintain the stability of soil structure, the need for continuous liming of the soil is apparent.

Soil Compaction. - In natural conditions, soil compaction occurs when the soil is easily dispersed and converted into a soil suspension. When dry, a dispersed soil will settle in a close packing. The soil particles settle down in an oriented fashion without forming pore spaces. The plate-like structures of clays are especially suited for formation of such compacted packages. Dispersion takes place on exposed soil surfaces because of the action of raindrops breaking down the soil structure.

In crop production, compaction occurs by soil particles being squeezed together by the use of heavy equipments, such as the soil underneath tractor tracks. Compacted soils are usually characterized by high bulk density values ($1.6 - 1.8$ Mg/m^3). Compaction will, of course, inhibit root growth. When the subsoil is compacted, roots grow laterally in the surface soil where compaction is less. Another impact of compaction is that during rainy periods the surface soil tends to become flooded, because water percolation through the subsoil is inhibited due to destruction of pore spaces caused by the compaction. However, the rules in soil physics suggest that as water content in soils increases, the density of soil increases, to decrease again in the presence of large amounts of water at saturation. Water increases swelling and reduces friction between soil particles. The soil particles are totally surrounded by thick water layers, and tend to slide and roll over one another more easily than under dry conditions. Therefore, at saturation, soil particles are separated by water, and this decreases compaction. However, the decrease is apparently only temporary, because upon drying the suspended soil particles resettle in an oriented close packing. The compaction is more pronounced in heavy-textured than in light-textured soils, because the plate-like structures of clays may stack up closely along the flat surfaces. Sand particles are, in contrast, coarse in size and irregular in shape, hence are less adapted for a close packing upon settling.

Surface Seals and Crust Formation. - Formation of soil crust and surface seals are also physical and chemical. Soil crusting and soil surface sealing are in fact similar phenomena. According to Singer and Munns (1996), the difference is that surface sealing takes place after dispersion of soil particles, especially clays, hence when the soil is wet. On the other hand, soil crusting occurs when the dispersed soil dries and becomes hard.

When the soil is exposed, the soil aggregates tend to be broken down easily by raindrops. Water drops from sprinkle irrigation may do the same damage on soil structure. The dispersed soil particles move with the percolating waters into the soil and clog the soil pores. Infiltration and percolation of water are drastically reduced and the dispersed soil particles will settle in a close dense packing, sealing the surface soil. When such a soil dries, the dry surface seal is then called a soil crust. Not only will soil crust inhibit infiltration and water percolation, but it will also restrict germination of seed and soil aeration needed for root growth. Runoff and soil erosion are increased by the presence of surface sealing and crust formation. Temporary control measures for surface sealing and crust formations are liming and the use of soil stabilizers, as discussed in Chapter 7.

8.2.2 Chemical Degradation

Two types of chemical soil degradation can be distinguished: (1) soil degradation due to leaching, and (2) soil degradation due to contamination or pollution.

Leaching. - Leaching has been defined as the loss of elements with percolating waters. Since plant nutrients are among the elements that are washed away through the soil profile, the process decreases soil fertility. The nutrients affected by leaching are mostly those that remain soluble in the soil solution, and the most significant are K^+, SO_4^{2-}, and NO_3^-. Ammonium, NH_4^+, and $H_2PO_4^-$ ions are not affected, because ammonium is a cation that will be adsorbed by the clay complex, whereas phosphate is usually subject to fixation. The exceptions to the preceding remarks may occur in sandy soils. Leaching of plant nutrients is very serious in humid region soils and is less significant in soils of the arid regions. The rate of leaching in humid regions is partly offset by the recycling process in soils under a forest cover. The nutrient elements that have been transported to the subsoil are taken up by plant roots and transported to the aboveground part of the plants. With leaffall and decomposition of the leaf residue, these nutrients are returned to the surface soils. However, with the destruction of the forest, the neutralizing effect of nutrient cycling on loss of nutrients by leaching is lost for ever. Since most of the nutrients are bases (K, Ca, and Mg), leaching of these elements will produce acid soils. Therefore, with the destruction of the forest, not only does the neutralizing effect of nutrient cycling on leaching disappear, but the tendency increases for formation of acid soils.

In general, leaching is a natural process that will take place in all soils, the most significant, of course, in humid region soils, and there is nothing that we can do about. A forest or other type of vegetative cover that can provide large amounts of plant residue for nutrient cycling can only offset part of the effect of leaching. Under a conifer forest, leaching is essential for formation of spodosols. In tropical regions, oxisols and ultisols will not be formed without adequate leaching for movement and removal of Si by a process earlier called *desilicification*. In crop production, losses of nutrients by leaching can be temporarily controlled by the use of fertilizers. Leaching losses of the nutrients from the fertilizers still take place, but can be reduced by applying the fertilizers at the right time and in the proper amounts.

Contamination or Pollution. - Many people often used the two terms interchangeably; however, some scientists suggest making a distinction between contamination and pollution (Wild, 1993). *Contamination* is defined as the introduction of hazardous chemical compounds or organisms into the soil without necessarily harmfully affecting the soil or the environment. On the other hand, *pollution* infers the introduction of harmful compounds or organisms into the soil that results in the degradation of the soil and the environment, aside from being harmful to organisms, animals, and humans. When present in appropriate amounts, all nutrient elements are beneficial for plant growth and are not harmful for other organisms. However, in relatively large concentrations these beneficial elements can become toxic. The accumulation of these elements below toxicity levels is an example of contamination. When the elements are accumulated at and above toxic levels, then this can be called pollution. The cationic types of these substances can be decontaminated by the soil's buffer capacity. Some of the anionic types, e.g., phosphate compounds, can be decontaminated through fixation with Al in acid soils, or with Ca in basic soils. However, other anions, such as NO_3^- and Cl^-, are more difficult to decontaminate, because they are usually subject to negative adsorption. When these ions escape, they may endanger the quality of groundwater, rivers, and lakes. The problem of eutrophication, as a result of accumulation of nitrates and phosphates, has been discussed in preceding chapters.

8.2.3 Biological Degradation

Biological degradation of soils can be distinguished into two categories. The first category is the degradation process brought about by the destruction of soil organic matter and the disappearance of microbial, plant, and animal life as a result of deforestation. The second category is related to crop production that introduces large amounts of chemical compounds into the soil. The latter may accelerate the disappearance of soil organisms that started with the destruction of the forest.

The substances entering the soil that may bring about biological degradation are organic waste and organic pesticides. Organic waste originates from agricultural, industrial, and domestic operations. Contamination with solid waste disposed of on soils present risk of fire, explosion, or chemical toxicity. Buried in landfills, anaerobic decomposition of organic waste may produce methane gas, a serious hazard for fire and explosion. Industrial and agricultural organic wastes may contain heavy metals that are toxic for soil organisms. They may also contain a number of pathogenic organisms that can cause serious illnesses, even death, to animals and humans. Recent reports from the press media indicated an outbreak of fish kill in the coastal waterways of Maryland (*New York Times*, 15 Sept., 1997). It was believed that the fish were dying because of infection with *Pfiesteria piscicida*, a single-cell pathogen, carried into the streams by the discharge of effluents from surrounding agricultural fields treated with manure. Such a harmful effect caused by the same microbe was also noticed in the hog-producing areas in North Carolina.

Pesticides are introduced into the soil by soil surface application or through sprays on plants dripping from the canopy or foliage. Their potential for biological soil degradation depends on their biodegradability and toxicity to non-target organisms. Although the assumption exists that used in low concentrations many of the pesticides are only destroying the target organisms, the fact is that fungicides used for control of soil-borne pathogens have often been noted to reduce drastically the population of many non-target soil organisms. Notorious among the pesticides is the fumigant methyl bromide, which is capable of making the soil completely sterile. The chances for biological degradation are increased by destroying many of the beneficial organisms, especially those that are required in nutrient cycling, organic cycling, and the like. However, pesticides are also subject to decontamination and/or detoxification through interaction with humic acids or adsorption by clay minerals, processes that have been discussed in preceding chapters. In summary, pesticides are subject to all kinds of reactions and transformation in the soil. They may be leached and volatilized. They can also be altered chemically into nontoxic form, but some can become even more potent than before. It is often noted that several pesticides have a residual effect that is harmful, or that they become even more harmful the next season. Aside from interactions with humic acids, interactions with microorganisms are considered the major pathways for transformation of pesticides. Though not much is known, reports have been presented that several microorganisms are capable of consuming pesticides through a process called *cometabolism*. This is a process by which the pesticide is taken up by the microorganism, but is not used as food or as a source of energy. Apparently, the pesticide is useful for cell growth and in assisting decomposition of compounds by the cell as part of cometabolism. More will be discussed about pesticide problems in Chapter 11.

8.3 ENVIRONMENTAL CHANGES

Because of population pressure and poor economic conditions, especially in Africa, South America, and Southeast Asia, large areas containing poor soils have been deforested for food and timber production. Forests have been cut at an alarming rate, and frequently insufficient time has been allowed after cultivation to return the cleared areas to forest. This appears to be more harmful to the environment than the benefits it brings of more food for the population. The environmental impact is expected to reach drastic proportions within the foreseeable future. However, such a view is not shared by many scientists in Brazil (Fearnside, 1987). Surveys conducted on deforestation of the Brazilian Amazon using *LANDSAT* satellite images indicated that only 1.55% of the area legally defined as Amazonia had been deforested in 1978. Although, on a regional scale the cleared area has increased from 0.5% to 3.12%, these increases are believed not to be widespread. Most of the increases in deforestation are related to cultivation efforts by large landholders, and transmigration because of improved road access. Nevertheless, environmental considerations necessitate careful deliberation before clearing rain forest areas, and a suitable balance must be found between the necessity of clearing the tropical rain forest for crop production and the rate or degree of destruction in environmental quality that it brings.

8.3.1 Regional Climatic Changes

Regional Changes in Temperature and Precipitation. - The destruction of the tropical rain forest causes permanent changes in the environment. Concern has been expressed in many countries of the world that the extensive slash-and-burn method produces large amounts of CO_2 and other pollutant gasses, contributing to the so-called *greenhouse effect*. The resulting *global warming* can bring drastic changes in the earth climate. Emission of CO_2 by industry and vehicular transportation, and other factors, related to the distance of the earth from the sun, are perhaps bigger contributors to the greenhouse effect. If global warming is to occur by 0.3° per decade, as predicted, the sea level will rise to such a level that extensive areas of coastal plains in the world, such as in Bangladesh, will disappear. In upland soils the increase in temperatures may increase the rate of rock and mineral weathering, and organic matter decomposition, accelerating soil degradation. However, recent studies indicate that increased CO_2 emission may not be the culprit for increasing the Earth temperature. The concept of global warming because of increased CO_2 content was disavowed as a *myth* by A. Robinson and Z. Robinson, chemists at the Oregon Institute of Science and Medicine. They wrote in the *Wall Street Journal*, December 4, 1997, that during the last 20 years when the Earth was experiencing the highest CO_2 levels, temperatures on Earth have actually decreased. The data presented indicate that the temperature of the Earth's atmosphere fluctuates corresponding to fluctuations of solar activities and other factors. Since the latest ice age, called the *Little Ice Age*, some 300 years ago, the temperature has been rising, and remains today at a level below a 3000-year average.

A more tangible effect of deforestation is perhaps the slight decrease in precipitation. Deforestation in the Amazon rain forest of Brazil results in a decrease in evapotranspiration, and hence in total cloud formation and cover. Therefore, rainfall also decreases by 0.5 to 0.7 mm/day, but nevertheless no regional temperature changes have been reported (Henderson-Sellers, 1987).

Desertification. – In arid regions, especially those bordering the deserts, clearing the savannah forest can turn the deforested areas into desert lands. The process is known as *desertification*. It was reported by the United Nations in 1984 that 3.5 billion hectares of the arid region croplands in Africa and Asia are affected by desertification (Singer and Munns, 1996). The reason for the change is still unknown but a number of scientists believe that desertification is a natural process caused by changes in climate due to large-scale deforestation of the region. It appears to happen only in arid regions, since in humid regions, cleared areas will turn into a forest again if sufficient time is allowed for the regrowth of the vegetation. As in Brazil, removal of forest cover may affect evapotranspiration, and hence affect cloud formation, which is very important for the arid regions at the time when the monsoonrains start. Overgrazing and overcultivation, together with removal of trees, shrubs, and even animal manure for fuel, common practices in Africa and India and other arid regions in Asia, are enhancing the process of desertification. The neighboring desert usually expands and /or moves with the wind. Blown by the wind, it is as if the sand dunes are migrating into the empty deforested wastelands. Attempts to restabilize the sand dunes have been made in Africa by planting trees, but many people believe that it is a lost cause and no human intervention will be successful unless drastic measures are taken to revegetate the land, as long as the rains are willing to cooperate.

8.3.2   Effect of Increased CO_2 Emission

The increase in CO_2 production by both agricultural and industrial operations may have a pronounced effect on photosynthesis. An increase in CO_2 content in soil air will not only accelerate soil-forming processes, as discussed above, but a doubling of the CO_2 content in atmospheric air is believed to increase the yield of C_3 plants, such as soybean, potato, oats, barley, wheat, and rice by 10 to 50% (Wild, 1993). C_4 plants, e.g., corn, sugar cane, and sorghum, appear to be less affected by an increase in CO_2 content in our air. C_3 plants fix CO_2 by the so-called *Calvin cycle*, whereas C_4 plants use another process for CO_2 fixation, sometimes called the *Hatch-Slack cycle*. This cycle was discovered by Hatch and Slack in sugar cane and enables the plants to store more CO_2 in the chloroplast. It is the reason why C_4 plants do not respond as well as C_3 plants to an increase in CO_2 content. The growth rate of C_4 plants is consequently greater than that of C_3 plants.

The trees and other plants in the rain forest are also the major suppliers of oxygen in the air we breathe. They also purify the air by absorbing excess carbon dioxide produced by modern industry and prevent it from accumulating in the atmosphere. With the help of sunlight, green plants produce carbohydrate from carbon dioxide and water. During this process, known as photosynthesis, oxygen is formed and released. When the forest disappears, these processes — so essential to our environment and health — cease.

8.3.3 The Issue of Biodiversity

The destruction of the tropical rain forest also destroys a variety of plants and animals. Such ecosystems are characterized by a highly diverse and interdependent population of plants and animals. The interaction between these plants and animals is so close that the extinction of one species may have direct and indirect effects on other species. For example, the Brazil nut (*B. exelsa*) is pollinated by bees. Any changes that destroy one of these organisms may harmfully affect the other (Mori and Prance, 1987). If the pollinators vanish, these nut trees will also disappear. If the trees are destroyed, then the pollinators will become extinct. Seeds must be dispersed and birds play an important role in this respect. Even if cleared areas have been allowed to return to forest, original habitats for plant and animal life have been destroyed. An example unique in the United States is the predator-prey relationship between the black-footed ferret and the prairie dog in the Great Plains. Since the habitat, the prairie, for prairie dogs has mostly been destroyed, the number of these animals has been decreased dramatically, and so does the number of the black-footed ferrets, who depend on prairie dogs for their main prey.

Every day and every moment wild animals are vanishing. Plants that are commonplace and animals that are pleasing to the eye are disappearing forever. It will be a global tragedy if most of the animals vanish, except for domestic cats and dogs, and our Earth becomes inhabited only by human beings. When the forest disappears, we also lose valuable sources of chemicals, medicine, and fiber. Many of the plants are known to have economic value due to their medicinal properties, and these will be lost forever with the disappearance of the rain forest.

8.4   PLANTATION AGRICULTURE AND AGROFORESTRY

In Africa and South America the increase in crop production comes by clearing the tropical rain forests. The soils are usually highly leached and low in plant nutrients. Nevertheless, natural rain forest ecosystems on tropical soils have a large standing biomass. Large amounts of nutrients are contained in the plant biomass (Table 8.1), which will be released upon decomposition of the litter. A dense vegetative cover will add substantially to the organic matter content in soils, which is the main source of soil nitrogen. It is a well-established fact that soil nitrogen content increases with increased organic matter content. A positive correlation exists between organic carbon and total nitrogen contents, which usually takes the form of a linear regression (Tan and Troth, 1981):

$$N = a + bC$$

where N = total %N and C = %C_{org}. The equation above indicates that nitrogen content increase linearly as organic carbon increases in the

Table 8.1   Plant Nutrient Content in Soil and Vegetation in Tropical
Forests (in % of Total Stock)

Region	N	P	K	Ca	Mg
Ivory Coast (Banco)	82.2	50.0	13.3	8.3	15.1
Ghana (Kade)	69.2	8.6	41.6	49.1	49.6
Venezuela (San Carlos					
de Rio Negro)	62.6	59.3	18.1	38.9	20.2
Panama	00.0	11.6	10.1	84.3	84.3
Puerto Rico	00.0	83.6	8.9	47.1	69.3
Zaire (Yangambi)	89.5	98.0	66.3	33.8	42.0

Sources: Odum and Pigeon (1970); Golley et al. (1975); Herrrera (1979);
Bernhard-Reversat et al. (1979); Lavelle (1984).

soil. Therefore, the destruction of soil organic matter due to deforestation may have serious implications for the nitrogen content in soils.

A variety of problems develop after deforestation. After deforestation, the surface soil, in which most of the nutrients are recycled,
can support crop production for a year or two. Soon, the cultivated
crops and leaching processes exhaust the soil's nutrient supply. Crop
yields decrease and the land is abandoned, leaving bare soils exposed
to the impact of tropical rains, and severe erosion becomes a problem.
When the surface soil has been stripped off by accelerated erosion,
the exposed subsoil, often rich in iron, hardens and forms an iron pan.
This pan formation inhibits further vegetative growth.

The disastrous effect of deforestation as discussed above can be
avoided, and sustainable economic yields are possible in tropical
agriculture by the application of a nutrient cycling system. Such a
system has been known in the tropics for decades under the name
plantation agriculture, and recently a variation of this method has
been introduced under the name *agroforestry*. In plantation
agriculture, the crops are composed of trees or shrubs (Figure 8.1),
and only the fruits or young shoots, such as coffee, cacao, and tea, are

Figure 8.1 Cacao (*Theobroma cacao*) grown in plantation agriculture with *Leucaena glauca* and *Glyricidae* sp. as shade trees (Courtesy: Didiek H. Goenadi, Research Institute for Biotechnology of Estate Crops, Bogor, Indonesia).

harvested. These crops are grown together with nitrogen-fixing legume trees that provide the crop with necessary shade, N_2, and other nutrients. A widely grown legume tree is *Leucaena leucocephala,* and another is *Leucaena glauca* (Figure 8.2). When used as mulch, the leaves of these legume trees may add 150 kg/ha of N per year. The roots of these shade trees penetrate deep into the subsoil and take up nutrients beyond the reach of the roots of the coffee, cacao or tea plants. These nutrients are recycled to the surface soil by way of litter- and leaf-fall.

Plantation agriculture has been successful in controlling part of the degradation of soils. The crops, such as coffee, cacao, tea, rubber, and in South America also bananas and sugar cane, are mainly grown for export. It is generally a huge operation using high technology and research input, and is sometimes accused of being an operation of little

Figure 8.2 (A) Agroforestry of *Leucaena* sp., (B) Closeup of *Leucaena glauca* tree (Courtesy: Didiek A. Goenadi, Research Institute for Biotechnology of Estate Crops, Bogor, Indonesia).

or no benefit at all to the indigenous small-scale farmer (Brady and Weil, 1996). Such a criticism stems more from the exceptions than from the general cases. It is true that considerable profit has been gained and exported overseas, but in Indonesia, even under the Dutch colonial rule in the past century, the big plantations have pumped

much money into the economy of the regions. In the author's experi-
ence, the regions surrounding a plantation were always better off, because of the many jobs, roads, schools, and other benefits that the
plantation provided to the region. In the old days, the management
team from outside the region was composed only of a few people when

compared to the huge staff and labor locally acquired. Even many of the research personnel were local people. Today, many of the plantations in Southeast Asia are owned and operated by the government, and the local farmers are encouraged to grow the crops also on their farms neighboring the plantation. For example, oil palm plantations in north Sumatra, Indonesia, are operating in this way with the believe that it is more economical to let local farmers also raise the crops in support of their own crops grown in the plantations. Their factories are then additionally supplied with products purchased from the small farmers, hence ensuring a steady supply of less expensive raw material for palm oil production. Apparently, such a relationship between plantation and small-scale agriculture is not limited to oil palm cultivation. Cacao, rubber, and tea are also enjoying a similar mutual benefit as that experienced in the oil palm industry. The small-scale rice farmers reap the benefits by selling their excess paddy to a nearby rice plantation, or by taking advantage of the rice mill in the plantation for processing their paddy into rice for their own consumption.

Agroforestry. – This appears to be an extension of plantation agriculture, because of its many similarities in the method of operations. *Agroforestry*, as defined today, is a system of crop production by growing trees in association with crops and pastures. The trees are grown in rows, providing the forest system, separated by strips of soils cultivated with food crops. Such a system originates from the indigenous, frequently considered primitive, system called *shifting cultivation*. It was developed by the native people in the tropics to restore soil fertility by nutrient cycling. After the soil has been cultivated for two or three years, it is then abandoned for 5-10 years, allowing the regrowth of the forest and sufficient accumulation of organic matter for recycling purposes. The topsoil is rejuvenated by a combination of effects due to nutrient recycling, accumulation, and conservation. However, large acreages of land are required to produce enough food with shifting cultivation. To minimize the damage to the environment, perhaps 10 to 20 hectares of land per person are required, which is a far bigger acreage than a farmer in the temperate region uses to cultivate.

8.5 INTENSIFYING SOIL PRODUCTIVITY

In Europe, where population pressures have always been strong, most of the good soils are already being used for food production, and little land remains for agricultural expansion. Approximately 80% of the arable land in Europe is already used for crop production, whereas in North America 60% of all arable land is cultivated (World Resources, 1987; FAO, 1987; President's Advisory Panel on World Food Supply, 1967). In Asia and Russia, 73% and 65%, respectively, of the arable land are in use. These figures may be higher today. Therefore, these countries have little opportunity for increasing food production by opening new lands. At the same time, they are already growing as many crops as possible during the year. Thus the third option, increasing cropping intensity, is also unfeasible. The only viable option for these countries is to intensify crop production on soils under cultivation. Complicating the situation is the fact that large areas of arable land are constantly being taken out of agricultural use for housing and urban development. In addition, younger generations are leaving the farms for the cities, and more food must be produced by fewer farmers on less land than ever before. Only through the combined efforts of scientists, professionals, and others loyal to agriculture can intensification of crop production be achieved.

8.5.1 Green Revolution

Ironically, the push for intensification of food production in the world was set off by the *green revolution* (Uribe, 1975) in Central America and Southeast Asia. The green revolution, a revolution in agricultural production, was initiated by a team of scientists from the Rockefeller Foundation. Under the leadership of George Harrar, these scientists were sent to Mexico in 1943 to improve and increase corn production. At the onset, it was an attempt to transfer US techniques in crop production to the farmers in Mexico. However, today the green revolution has developed into a global effort to improve food and fiber production.

The innovations that George Harrar's team introduced — genetic improvement of local corn, proper soil management, use of fertilizers, and weed, pest, and disease control — enabled Mexico to become self-sufficient in this important staple food. The success of the Mexican project led to expansion of the program toward improving wheat production in the Punjab, India, and Mexico (Reitz, 1968). Results of this team effort enabled farmers in India and Mexico to double their wheat production and boost their total tonnage in food production. Norman Borlaug, who played an important role in the wheat improvement program, was later awarded the Nobel Prize in agriculture and food production. The effort in improving world food production was continued in 1961 by the establishment of the International Rice Research Institute (IRRI) at Los Banos, the Philippines, under the combined sponsorship of the Rockefeller and Ford Foundations (Uribe, 1975).

Rice is the major staple food grain in Asia, and attempts to increase local rice yields in Southeast Asia were at first disappointing. The use of fertilizers and proper control of weeds, pests, and diseases caused the local varieties to grow excessively tall, hence caused the plants to lodge or fall over easily. At the IRRI, crossings were made between a tall tropical 'indica' variety from Indonesia and the 'ponlai' variety from Taiwan. The 'ponlai' rice variety was a small 'japonica' variety developed for tropical and subtropical conditions. The result of the crossing was a dwarf, stiff-straw, high-yield variety of rice, introduced for the first time under the name *IR8-288-3*, or better known as *IR-8* (Chandler, 1968). This new variety had a grain-straw ratio of 1.2, instead of 0.6 as in the local varieties of rice. It had, therefore, more grain than straw and it matured early and was photoperiod insensitive. Not only could the rice yield be doubled by planting IR-8, but the farmers could also have three crops a year.

Although there is no doubt that food production has been increased considerably, several scientists remain skeptical about the green revolution. Among their major concerns are the social and economic factors indicating that too high a price is paid for the increase in crop yields (Moore and Moore, 1976). The new crops need more fertilizers and pesticides, affordable only to rich farmers and big landowners. More energy-intensive cultivation methods must be applied, which require mechanization. The latter may result in unemployment and increased migration of rural folks to the cities. Another question raised about the green revolution is the decrease in variability in the gene pool. Breeding of high-yielding varieties has resulted in genetic uniformity, and cultivation of these new plants as monocultures increases the chances for the development of pests and diseases. The control of these pests and diseases will then be very difficult, since breeding new resistant plants would be impossible due to the loss of variant genes. This problem will be discussed in the next section in relation with the southern corn blight disease in the United States.

Although urban migration has occurred in Southeast Asia, the reason for such a migration can be traced to a multitude of factors, and it is incorrect to blame it solely on the green revolution. The vulnerability of the high-yielding crops to pests and diseases is perhaps also true, but this is a general problem with plant breeding and not necessarily the effect of the green revolution. The rice tungro disease reported during 1970–1971 (Moore and Moore, 1976) did not affect the millions of hectares of rice, but was limited to a few thousands hectares, and was controlled rather quickly.

It is apparent that the positive effects of the green revolution outweigh its negative effects. The effort started by the IRRI has spread to neighboring countries, especially Indonesia, where a long-standing project from the time of Dutch colonialism, to improve rice yields, is revitalized. Due to the introduction of its own high yielding varieties, Indonesia is currently reported not only to be self-sufficient in rice production, but to have gone from being a rice-importing country to being an important rice-exporting country.

8.5.2 Hybrid Corn

In the United States, intensification of crop production preceded the green revolution. It started perhaps in the period 1812–1877 with the discovery of open-pollinated corn, which became known later as *hybrid corn* (Harpstead, 1975). The practical seed production of hybrid corn and the use of the *double cross* led to the development of the *cornbelt* in the Midwest. Today we find that increasing the yield of corn to 10,000–20,000 kg/ha (equivalent to 200 to 300 bushels/acre) by the use of high-yielding hybrid corn and proper management of the soil is not an impossible task (Aldrich et al., 1975).

As is the case with intensification of rice production, the efforts in breeding new high-yielding corn varieties and their cultivation in monocultures create some concern about the vulnerability of the crops to pests and diseases. As indicated earlier, many scientists were alarmed about a possible loss of beneficial genes that can be used to develop new resistant plants. However, the sudden explosion of the southern corn blight in the United States, destroying in 1970 almost one-fifth of the US crops, has not supported the negative concerns. The corn blight disaster was controlled within the year by the production of new varieties by using plant breeding.

Such intensive production methods, producing high yields, require a large investment of capital. Heavier application of fertilizers than usual, and larger amounts of chemicals for controlling weeds, insects, and diseases must be used. This need for such capital expenditures is a major obstacle in Africa, South America, and Southeast Asia, where economic conditions present major problems. Some observers also feel that the modern agriculture of the Northern Hemisphere is too energy–intensive, and that methods requiring less energy input per unit area are preferable, especially in the third world countries. Less intensive methods are less expensive, but more land is then required for farming to increase food production. The consequence of cultivating more land at lower yield levels is that less desirable soils will be used for agriculture. Not only will this result in lower output per unit energy input, but it will also increase the danger of destroying the environment.

8.5.3 Use of Fertilizers

Fertilizers are defined as any material applied to soils to increase yield, and to improve the quality and nutritive value of crops. The bulk of fertilizers is artificially produced by chemical companies and the oil industry. These fertilizers contain one or more of the elements essential for plant growth. The elements are carried in forms available to plants. Among the many nutrients required by plants, three elements, N, P, and K, are the major components of inorganic fertilizers. Based on these three major *fertilizer elements*, the artificial fertilizers can be distinguished into (1) nitrogen fertilizers, (2) phosphate fertilizers, and (3) potash (potassium) fertilizers. The three groups of fertilizers are considered single-element fertilizers, in contrast to a fourth group called mixed fertilizers. The mixed fertilizers contain two or all three of the major fertilizer elements, and may also have other plant nutrients, such as the micronutrients.

The N, P, or K content, or quality of the fertilizers, is indicated by the *fertilizer grade*. This fertilizer grade is usually expressed in % total N, % available P_2O_5, or % soluble K_2O. The grade of mixed fertilizers is composed of three figures, e.g., 10-10-10, 4-12-12, etc. A 4-12-12 fertilizer mixture contains 4% total N, 12% available P_2O_5, and 12% soluble K_2O. Currently, there is a trend toward expressing fertilizer grades in the elemental percentage for scientific purposes, and in the oxide percentage for fertilizer sales. Some of the common artificial fertilizers and their grades are listed in Table 8.2. The equivalent acidity listed in the table is the amount of lime required to bring the soil pH back to pre-fertilizer application levels. For example, the use of 1 Mg (1000 kg)/ha of ammonium nitrate will decrease the soil pH enough so that 599 kg/ha of limestone is needed to restore the pH to the original level before fertilization. Therefore, the equivalent (or potential) acidity of 1000 kg ammonium nitrate equals 599. Fertilizers that decrease soil pH are called *acid-forming fertilizers*. All ammonium fertilizers are acid-forming fertilizers. In contrast, nitrate fertilizers, such as $NaNO_3$, are *basic-forming fertilizers*. The equivalent acidity of $NaNO_3$ is, therefore, -294.

The importance of artificial fertilizers in crop production dates back to 1842 when J.B. Lawes and J.H. Gilbert invented the production of superphosphate at the Rothamsted Experiment Station in England (Brady, 1990; Tisdale et al., 1993). However, the growth of the fertilizer industry was propelled to its current dimension after the discovery of the production of ammonia, NH_3, by reacting N_2 gas with H_2 gas, according to the so-called *Haber-Bosch process* (Tisdale et al., 1993). Intensifying crop production requires higher yields per unit area of soil. This increases the demand placed on soils to provide sufficient amounts of nutrients, and fertilizers are applied to remove

Table 8.2 Major Inorganic Fertilizers in Agriculture

Fertilizer	N* %	P$_2$O$_5$ %	K$_2$O %	Equiv. acidity
Nitrogen fertilizers				kg/ha
Ammonium nitrate, NH$_4$NO$_3$	33.5			599
Anhydrous ammonia, NH$_3$	82.0			1492
Sulfate of ammonia, (NH$_4$)$_2$SO$_4$	20.0			1109
Urea, CO(NH$_2$)$_2$	45.0			756
Phosphorus fertilizers				
Superphosphate		20.0		
Triple superphosphate		48.0		
Potassium fertilizers				
Muriate of potash, KCl			60	
Sulfate of potash, K$_2$SO$_4$			50	

*Total N, available P$_2$O$_5$, and water soluble K$_2$O.

the limitation of crop production caused by an inadequate supply of plant nutrients in soils.

The use of fertilizers worldwide has, therefore, increased considerably since 1950. In North America and Europe, the use of fertilizers accounts for 30 to 50% of the expenditures in agricultural operations. The use of nitrogen fertilizers has increased from 10 million tons in 1950 to 80 million tons in 1990 worldwide. The use of P and K fertilizers did not increase dramatically during the same period. According to Brady and Weil (1996), fertilizer use in North America and Europe decreased or leveled off in 1990 because optimum yields had apparently been reached, whereas additional increases in yield would prove to be too expensive. In contrast, the use of fertilizers in third-world countries continues to increase in their efforts to reach optimum yields.

Because of fertilizer application, the yield of wheat improved steadily from 800 kg/ha in 1950 to 2382 kg/ha in 1987. During the same period corn yield increased from 2272 kg/ha to 7250 kg/ha in 1987. Corn yields exceeding 300 bu/a (18142 kg/ha) are not uncommon today by the application of fertilizers in combination with the use of hybrid corn and the proper management practices (Tisdale et al., 1993). However, as the use of inorganic fertilizers increases, the potential hazard for contamination of surface and groundwater increases considerably. Therefore, new techniques have to be developed to control the harmful effect on environmental quality. The use of fertilizers has been shown to increase yield and prolong the period of a good yield. However, even fertilizers have been known to fail in halting yield decline over a period of years in upland rice fertilization trials in Peru (Jordan, 1985).

8.6 RESIDUE- OR NO-TILL FARMING

To reduce the hazard of decreasing the quality of the environment by intensification of crop production, a new method has currently been introduced to the United States.

With the development of the moldboard plow in 1837, farmers settling in the Great Plains of the United States plowed through the tough root system of prairie grasses. Exposing the black topsoil of the mollisols, rich in organic matter and plant nutrients, they planted corn and wheat. Despite several years of ravaging dust bowls when drought-stricken soil was lifted by storms and carried as far east as the eastern seaboard, America became the breadbasket of the world. During 1950-1980, more plows and more heavy machinery of mounting complexity were developed. Notwithstanding the avai-lability of modern equipment, fertilizers, and the use of high-yielding crops, farmers today appear to be making little economic headway. They are financially strapped and overburdened by overhead expenses from the monstrous machines needed to plant and harvest corn, wheat, and soybeans. Additionally, farm legislation in 1955 forced them to adopt conservation methods in order to receive crop support payments. This factor, together with a strong desire to reduce the overhead in machinery and to convert from an intensive into a less labor-intensive operation, led farmers to investigate more efficient farming methods. Recently, a new method called *residue farming* by the USDA Soil Conservation Service, popularly known as *no-till farming*, seems to have taken hold. This technique has been known for some time, but has not attracted too much attention because of the lack of economic and environmental urgency. However, following recent years of severe economic depression and disastrous farm surpluses, farmers permitted millions of acres of farmlands to lay fallow. This compelled the farmers to adopt better and more efficient farming methods on the lands still under cultivation. One of these methods is residue farming, which is now being applied on more than 70 million acres of US croplands. This method is spreading as a renewed burst of the green revolution from the coastal plains of the Atlantic seaboard to the rich floodplains of the Mississippi River valleys and onto the great plains in Nebraska and the highlands in Oregon, where corn, soybeans, wheat, and other crops are grown.

In residue farming, the residue or stubble from the previous year's crop is not plowed under. Instead, it is left undisturbed (in place) to hold soil and conserve soil moisture. The plows and machines are retired, and the seeds for the new crop are chiseled in between the stubbles. The seedlings are allowed to sprout through the decomposing vegetative residue, hence, the name residue or no-till farming (Figure 8.3). It is sometimes also referred to as *mulch-till farming*. Weeds are controlled by newly developed biodegradable herbicides. Insects are also controlled with biodegradable insecticides or by adoption of crop rotation to break the insect life cycles. Unconfirmed reports indicate that the application of residue farming has enabled farmers to decrease their labor input considerably, allowing them to cut expenses of producing crops by 30% to 40%. However, some caution should be executed to prevent the occurrence of allelophaty as discussed in Chapter 3 and elsewhere in this book.

Figure 8.3 Residue farming of soybean (*Glycine max*) (Courtesy: D.L. Armstrong, editor, Potash & Phosphate Institute, Atlanta, GA).

8.7 ORGANIC FARMING

Organic farming does not fall in the category of intensifying crop production, but is somewhat related to residue farming. Organic farming is also a system of low external input of crop production, but without using artificially produced agricultural chemicals. It has as a main objective the production of food void of chemical contaminants and relatively safe for human consumption. The use of commercial fertilizers, insecticides, and pesticides is avoided. Artificial fertilizers are replaced by organic manures and compost, whereas soil organic matter and nitrogen content are maintained by growing leguminous green manures. Pests and diseases are controlled by the proper selection of resistant crops, and use of the appropriate agronomic practices, such as crop rotations and *buffer plants*. The latter may kill or expel insects from the main crops. Mixed reports are available on the yield and quality of the crop produced by organic farming. A report from the Council for Agricultural Science and Technology (CAST, 1980) shows a wide range in crop yields as a result of organic farming, from 56% to 107% of the yield of *conventional farming.* Similar examples of yields were also reported in England (Wild, 1993). The average yields over a 14-year period of experiments in eastern England, as compared with those grown conventionally with fertilizers, were 93, 80, and 75% for organically grown wheat, barley, and beans, respectively. Therefore, the opinion exists that organic farming produces less than conventional farming, and data presented by Miller and Gardiner (1998) suggest that organic farming on a global scale, replacing conventional farming, would result in the production of less than half the amount of food. The crops, especially fruits and vegetables, are frequently deformed and display some blemishes or imperfections due to slight pathogenic infection. They are free of chemical contaminants and are usually more expensive. Their market value depends on the willingness of the customer to accept a less perfect shape of fruit or vegetable at a higher price. Nevertheless, organic farming today seems to be making headway, since many supermarkets in the United States are displaying large sections of a variety of organic produce. The quality of the products has apparently been increased substantially and fewer blemishes are noticed. Moreover, the price of organically grown produce has been leveling off.

8.8 LOW-INPUT SUSTAINABLE AGRICULTURE

Low-input sustainable agriculture or *LISA* refers to agricultural operations with minimal usage of energy and artificial chemicals. As with residue tillage, it is aimed at reducing the costly overhead of heavy equipments and the cost of energy intensive operations. The use of artificial chemicals is reduced, but is not totally avoided, as in organic farming, which is totally dependent on natural organics. The difference with residue farming or no-till farming is that LISA allows the soil to be tilled and cultivated. Miller and Gardiner (1998) equate

LISA with the farming operations of the *Amish people* in the midwestern states of the United States, who even today are tilling and cultivating the soil with a traditional plow pulled by horses or mules. The authors above believe that to practice LISA it is essential to combine the best practices of organic farming with methods that are inexpensive and less energy dependent. This includes the use of lower amounts of fertilizers and pesticides, and a decrease in using machinery and/or heavy equipment, all practices that will result in lower crop yields. Hence, as with organic farming, LISA does not apply to the category of intensifying crop production, but does correspond to the theme of the book as reflected by the title. LISA is just a method for making crop production less expensive, and consequently it can be argued that because of the lower operational expenditures, lower yields do not necessarily result in profits going down. This is perhaps true for subsistence agriculture, but in crops grown for the market, lower yields can be a disadvantage in today's strongly competitive global economy. However, more research is required to address the economical implications of LISA. Another equally important factor is that agricultural operations which are dependent on more manual labor, involve back-breaking work, and human power is also a very precious commodity.

CHAPTER 9

SOILLESS AGRICULTURE

9.1 HYDROPONICS

Hydroponics, from the Latin words *hydro* (water) and *ponos* (labor), is by definition "growing plants in water." It is also referred to as soilless culture of plants (Jones, 1983), *water culture*, or *nutriculture* (Hoagland and Arnon, 1950). Water culture or hydroponics is a method of growing plants with their roots in a solution containing a balanced proportion of nutrient elements essential for plant growth. In the past, the names *tray agriculture* and *tank farming* were also used for this method. Although the technique of growing plants in water has been known for centuries, hydroponics has only recently gained its present popularity for growing vegetables and other plants. Early in the 17th century, Jan Baptista Van Helmont indicated, as a result of his now famous willow tree experiments, that water was the most important nutrient for growing plants (Brady, 1990). The importance of water was contended later by John Woodward, who experimented with growing spearmints in river water, rain water, and muddy water. He argued that the mud, not water, was the substance for increased plant growth. Although others indicated that organic matter or *humus* was consumed by plants instead of mud, the concept of mud as a plant food was supported by Jethro Tull's discovery in the 18th century. Tull noticed that increased plant growth was obtained by tilling the soil, and believed that fine soil particles, produced by cultivation, were taken up by the growing plants. In the beginning of the 19th century, De Saussurre discovered the process known today as *photosynthesis*. This notion that plants absorbed CO_2 from the air and at the same time released O_2 with the help of sunlight was not well understood at that time. However, this new approach brought a change in the concept of plant nutrition. It was finally realized that plants obtain nutrients from both the air and water.

Today the idea of water culture or hydroponics has been revised and expanded to include all methods of growing plants in an artificial medium supplied with nutrient solutions. Although it is frequently assumed that hydroponics is a soilless culture, several of the growth media today often include *artificial soil*, which is made up of a mixture of sand, or vermiculite, and organic compounds such as peat and the like, especially in the aggregate, and adsorbed nutrient techniques. All these materials are soil constituents, and strictly speaking methods using artificial soil media do not qualify as soilless agriculture. However, since growing plants in artificial media has gained considerable importance today, this method will also be addressed in this chapter. All these methods allow intensification of agricultural production without using natural soils or destroying valuable areas of rain forest.

9.1.1 Water Requirements

As indicated above, in hydroponics plants are grown in water to which nutrient mixtures have been added. Therefore, this method requires a lot of water, considerably more than methods using soil for growing plants. Most of this water serves only as the medium for nutrients and root growth and will often be discarded after harvest of the crops. However, at present, attempts are being made to recycle the used water. In this case care should be taken to prevent allelophaty, since it is known that roots may secrete a variety of organic substances that can be harmful to other crops. In addition, some eutrophication may also have taken place, making the purification of used water necessary.

The amount of water taken up by plants grown by hydroponics is expected not to differ significantly from that taken up by plants grown in the soil. Data on water consumption of tomatoes grown side by side in water culture and soil showed some evidence of this. The amount of water used to produce 45 kg (100 lb) of tomatoes was 971 L (257 gallons) in water culture as opposed to 839 L (222 gallons) in soil as reported by Hoagland and Arnon (1950). A difference of 132 L or 35 gallons of water could be accepted as a large difference by some authors. Scientifically there should not be any difference present in water uptake of plants within the same species. The water requirement of a growing plant, as expressed in terms of *transpiration ratio*, is a plant characteristic, and should be relatively constant within one plant species. It shouldn't differ much whether the plant is grown in water or in soil. By definition, the transpiration ratio is the unit of water needed to produce a unit of dry matter. It is calculated from the weight of water lost by evapotranspiration divided by the weight of dry plant material. Hence, its value may vary according to differences in climatic conditions. Generally, it is much higher for crops grown in arid regions than for similar crops grown in humid regions. Intense solar radiation, higher temperatures, and stronger wind action tend to raise the value of the transpiration ratio. The value of the transpiration ratio is also known to differ between plant species. For example, it is low for sorghum (277), but high for alfalfa plants (853). The figures between parentheses mean that sorghum needs 277 kg of water for the production of 1 kg of dry matter, versus alfalfa, which requires 853 kg of water for the production of 1 kg of dry matter. Rice plants exhibit transpiration ratios around 682, which is less than the water requirement of alfalfa, yet rice grows in water or inundated soil.

9.1.2 Nutrient Solutions

A number of nutrient solutions have been developed for hydroponic cultures, such as the *Hoagland, Sach's, Knop's, Pfeffer's*, and *Crone's* solutions (Jones, Jr., 1983; Douglas, 1976; Hewitt, 1966; Hoagland and Arnon, 1950). One of the nutrient solutions frequently used in the

Table 9.1 Composition of a Hoagland and Arnon and Knop's
Nutrient Solution

Compound	Hoagland	Knop's
	Moles/L	g/L
$NH_4H_2PO_4$	0.001	-
KNO_3	0.006	0.2
$Ca(NO_3)_2$	0.004	0.8
$MgSO_4$	0.002	0.2
KH_2PO_4	-	0.2
$FePO_4$	-	0.1
	mg element/L	
H_3BO_3	0.5	
$MnCl_2.4H_2O$	0.5	
$ZnSO_4.4H_2O$	0.05	
$CuSO_4.5H_2O$	0.02	
$H_2MoO_4.H_2O$	0.05	
Fe tartrate	1.37	

Sources: Hoagland and Arnon (1950); Jones, Jr.(1997), and Hewitt
(1966).

United States is the so-called *Hoagland and Arnon solution*, often
called *Hoagland solution,* which has a composition as listed in Table
9.1. According to the general opinion the other types of nutrient
solutions are equally as effective for water cultures as the Hoagland
solution. As correctly indicated by Hoagland and Arnon (1950), there
is no one composition superior to another. Plants are assumed to be
able to adapt remarkably to different chemical environments; otherwise
they would not be growing in all kinds of soils in nature. This opinion
is somewhat confusing, since many of the other nutrient solutions
carry relatively less nitrogen than the Hoagland's, as shown in Table
9.1 for the Knop's nutrient solution, and yet N_2 is an essential nutrient

element for plant growth.

Today, it is suggested to use different solution compositions depending on the stage of growth and also on plant species. A so-called *starter solution*, composed of a half-strength Hoagland solution, is sometimes suggested in the United States for the growth of young plants. This is followed by a *vegetative* (or *pre-anthesis*) and a *fruit* or *seed* (or *post-anthesis*) *refill* at the more advanced stages of plant growth. Different plant species also have different requirements especially in the micronutrients. Too little can inhibit the growth of plants, whereas too much of a specific micronutrient can become toxic. Therefore, nutrient solutions for sensitive plants should contain only very small amounts, not exceeding the *threshold* or *critical value* of that particular micronutrient. Each type of plant species exhibits a specific critical value for micronutrient deficiency or toxicity. According to the experts in micronutrients, the requirement for a micronutrient is generally < 50 mg/L, and sensitive plants exhibit only a small range of tolerance to deviations from the required amount (Ashworth, 1991). Several countries in Europe have established a *maximum acceptable concentration (MAC)* limit and a *maximum application limit (MAL)* of trace elements in soils (Pais and Jones, 1997). Perhaps such a concept can also be worked out for hydroponic operations.

Respiration of plant roots requires the presence of O_2, but since the medium is water, only limited amounts of O_2 are available. As discussed in Chapter 5, the maximum concentration of dissolved oxygen amounts to only 8.3 mg/L, and this concentration decreases with increased temperature. As a rule, respiration also increases with increased temperature, hence the demand for O_2 by plant roots increases rapidly with the rise in temperature. Fortunately, considerable fluctuation in temperature is not too serious in water media. Nevertheless, to maintain a constant supply of this small amount of dissolved O_2, *aeration* is necessary, which is usually attained by bubbling air into the nutrient solution. Algae growth can be a problem, since algae compete for nutrients and tend to enhance eutrophication. Algae can be controlled, though not totally, by keeping the rooting medium in the dark.

The nutrient solutions, as discussed above, have to be constantly monitored for nutrient content and changes in chemical reactions. An imbalance in nutrient composition occurs during the growth of plants due to a differential rate in uptake among the various nutrient elements. Respiration of roots also produces CO_2, which reacts with water to form H_2CO_3, a weak acid that causes the solution pH to decrease. Consequently the nutrient solutions have to be refilled, changed or renewed whenever nutrient deficiencies and pH changes are becoming limiting factors. The return or used solutions can be monitored automatically by computers, and by using computer-controlled injectors, deficient nutrient contents can be adjusted and the solution composition re-balanced. If monitoring of nutrient imbalances cannot be conducted, the nutrient solution must either be purified by other methods or discarded, which is wasting precious amounts of water.

9.1.3 The Issue of Crop Yield

In the opinion of Hoagland and Arnon (1950), water culture is not superior to growing plants in natural soil. The yields are not strikingly different, and growing plants in water cultures does not guarantee the production of food safer for human consumption than food grown in the traditional manner in soil. Regardless of the method used for food production, rigid sanitation measures must be observed to prevent contamination of the products by insecticides and/or other chemical residues and pathogenic infections. Another issue in water culture is plant spacing for maximum yields. Plants can perhaps be more closely spaced together in water cultures than in soils. However, the rules of plant physiology dictate that the density of stand giving the highest yield is determined by the amount of light received by plants, when all other growth factors are optimal.

Several scientists today seem to disagree with the yield potential of crops grown in hydroponics as stated above, and are of the opinion that hydroponic crops outyield those grown in the field. This is completely justified because in hydroponics all the factors affecting growth and yield are completely under control as compared to field conditions. On a per plant basis, crop yields are expected to be higher in hydroponics than in the field. When this yield per plant is extrapolated for the yield per hectare, a common practice in greenhouse experiments, the difference is usually highly amplified, and plants grown in hydroponics are outyielding those in the field by a factor of 2 to 30 times or more. For example, Miller and Gardiner (1998), citing Resh (1995), presented yields of tomatoes grown in soils of only 11,000 - 22,000 kg/ha as compared to the yields of 135,000 - 674,000 kg/ha of tomatoes grown in hydroponics. Comparisons of this kind should be made very carefully, since yield factors for plants growing on a large scale in the field are often completely different from those of plants in hydroponics. Extrapolation of data from single plants is an unreliable method. Because of the process of amplification, conversion of the yields of a few plants to yields of plants per hectare will result in extremely high values. Potatoes were also listed by the authors above as yielding 8.9 times higher or more when grown in hydroponics than in soils. These potatoes were presumably grown in pots containing artificial soil, since it is very difficult to understand how tubers can grow in water.

9.1.4 Significance in Food Production

Hydroponics, used in combination with biotechnology, allows for immense possibilities in intensification of crop production. However, the method is considered more difficult to apply for commercial crop production than the conventional method using soil. Not only does the commercial grower require a thorough knowledge of plant nutrition, but the grower must perfectly design and manage a routine procedure of growing, which involves the diagnosis and testing of plants, rooting media, and nutrient solution (Jones, 1983, 1997). However, hydroponics is perhaps the best alternative for crop production in the desert regions of the world. It may one day become the method for growing food and fiber on the moon (Ming and Henninger, 1989). Currently ongoing research is reported with Utah Space Wheat solutions for use in space flight (Miller and Gardiner, 1998). Jones (1983) also indicated that in cases of space and soil limitations, such as in many urban areas, the individual can apply hydroponics to grow food for his family. However, hydroponics does not qualify to be grouped with LISA. It is an operation that is very expensive in time, energy and effort, and not all crops can be grown using hydroponics. Highly skilled personnel must also be available, as indicated above. All these make the product generally more expensive in the market than that grown in the field. An exception to this is perhaps the cultivation of algae by hydroponics.

Hydroponic Cultivation of Algae. - A less widely known application of hydroponics is the cultivation of algae, which can be carried out indoors or outdoors in ponds (Figure 9.1). Algae are lower plants that contain chlorophyll, enabling them to produce carbohydrates by photosynthesis. Blue green algae are also known to be capable of fixing nitrogen from the air. Therefore, algae are also important as a protein source. Their capacity for protein production far exceeds that of the peanut, pea, or legume plants, and their protein content is believed to be greater than that of beef (Smith and Lewis, 1991). Some types of algae are high in vitamin B12, whereas others may contain ß-carotene.

Algae grow best in the tropics and subtropics, areas characterized by high temperatures and abundant sunlight. Today, large-scale pro-

Figure 9.1 Indoor cultivation of algae. (Courtesy: Didiek H. Goenadi, Biotechnology Research Institute for Estate Crops, Bogor, Indonesia.)

duction of algae in outdoor ponds is found in California and Israel. In northern Israel, liquid municipal sewage is treated with algae to produce water suitable for irrigation. The algae are periodically harvested and pelletized for use as chicken and fish feed (Smith and Lewis, 1991). It was estimated by these authors that algae production used only 2% of the land area required in conventional agriculture to produce an equal amount of organic matter.

In several other countries, algae are important food sources for humans. They are a major component of the diet in Japan, where a variety of seaweed is used for human consumption. Kelp, *Fucus* sp., one of the many forms of seaweed, is also known to have medicinal properties. It is used in Asia as an effective remedy for goiter, because of its high iodine content. Practitioners in alternative medicine believe that kelp is very beneficial against thyroid disorders. *Sodium alginate*, a component of kelp leaves, is said to have the capacity to prevent the human body from absorbing heavy metals and radioactive elements, such as strontium 90. Therefore, algae (kelp and seaweed) farming in the coastal waters of Japan is a booming business today. In addition, algae is a source for the manufacture of agar and gelatin, ingredients for making desert and cake in the United States. Not only is agar used for cake and dietary purposes, but consumption of agar (jel-o) is believed to control brittleness of your nails.

9.2 AEROPONICS

Related to hydroponics is *aeroponics*, which involves growing plants in the air (aero = air). In hydroponics, plants are grown in water to which nutrient mixtures have been added. On the other hand, in aeroponics plants are grown in the air, and dilute nutrient solutions are sprayed onto the hanging roots (Figure 9.2). The idea with aeroponics is to reduce the amount of water and nutrients used in growing plants to a bare minimum for maximum yields. Similar nutrient solutions as in hydroponics are generally employed in aeroponics. The nutrient solution can be sprayed or can be applied as a fine mist. Jones (1997) believes that continuous exposure of the roots

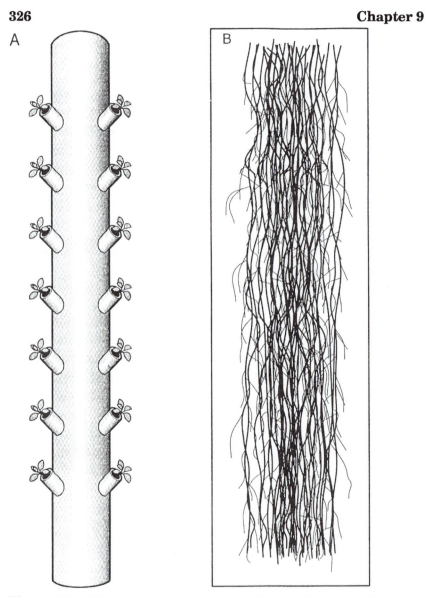

Figure 9.2   (A) Aeroponics, or the culture of plants in the air. The plants are allowed to grow in hollow plastic tubes. (B) Roots of plants, grown in aeroponic cultures, hanging vertically in the tubes. They are sprayed periodically with nutrient solutions. (Drawings by W.G. Reeves, Art Coordinator, University of Georgia.)

to a fine mist is superior to intermittent spraying or misting. Excess nutrient solutions dripping down from the hanging roots are collected and recycled. The chances for allopathy are small since the solutions are sprayed again onto the same roots. Monitoring of nutrient imbalances and changes in chemistry is, however, necessary. Aeration seems not to be a big problem, but the need for keeping the roots in the dark is more apparent in aeroponic than in hydroponic cultures. Another problem is the need of support for plants, which is more serious in aeroponics than in hydroponics. These conditions are perhaps the reasons why only a limited variety of plants can be grown. Aeroponics are suitable only for growing certain plants. Leaf vegetables, such as spinach and lettuce, are adaptable and will lend themselves to cultivation by aeroponics. Yields of the crops have been reported similar to or even above those that have been grown by conventional methods in soils. However, aeroponics is, no doubt, not a system within the category of LISA, and the operational expenditures appear to be far in excess of those of ordinary hydroponics.

9.3 NUTRIENT FILM TECHNIQUE

This technique, also called NFT, was introduced by Cooper (1976) in England. The relationship with aeroponics is that it also has as a purpose the efficient use of water, or in other words preventing waste of water. With this method, the plants are grown in a narrow tube or trough through which a small amount of a solution is passed continuously so that the roots are coated only with a very thin layer of solution, hence the name *film*. Usually, the trough is arranged on a slope so that the nutrient solution can flow from the top to the lower parts by gravity. However, serious problems in nutrient flow and aeration often arise with plants that develop large root masses, resulting in root death. As indicated by Jones (1997), the nutrient and oxygen flow downwards will be intercepted at the upper end of the slope by the root system, which grows considerably in size. The author above suggests using plants that will develop less dense root systems.

9.4 ARTIFICIAL SOIL

Many authors consider cultivation of crops in pots containing soil constituents, such as sand, manure, peat, bark, sawdust, straw, perlite, vermiculite and other artificially prepared rocks, to also be soilless or hydroponic cultivation. However, since the rooting medium is composed of a mixture of soil constituents, the growth medium is more correctly called *artificial soil* than a *soilless* medium. It is composed of solid soil constituents or soil material, and not water (hydro), selected for the purpose of providing mechanical support to the growing plants (Ellis and Swaney, 1947). In hydroponics, support for plants has to be built in the form of a solid frame above the water or the plants have to be attached to a floating frame. The solid growth media, or artificial soil, is similar to a natural soil in its mechanical support. However, the main reason for creating an artificial soil is its versatility and adaptability for specific requirements of plants. An artificial soil is a soil made artificially so that its properties can be manipulated to suit the growth of specific plants for maximum performance.

Many types of mixtures are possible and they can perhaps be distinguished into *inorganic* and *organic mixtures*. The inorganic mixtures are composed of mixtures of sand, gravel, and/or vermiculite. This type of mixture is operated in a similar way as hydroponics, with the difference that nutrient solutions are employed for growing the plants to field capacity only. In this way, aeration, by bubbling air in the medium, is not necessary. The sandy medium contains a considerable amount of macropores and acts similarly as a sandy soil. The organic mixtures are mixtures of sands, gravel, vermiculite, and organic matter. A natural soil in optimal condition for plant growth is made up of approximately 45% mineral matter, 5% organic matter, 25% water, and 25% air. The ingredients sand, gravel, vermiculite, and organic matter are then mixed together to approach this natural composition. A basic composition of a potting soil contains sand, peat, and topsoil in a ratio of 1:1:1 (Foth, 1990; Brady, 1984). To make an organic mixture the proportion of topsoil can be replaced by peat so that the ratio becomes sand : peat = 1 : 2. Sand can be replaced by a mixture of sand and gravel, whereas peat can be replaced by other types of organic matter, e.g., pine bark or coconut husk. Perlite can also be used instead of sand, and the variations are endless. Jones (1997) considers the *Cornell Peat-lite mixes* and the *California basic mix* as the two basics for preparing other types of organic mixes. These artificial soil mixtures are handled differently than the hydroponics. They are not supplied with nutrient solutions, but the nutrients are provided in the form of fertilizers. For example, the Cornell Peat-lite mix is a mixture of sphagnum peat moss and horticulture vermiculite No. 2 to which is added ground limestone, superphosphate, and a 5-10-5 fertilizer. By adding, withholding, or replacing specific ingredients, the rate of growth and yield of plants can be easily manipulated. The behavior of such a potting soil is usually quite different from that of soils in the field. For example, the concept of field capacity in natural soils does not hold for potting soil. The water content in a potting soil for optimum plant growth is considered too wet for the soil in the field. Often, the bottom part is in a more reduced state than the surface of the potting soil where oxidation condition prevails. Though hydroponic experts consider a potting soil medium similar to a hydroponic medium, it is in fact two completely different systems, in composition as well as in operation. One contains artificial soil, whereas the other contains water. One is supplied with solid fertilizers and limestone, but the other is supplied with nutrient solutions. Nevertheless, some of the problems in hydroponics also surface in artificial soils as a result of growing plants, such as accumulation of soluble salt.

9.4.1 Orchid Medium

The cultivation of grown orchids is perhaps the oldest example of using an artificial soil. Germination and seedling growth of orchids are, however, conducted on solid (agar) nutrient media, a procedure that is in the area of tissue cultures that will be discussed in Chapter 10, on Biotechnology in Soil Science and Agriculture. At a certain age, the orchids seedlings are then transplanted into pots containing artificial soil, composed of organic mixes. The best time to transplant the seedlings into pots is when a healthy root system has developed. These seedlings are not transplanted and allowed to grow in hydroponics, because this will be very harmful for the development of mycorrhizae in the roots, essential for the symbiotic relationship of the orchid plants. This is another reason why the name *artificial soil*, or just potting soil is used in this book, instead of using the terms hydroponics or soilless medium as is the case with other authors.

Orchid growing is more an art than a science, and it is big business. The orchids are available in all kinds of varieties, e.g., cattleyas, dendrobiums, cymbidiums, vandas, and phalaenopsis. Each of them requires a specific medium for root growth, though they will also grow in a standard organic mix provided for the amateur grower by commercial orchids farms. An example of a standard mix for cattleyas and dendrobiums is 3 parts of osmunda, 1 part of sphagnum moss, and 1 part of artificial pumice or volcanic gravel. As discussed earlier, replacements and adjustments of composition are always possible, which provide for an endless variation in potting mixtures. In Thailand, coconut husk, replacing osmunda and sphagnum, is used and appears to serve the purpose well. Chopped fern bark is also frequently used in the mixture. Phalaenopsis is often grown on just a piece of black fern bark or sawed-off fern trunk. Osmunda fiber is considered the best material for making the mixes. However, any other vegetable fiber can be used, provided it allows for good drainage and has the proper acidity (Sander, 1969). Sphagnum moss is an essential constituent to ensure a certain degree of acidity. Sometimes small pieces of charcoal are added to the mixture. For transplanting large plants, broken pieces of brick, the size of coarse gravel, are often used. The use of fine dust particles should be avoided because these will impede rapid drainage, which is an important factor for the growth of the orchids.

9.5 AQUACULTURE

Aquaculture is by definition the culture and husbandry of aquatic organisms. It encompasses the control and management of aquatic plants and animals in controlled or selected environments for economic or social benefits (Bell and Canterbery, 1976). In a narrow sense it is *fish farming*, which is considered more of an art than a science. Fish farming has been practiced since ancient times. The cultivation of fish in ponds today originates from the practice of prehistoric people living in coastal regions keeping live fish in baskets submerged in the river or sea (Tiddens, 1990). Fish farming can be traced back to the Egyptians in 2000 B.C., who grew *tilapia* (*Tilapia nilotica*), and to the ancient Chinese people, who were known for their carp (*Cyprinus carpo*) culture. From these ancient methods, aquaculture has developed into its present modern science in the same way as the development of agriculture. Some consider it a type of agriculture, with the main difference being that aquaculture is aquatic, whereas agriculture is terrestrial. However, its progress to the present level has been much slower than that of agricultural science. In their quest for food, human beings have directed their attention to the cultivation of soils. The oceans and lakes were until recently a mysterious and impenetrable domain.

In the early days, aquaculture was practiced for the simple production of a cheap source of animal protein. It was conducted at first by growing fish, which were hardy and easy to grow and popular as a food source, such as carp, catfish, tilapia, milkfish, and mullets (Bell and Canterbery, 1976). This has expanded in several countries, such as in Japan, into a modern operation to include growing aquatic organisms which command high prices in the market. The "inexpensive" fish is grown and processed into feed for cultivating the higher priced fish and shellfish (Kafuku and Ikenoue, 1983). Aquaculture is a highly specialized operation in Japan, and can be distinguished into *freshwater aquaculture* and *marine aquaculture*.

9.5.1 Freshwater Aquaculture

Freshwater aquaculture in Japan involves the cultivation of highly priced freshwater fish, such as chum salmon (*Oncorhynchus keta*), ayu (*Plecoglossus altivelis*), eel (*Anguilla japonica*), and several species of carp. It has been expanded recently into growing terrapin (*Tryonix sinensis*), a softshell turtle considered an expensive delicacy, and freshwater oyster (*Hyriopsis schlegelii*). This type of oyster is grown not for food but for the pearls, which are of high quality, and have been compared to the once world-famous, now extinct, Persian Gulf natural pearl (Kafuku and Ikenoue, 1983).

9.5.2 Marine Aquaculture

In Japan, marine aquaculture includes growing valuable sea items, such as green turtle (*Chelonia mydas*), nori (*Porphyra yezoensis*), and wakame (*Undaria pinnatifida*). Kuruma prawn (*Penaeus japonicus*), scallop (*Patinopecten yessoenensis*), oyster (*Crassostrea gigas*), and abalone (*Haliotis (Nordotis) discus*) are also favored sea crops, because of the high prices they bring in the marketplace. Nori and wakame are seaweed or algae, and are important foodstuffs in Japan. Wakame is used as an ingredient in soybean paste (miso) soup, whereas nori sheets are employed for making sushi and rice balls. Numerous kelp and seaweed farms are thriving at the coastal regions in Japan. In Southeast Asia, seaweed is an important source for agar production. Some seaweeds are claimed to have medicinal properties.

Japan has also intensified its cultivation of pearl oyster (*Pinctada fucata*) in the sea; since 1919 cultured pearls have been acclaimed in the London pearl market (Kafuku and Ikenoue, 1983). A high tech center for the cultured pearl is located at Mikimoto Pearl Island, where oysters are surgically implanted with a small pellet of ground oyster shells that serve as seed for the growth of pearl. The implanted oysters are returned to the bay to grow. Mikimoto pearls are famous for their beauty, and Mikimoto and similar companies have given Japan a monopoly in today's world pearl market. Amazingly, the oyster shells that are ground to serve as seed are imported oyster shells from the Mississippi River delta, Louisiana.

9.6 AQUACULTURE IN THE UNITED STATES

9.6.1 Salmon Ranching

Aquaculture has not flourished in the United States as it has in

Japan. In contrast to Japanese people, Americans prefer to consume beef and chicken rather than fish. At present, fishery science in the United States has been largely limited to managing and improving the population of wild fish. Fish hatcheries have been established to produce young fish for restocking streams and lakes, such as trout. Perhaps the closest concept in the United States to modern aquaculture is an operation called *salmon ranching*. It is usually conducted by large corporations, such as the Weyerhaeuser Timber Company and British Petroleum Anadromous, Inc. (Tiddens, 1990). Salmon are born in fresh water, but spend most of their adult life in the oceans, hence the salmon is an *anadromous* fish. They may return to their original breeding grounds to spawn and die. In the salmon ranching operation, wild salmon are caught during the annual salmon run, and their eggs are milked and hatched artificially. The young fish are reared in artificial ponds to be released later in the rivers. It is hoped that this process of releasing young fish, called *imprinting* to the surroundings, will encourage the adult fish to return in the future to the same rivers and breeding grounds to be harvested.

A similar operation, but on a smaller scale, is trout farming. The trout eggs are also hatched artificially, and the young fish are released to restock mountain streams and rivers for sport fishing. Most of the hatcheries are government owned. The few commercially owned trout farms are producing trout for the food market.

9.6.2 Channel Catfish Farming

Another development towards modern aquaculture in the United States is the cultivation of freshwater fish in ponds. This practice started in 1927, when scientists from the Agricultural Experiment Station, University of Alabama, under the leadership of H.S. Swingle, raised bluegills, bass, and other freshwater fish in ponds (Shell, 1975). At first, the fish were grown for sport fishing, but the experiment was later expanded for food production. Swingle and his group demonstrated that fertilizing ponds with N, P, and K increased the growth of algae and other aquatic organisms, which are major food sources for many types of fish. An optimum NPK:fish ratio was developed for optimum fish production. Later, Swingle added channel catfish to his experiments, and his report in 1959 became the basis for much of the US channel catfish industry today. What Swingle was doing, was in fact taking advantage of an eutrophication process, called earlier *cultural eutrophication*. This is one example that a degradation process is turned around to becoming useful.

Finally worth mentioning is the intensive shrimp farming in ponds conducted by the Marine Culture Enterprises in Hawaii. Shrimp are also grown artificially in Panama by the Purina Company, as a joint venture between Panamanians and Americans. Shrimp and catfish are favored delicacies, hence are in high demand in the food market of especially the southern United States, famous for its fish fries.

9.7   THE LAW OF THE SEA

Several factors have contributed to the current heightened interest in aquaculture. Over-fishing of oceans and lakes by huge fishing vessels, using improved netting techniques that act literally as gigantic vacuum cleaners, has markedly decreased the worldwide catch of wild fish. Salmon, herring, and halibut populations are decreasing at an alarming rate. Discharge of pollutants from coastal towns, maritime shipping, and offshore oil production is compounding the problem. Not only do these contaminants threaten the safety of human consumption of fish, they also play an important role in the decline of reproduction of several species of fish, such as salmon. As one response to these concerns, the United Nations Conference in 1987 formulated the *Law of the Sea*. Many nations worldwide sought to protect their fishing rights by claiming a 200-mile zone extending seaward from the coast of each nation as their exclusive property. Consequently, such a law means the end of the freedom of fishing in the high seas (Tiddens, 1990).

CHAPTER 10

BIOTECHNOLOGY IN SOIL SCIENCE AND AGRICULTURE

The advancements in *biotechnology* have created new frontiers, and new vistas of opportunity for increasing and improving food and fiber production. Biotechnology involves the application of organisms or their subcellular components to create new substances or organisms (Smith and Lewis, 1991). It can be distinguished into *food biotechnology* and *agricultural biotechnology*. In the author's opinion, there is also room for the division *soil biotechnology*, a science which has yet to be developed. The manufacture of soil organic compounds, e.g., humic acids, and the application of soil-borne organisms in the control of pests and diseases fall in this category. Today, attempts have been made to use soil organisms in the synthesis of new minerals and fertilizers, and in the cleanup of contaminated or polluted soils. In many cases soil biotechnology is closely related to agricultural biotechnology, making a sharp distinction between the two sciences almost impossible.

Another equally important branch is biotechnology in medical science. This branch is mostly involved with genetic engineering and has recently developed engineered microorganisms capable of producing insulin, a hormone essential for the metabolism of carbohydrates. However, this topic is beyond the scope of the book.

10.1 FOOD BIOTECHNOLOGY

Food biotechnology involves techniques for the production and processing of food. Modern food biotechnology finds its origin in the fermentation of food and beverages, techniques practiced since ancient times. Beer was produced by Sumerians and Babylonians as far back as 6000 BC, whereas bread making was already known by Egyptians in 4000 BC. Cheese production was also known in ancient times. Although these ancient people can be considered pioneers in biotechnology, it was Louis Pasteur (1857–1876) who is commonly regarded as the founder of modern biotechnology.

Today the techniques in biotechnology are applied to improve the fermentation of beverages and the production of a variety of food products from cereal and soybean, such as bread, soy sauce, tofu, and tempeh. Soy sauce and tofu were introduced in Europe and the United States from China via Japan. Because of their high protein, low fat, and low cholesterol content, they are important additions to vegetarian diets. In Japan, sophisticated biotechnological operations are used to make these products. Less well known in the United States is *tempeh*, developed centuries ago in Indonesia from soybean. Tempeh is produced by fermentation of soaked soybeans with a fungus, *Rhizopus oligosporus* (Smith and Lewis, 1991). The Netherlands, which has close historical ties with Indonesia, is an active center for tempeh production, using advanced biotechnological methods.

In the United States, tempeh enjoys increasing popularity as a healthy vegetarian food with meat-like texture. Today, both tempeh and tofu are offered in several varieties in many supermarkets. The tempeh, produced only from soybean, called the single-grain tempeh, is a traditional Indonesian version of *soyfood*. Prepared as spicy fried chips, it is a delicious item often garnishing a usually elaborate Indonesian *rijst tafel*. Several kinds of tempeh cultured from a mixture of grains are now available in the United States. Soybean, sometimes called nature's most perfect food, is mixed and cultured with brown rice, millet, and/or barley to yield two-, three-, or four-grain tempeh. To make the product more attractive all the grains can be grown organically. Tofu, on the other hand, is made only from soybeans but can be found in different textures. It is offered in a firm, soft, or silken texture. Both tempeh and tofu should not be consumed prior to cooking, because eaten raw they are tasteless. They can be sauteed, stirfried, or fried as chips to be mixed with a variety of main courses from salads to soups and stirfries. Stuffed with ground shrimp, pork, turkey, chicken or beef, stuffed tofu served steamed or fried is a favored delicacy for lunch or dinner in restaurants from Hong Kong, to Bangkok, Singapore, and Jakarta.

10.2 AGRICULTURAL BIOTECHNOLOGY

10.2.1 Crop Production

Agricultural biotechnology is the science of creating new strains of plants and animals and improving agricultural products and chemicals by *genetic engineering*. The discovery and advent of new recombinant DNA technology open new possibilities for the development of superior crops and animals. The green revolution, started by George Harrar and his team, is hereby propelled into new areas and dimensions. Breeding programs can be accelerated. The application of genetic engineering to plants can confer novel properties in the recipient crops. Crop improvements which were previously not possible in a traditional way are now made possible by the application of recombinant DNA techniques. The shelf life of food and vegetables has been lengthened considerably by the use of biotechnological methods, and several plants have been made more resistant to pests and diseases. Genetically engineered tomatoes have recently reached the US markets, causing public and media uproar. The general belief of consumer advocates is that genetically engineered crops are a threat to consumers and the environment. Suspicions about genetic side effects and contamination in humans as a result of consumption of genetically engineered products almost borders on superstition. This issue was escalated when European markets blocked the import of beef tainted by hormones, which was followed by barring the import of genetically altered crops grown in the United States. According to FDA experts, there is no scientific evidence that genetically engineered food crops are harmful to consumers. Nevertheless, public opinion is the reason why the transgenic tomatoes disappeared from the supermarket shelves as quickly as they appeared. More recently organic tortilla chips, made of genetically engineered corn, have had to be recalled. This was not the end, since lately baby food, containing genetically engineered corn, soybean, and a variety of fruit, was recalled by the Gerber company. Apparently, the FDA public relations information or extension services lack the data or the efficiency that might convince the general consumer of the safety and sanitation of genetically engineered products. Public's apprehension is apparently not limited to genetically altered food products, but also in growing plants that have been doctored by genetic engineering. Recently, plants have been developed that are capable of producing the Bt toxin giving them protection against the caterpillar of the diamond back moth (see page 99). However, their release for widespread cultivation has triggered fears for development of insect resistance to the Bt-toxin. The news media as well as several scientists have scared potential growers even more by spreading reports that in laboratory tests, pollen from Bt-transgenic corn plants was deadly to non-target organisms. Rebellion against growing genetically altered plants seems to increase with the development of crops producing sterile seed. To the companies this is big business, but to the growers this is a potential threat since they are compelled to buy seed every year.

10.2.2 Environmentally Friendly Pesticides

As has been known for years, plants can produce chemical compounds of importance for pesticides as well as for pharmaceutical purposes. *Pyrethrum* is one example of an insecticide produced by chrysanthemums. The production of such a chemical compound is the plant's natural defense mechanism to combat attack by insects and microorganisms. Another example is the oily aromatic substance in citrus skin which has a toxic effect on many insects. Other plants can produce a proteinase inhibitor to attack an invading pathogen. Such a chemical is *trypsin* inhibitor, which is carried by cowpea seed (Smith and Lewis, 1991). When trypsin inhibitor is introduced into tobacco plants, it appears to increase the resistance of these plants to infestation by the tobacco budworm, *Heliothis virescens*. Such bioinsecticides and other engineered microorganisms are being developed currently, with promising results for the biological control of pests and diseases. A more novel example is the development of plants capable of producing the Bt toxin, nicknamed *Bt plants,* which has created a lot of concern as discussed above.

From the above, it is clear that the extraction or production of allelophatical substances as discussed earlier falls in this category of biotechnology. Many of the plants, especially weeds, are now known to exhibit some type of allelophaty. Attempts are underway to use their residue for a so-called *selective pest control*. An example has been given earlier of the commercial production of solid sticks by compressing plant residue containing pyrethrum. When burned as incense, the smoke from the sticks kills mosquitos. Today, these sticks or coils are popular as cheap but efficient mosquito repellents in Southeast Asia, and seem making headway in US markets. However, when the allelochemicals are extracted and identified, they can perhaps be synthetically produced and used more efficiently as an environmentally friendly pesticide. The potential is enormous, since a whole array of allelochemicals, from acetic acid in wheat straw to alkaloids, flavanoids, quinones, and cyanogenic compounds in sorghum, peach, and Johnson grass, are waiting to be tapped for this purpose. Some allelophatic chemicals have to be yet identified, such as the toxin in the leachate of ryegrass that is reported to inhibit the growth of many woody ornamental plants (Miller and Gardiner, 1998).

10.3 TISSUE CULTURE

Another important aspect of agricultural biotechnology is *plant tissue culture*. Although it has not received as much publicity as genetic engineering, tissue culture has, nonetheless, enjoyed practical applications in agriculture. Long before the term biotechnology came into existence, tissue culture was used extensively in the propagation of orchids. This method of growing orchids, called in the past *asymbiotic orchid culture*, was used at first only for the production of orchid plants from seed. This technique accelerates the time for germination and growth of seedlings by at least 10 to 20 times that of plants produced in the conventional way, sowing seeds on well-rooted plants. However, some growers believe that stronger growing plants are usually produced in the conventional way than those from the same seed pod raised asymbiotically in the laboratory.

10.3.1 Nutrient Solutions

The asymbiotic propagation of orchids from seed succeeded after Lewis Knudson in 1918 made the discovery that orchid seedlings would grow only in a nutrient medium containing an organic compound in the form of sucrose (Sander, 1969). His nutrient formula, known as *Knudson's formula*, has enjoyed great popularity and was for some time the most used formula for the asymbiotic culture of orchids. As is the case in hydroponics, a great number of nutrient solutions have since been formulated by different authors. Sander stated that at least no less than 35 formulas have been presented, and all of them appear to be equally sufficient for the asymbiotic tissue culture of orchids. The standard Knudson's formula, which is still the most widely followed used formula, is composed as follows:

KH_2PO_4	0.25 g
$CaNO_3$	1.00
$(NH_4)_2SO_4$	0.50
$MgSO_4$	0.25
$FePO_4$	0.05
Sucrose	20.00
Agar	17.50
Distilled water	1000 mL

The pH of this nutrient mixture is adjusted to 5.0 before autoclaving. Today a variety of organic additives are used in the nutrient media of orchids.

10.3.2 Meristem Culture of Orchids

The technique developed for asymbiotic germination of orchid seeds has also been expanded to grow orchids from meristematic tissues or orchid buds. Such a technique allows for the preservation of superior plants by the production of genetically identical plants. It also has an immense potential for mass propagation of plants.

All plants contain within their buds the embryonic tissues that can give rise to similar cells or cells that differentiate to produce the definitive organs. This is what we call the meristematic tissue or the *meristem*. The bud is cut and peeled and placed aseptically into a nutrient solution sealed in an Erlenmeyer flask. The flask is then shaken in a mechanical shaker to keep the embryonic nuclei agitated in the liquid. Within days tiny seedling-like growths develop that are dissected into several sections, each of which is separately placed again in nutrient solutions and shaken again. The procedure can be repeated as many times as necessary until the desired number of seedlings has been obtained. This is then the method for the rapid production of a large number of plants with similar genetic properties, called *clones*. It can find application with other plants, where propagation through germination of seed is sometimes very difficult, such as with a number of forest trees in the tropics. To reforest a tropical region with, for example, ironwood trees in Southeast Asia is an almost unsurmountable task, since ironwood seeds are very difficult to germinate. Meristem cultures of such plants open the possibilities for producing rapidly the amount of planting stock needed for reforestation on a large scale. This issue will be discussed in more detail in the next section.

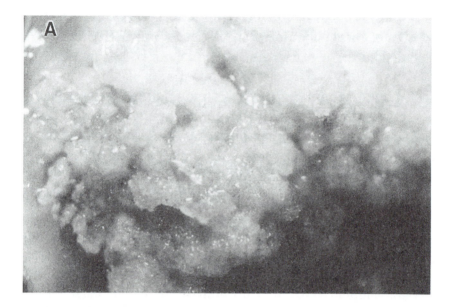

Figure 10.1 (A) Callus culture of slash pine (*Pinus elliottii* Engelm) seen with the light microscope. (B) Scanning electron micrograph of individual callus cells. Magnification: white bar = 0.1 mm (Irianto, 1991; Irianto and Tan, 1993).

10.4 ENVIRONMENTAL SIGNIFICANCE OF TISSUE CULTURES

The term tissue culture is considered a misnomer, because when it was introduced, it was restricted mainly to the culture of pieces of plant tissue *in vitro* (Bonga and Durzan, 1982). As indicated above, currently meristematic cells, parenchyma cells, embryos, callus, and/or other plant organs can be used in tissue culture (Kyte, 1983). Hence two groups of tissue culture methods can be distinguished: (1) methods using organized cell masses, such as meristems and embryos, which is frequently referred to as *micropropagation*, and (2) methods using unorganized cell masses, such as callus (see Figure 10.1).

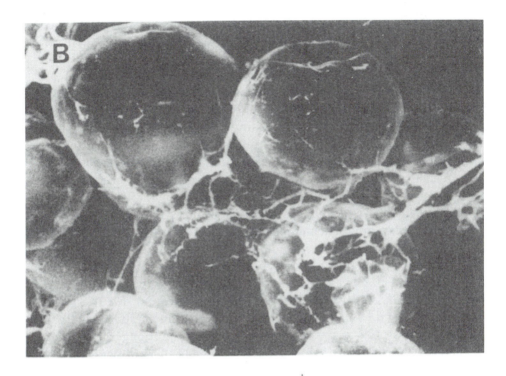

.1 mm

The first attempt to culture plant cells was conducted by Gottlieb Haberlandt. Although his experiment in growing aseptically palisade cells isolated from *Lamium purpureum* and *Pulmonaria mollisima*, was unsuccessful, Haberlandt is regarded today as the *founder* of plant tissue culture. The application of tissue cultures became successful only after the right formula and additives, including growth hormones, were discovered. In the early attempts, it was noticed that many plants cultivated *in vitro* behaved *heterotrophically* rather than *autotrophically*. Plants grown in cultures were unable to manufacture carbohydrates on their own from inorganic nutrients only. The usefulness of cold filtered or autoclaved organic supplements, such as sugar, malt extract, yeast, fruit juices, coconut milk, thiamine, and growth hormones has been examined with mixed results. After numerous attempts, the first successful growth medium was formulated by Murashige and Skoog (1962). This medium, called the *MS medium* after the names of the developers, was for some time a popular medium in tissue cultures. It is used in its original or modified form in the aseptic propagation of plant tissue, and a variety of culture media is now available.

Micropropagation enables the multiplication of clones and hence the production of plants of high uniformity and quality. On the other hand, tissue culture using callus or unorganized cell masses is often done to create variations in the regenerated plants, a phenomenon known as *somaclonal* variation. The regenerated plants differ from the parent plants. In general, tissue culture is practical because it permits mass production of plants. Large numbers of planting stock can be produced in a very small area at a reasonable cost within a relatively short time. Moreover, this technique is applicable to ornamental plants, horticultural and agronomic crops, as well as commercially important fruit and forest crops (Bajaj, 1989). Tissue culture may provide the answer to the cultivation of plants that are very difficult to propagate by conventional or traditional means. For example, many forest trees valuable for timber in the humid tropics, such as teak (*Tectona grandis L.*), ironwood (*Mesua ferrea* L.), and mahogany (*Swietenia mahogany* Jacq.), grow extremely slowly, and are very difficult to propagate by conventional vegetative methods or through seeds (Irianto, 1991). Tissue culture opens new possibilities for a rapid propagation of these trees. This method also enables the mass production of new tree stock for a rapid reestablishment of the tropical rain forest that is disappearing at an alarming rate due to increased demand for food, timber, pulpwood, fuel, fodder, and other forest products, as discussed previously.

10.5 SOIL BIOTECHNOLOGY

Soil biotechnology as a science is still in its infancy, but is rapidly growing in importance. It is already visible on the horizon and data are starting to accumulate. Soil biotechnology is here, and is expected to gain momentum in the foreseeable future as was the case with the science of humic acids. In some aspects, soil biotechnology overlaps with agricultural biotechnology, and a sharp separation is often not possible. For example, the extraction and production of allelochemicals are as much the topic of soil biotechnology as that of agricultural biotechnology. The most visible advances in soil biotechnology today are the production of humic acids, fertilizers, and pharmaceutical chemicals. Cultures of soil-borne organisms capable of destroying a variety of organic pollutants are also important aspects of soil biotechnology. Genetic engineering of soil organisms to fulfil such a function may well be underway.

10.5.1 Biofertilizers

The production of biofertilizers has attracted increasing attention in Europe and Southeast Asia. On the other hand, many of our US scientists have disregarded this important topic. The concept of biofertilizers stems from the advancing science in humic acids indicating that efficiency of most fertilizers and pesticides seems to increase in the presence of humic acids. Regardless of the disbelief or skepticism shown by many US scientists, the fact remains that fertilizers mixed with manure, as commonly practiced in the old days, are more effective than the use of fertilizers alone. In Europe, NPK fertilizers, mixed with humic acids to form coatings around the fertilizer's grains or pellets, are claimed to increase substantially their activity and nutrient efficiency. Even simple mixing of the fertilizers with humic acids is reported to enhance their chemical reactivity (Savoini, 1986). The significance of adding humic acids is especially of importance in the use of liquid fertilizers, or better fertilizers that have to be dispensed in suspension form. The fertilizer particles have to be kept in suspension by intermittent agitation and/or by adding a dispersion agent, such as attapulgite or bentonite. Humic acid appears now to be a better reagent than bentonite. The larger surface area and higher electric charges offered by humic acids make them more suitable in their functions as protective colloids, while in addition smaller amounts of humic acids are required than bentonite or attapulgite to enhance dispersion.

Advantage has also been taken of the beneficial effect of the chelating capacity of humic acids to produce humophosphates and a variety of humic acid micronutrient fertilizers, such as humoferrates. Unlike other chelates, such as EDTA and the like, with their high solubility, metal humates are relatively less soluble, but exhibit a gradual solubility and a higher resistance to degradation by light and microorganisms. Consequently, the chelates act as a slow release fertilizer, a factor of considerable advantage in many soils. Commercially produced iron humates in microgranule form or in capsules are now available in many European countries. The microgranule iron humate is compatible for mixing with other fertilizers, including phosphate fertilizers.

In Southeast Asia, attempts are made to produce superphospate from rock phosphate by treatments with soil microorganisms. In the conventional method sulfuric or phosphoric acids are added to rock phosphate to solubilize the phosphate. Since microbial secretions are known to exhibit a similar or stronger solubilizing effect on insoluble soil phosphates, the idea is to replace the use of inorganic acids, and mix microorganisms with rock phosphate for achieving the same degree of solubility as in superphosphate. The microorganisms, tested by several of the state universities in Indonesia, and by the Biotechnology Research Institute for Estate Crops, Bogor, Indonesia, are *Bradyrhizobium japonicum*, and *Aeromonas punctata*. Added as mixtures and incubated with rock phosphates, the resulting *biofertilizer* is claimed to be effective, especially, for use in marginal soils. The addition of Bradyrhizobium is believed to provide an additional advantage in the infertile soils because it increases nodulation in soybean plants.

Though a similar concept can be applied to pesticides, efforts in producing and testing *biopesticides* are still on the drawing board. Conjugated herbicide-humic acid compounds are produced commercially and are still being tested by Eniricerche, Monterotondo, Italy. Diuron-humic acid and linuron-humic acid complexes are reported to exhibit higher toxicity at lower concentrations than the herbicides alone (Genevini et al., 1992).

10.5.2 Biomedicines

The antiseptic properties of peat have been known for quite some time in Europe and Asia. Mud baths were common in the old days to control a variety of diseases, especially skin diseases. Even today a great number of European health spas are offering mud baths, such as in Lourdes, France. Face masks of muds are being promoted by major cosmetic companies in the United States. The American Red Cross applied in World War I a great number of peat moss pads as surgical dressings. The *bacteriostatic* function of peat was later traced to the humic fraction in peat (Gorniok et al., 1972).

Today, humic acids and related substances have found increasing application in the production of pharmaceutical chemicals. Especially in Eastern Europe, research to test the usefulness of humic acids is receiving considerable attention. In their overview MacCarthy and Rice (1994) reported that fulvic acid was found to exhibit *hemostatic* properties, and appeared effective in controlling ulcers in rats. The anticoagulant effect of humic acids was supported by tests at the Institute of Pharmacology and Toxicology of the Medical School at Erfurt, Germany (Kloecking, 1994). Used in small concentrations, NH_4-humate is also noted in cell cultures to inhibit infection by the herpes simplex virus. In Chapter 3, the beneficial effect of torfot, a chemical prepared from peat by the Ukranian Research Institute of Eye Disease at Odessa, was discussed as a topical treatment against myopia and opacification of the eye. Perhaps, one day a torfot-like chemical can be produced, useful in preventing or controlling cataracts, a form of opacification of the eye lens of elderly people. Possibilities for the use of humic acids in the rapid healing of wounds caused by cuts, crushing, and burning are being explored. Positive results have been reported on these aspects in Hungary by Jurscik (1994).

10.5.3 Microbial Genetic Engineering

This is a very wide area and covers not only medical science, but also food science, agricultural science, and soil science. For example in food science, superior organisms can be developed for a better fermentation process in bread, beer, tempeh production and the like. In crop science, genetic engineering of tomatoes to create new better tomatoes was discussed earlier as an example. Another example is tossed out here for contemplation is the possible use of genetic engineering to produce new, more efficient rhizobium strains, or a mycorrhiza capable of fixing nitrogen. The possibilities are endless and are limited only by our lack of imagination.

One aspect not focused on yet is culturing or creating organisms with better or substantially greater efficiencies in combatting organic pollutants in soils. Microorganisms already present in soils are capable of degrading compounds with a composition related to their food sources. When an unrelated substance enters the soil, often nothing seems to happen, and because of this many assume the pollutant to be nonbiodegradable. However, this period of apparent nonactivity is often caused because the organisms must first "get acquainted" with the foreign material. They have to multiply and produce sufficient amounts of enzymes before they can start the degradation process. This period of inactivity is called the *lag period*. This lag period may vary from a few hours to a few days. Other times, the organisms have to be seeded into the soil. Since they are entering a wholly new environment, often hostile to them, a similar lag period will be noted for the introduced microorganisms before they become active. The chances that they may not succeed and die are also possible. Many microbes are also very specific in their food source, and genetic engineering involving gene transfer can create a new species that is capable of attacking a variety of pollutants, called the *superbug* by Miller and Gardiner (1998). These authors note that multiple gene transfer in *Pseudomonas* sp. enables it to attack several hydrocarbons, such as octane and naphthalene.

CHAPTER 11

SOIL AND POLLUTION

11.1 DEFINITION AND CONCEPT OF WASTE

Waste is defined as useless, unwanted, or discarded material (U.S. Environmental Protection Agency, 1972). It can be distinguished into solid and liquid waste. Liquid waste is often referred to as *effluent*.

Agricultural, industrial, and other operations in our modern society produce large amounts of waste. The soil is traditionally the site for disposal of all these wastes, and people have been discarding waste since prehistoric times. However, there was little concern in the old days about pollution because the human population was not that large and there was plenty of space on earth for the amount of waste produced. However, with population growth and the revolutions in industry and agriculture, huge amounts of waste and a variety of new types of pollutants have been produced. For example, beef cattle in the United States were estimated to produce 92 million and dairy cattle 27 million metric tons/yr of manure (Donahue et al., 1983). Some of these types of wastes are difficult to deal with because of their diffuse or *nonpoint* nature, such as manure deposited at random by livestock, which may be washed into nearby streams (Troeh et al., 1980). The introduction of waste into the soil may result in *soil contamination* or *pollution*. Many people do not make a distinction between contamination and pollution, and use the terms interchangeably. However, some believe that these are two distinct phenomena, though related to each other (Wild, 1993). To them, soil contamination is the result of the addition of hazardous substances without endangering the environment. The mere present of a toxic substance may not be a problem, though it may create some concern for pollution upon further accumulation of the material. Therefore, a high degree of contamination may cause pollution. Pollution, then, refers to the addition of substances resulting in degradation of soils and the environment. For example, manure deposited randomly by livestock in the field is a form of contamination, but manure accumulated in feedlots is distinctly polluting the area. Because of concern about contamination and especially the degradation in environmental quality starting in 1970, a number of methods of waste disposal were explored. Today waste is buried in landfills, incinerated, dumped in the sea, or used again on farmlands. When material called waste is used again as soil amendment or for any other useful purpose, such as the commercial production of methane, it is no longer considered waste but rather a valuable resource material. If the demand for waste in energy and crop production increases, reclamation of waste may well become an important process in the future (McCalla et al., 1977).

The major types of waste reaching the soil can be classified as: (a) agricultural waste, (b) industrial waste, (c) municipal waste, and (d) nuclear waste.

11.2 AGRICULTURAL WASTE

The common perception of agricultural waste is organic waste, especially that pertaining to stable or barnyard manure. However, according to the definition of waste as given above, this is not adequate anymore. By such a definition then, agricultural wastes include the many forms of fertilizers, pesticides, plant residues, animal waste, and forest residues. Many of them are very beneficial when returned to the soil, such as plant residues and cattle and barnyard manure. Some of them, such as plant and forest residues, are important sources today for the production of methane and ethanol, previously called *biofuel*. However, the improper handling and disposal of the many types of agricultural wastes may cause contamination and pollution.

11.2.1 Fertilizers

The use of inorganic fertilizers, though essential to increasing crop production, can also prove to be hazardous to the environment. Of much concern is, for example, nitrate (NO_3^-) fertilizers or fertilizers that can be converted into nitrates. Ammonium (NH_4^+) fertilizers, for example, when used in well-drained soils may be converted into nitrates. This conversion of NH_4^+ occurs in two steps and is conducted by two types of bacteria: *Nitrosomonas* sp., responsible for the conversion of NH_4^+ into nitrite (NO_2^-), and *Nitrobacter* sp., responsible for the conversion of nitrite into nitrate (NO_3^-). The biochemical reactions, called *nitrification*, have been discussed in Chapter 2 (see reactions 2.4 and 2.5) and Chapter 3, section 3.4.6.

Nitrates are taken up by plants, but high nitrate contents in plants are considered unhealthy, especially when the crops are used for the manufacture of baby food. When the nitrates are not taken up by plants, concern is expressed for contamination of groundwater by nitrate ions. Nitrate ions, being negatively charged, are not adsorbed by the also negatively charged clay colloids, and hence are subject to leaching. By EPA standards, water containing >10% NO_3-N is unfit as a source of drinking water. Nitrate is sometimes reported also to be harmful to animals. The bacteria that reduce NO_3^- into NO_2^- are also present in the anaerobic environment of the cattle's stomach, pigs, sheep, and chickens and in the horse's colon, and may also cause some type of *methemoglobinemia* as in infants. As discussed earlier, the hemoglobin is oxidized into methemoglobin by the nitrite ions, and young animals especially may suffer due to lack of oxygen. A consumption of 50 mg of nitrate-N per kg of animal weight is reported very harmful for cattle (Miller and Gardiner, 1998). In addition to these hazards, the nitrification process produces protons (see reaction 2.4), which may increase soil acidity. Because of this and because the nitrate is in the form of nitric acid (HNO_3), ammonium-carrying fertilizers are frequently called *acid-forming fertilizers*.

Another fertilizer element considered a hazard in soils and the environment is phosphorus. Phosphorus is a plant nutrient and is required for plant growth. However, by the excessive use of phosphate fertilizers, large amounts of the phosphate may leach into streams and lakes. An overenrichment of lake water with phosphate and nitrate ions causes excessive growth of unwanted aquatic plants, a process called *eutrophication*. Whether phosphogypsum may contribute toward eutrophication problems is not known. This material, a by-product of the phosphoric acid industry, is mentioned earlier as a soil stabilizer. It is sometimes also used in peanut crops as a Ca source. However, it was declared a harmful material in 1990 because of its *radionuclide* content, a source of the harmful *radon gas* (Anonymous, 1990).

Environmental concerns, therefore, caution against the indiscriminate use of fertilizers or liming materials. An overuse, causing contamination and pollution, may impose restrictions by the government, that may prove to be very costly to agricultural operations.

11.2.2 Pesticides

Pesticides, another group of necessary agricultural chemicals in crop production, are used for the control of pests and diseases in crops. They include insecticides, fungicides, nematocides, rodenticides, and herbicides. All the names end with *-cides*, plural for -cide, meaning to cut or kill. According to Miller and Gardiner (1998), an environmentally friendly pesticide must have at least the following attributes: (1) must be safe to handle, (2) must be effective but short-lived, and (3) must not be harmful to human health (noncarcinogenic). Most of them are aromatic compounds, and chemical structures for some of the most widely used pesticides are shown in Figure 11.1. Their potential as a pollutant depends on their biodegradability and toxicity to animals and people. Pesticides that can persist in soils for a long time affect the food chain by a process called *biological magnification*,

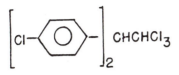

DDT , Dichlorodiphenyl-
trichloroethane

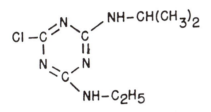

Atrazine

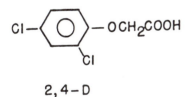

2,4-D

Figure 11.1 Chemical structure of DDT, triazine (atrazine), and 2,4-D.

meaning accumulation and subsequent concentration in the food chain. DDT may persist in soils for years, whereas 2,4-D is known to be a short-lived herbicide. It is reported to disappear in two to four weeks. Persistent insecticides can cause problems to non-target organisms. They may cause severe damage to non-target animals, or they can be consumed through a process earlier called *cometabolism*.

Persistence of pesticides is often expressed in terms of *half-life*. The half-life of a pesticide is defined as the length of time required for one-half of a given amount of the pesticide to disappear or decompose to other compounds.

The possibility arises that the decomposition products can be equally or even more harmful than the original compound. DDT, a chlorinated hydrocarbon developed by Paul Muller in 1938, was widely used during and after World War II as an insecticide, because of its usefulness in killing almost any insect. It was especially effective in controlling the malaria mosquitoes, afflicting the Allied Forces in the Pacific. For this important invention, Paul Muller was awarded the Nobel Prize in 1948. However, DDT is banned today because of its persistence in soils and its toxicity to animals and humans. The principal degradation product of DDT, called DDE, is even more persistent, and has been implicated in the development of thin eggshells in birds. Aldrin, another common pesticide at the time of DDT use, yields a decomposition product called dieldrin. Dieldrin is reported to be more toxic than aldrin. Triazines and 2,4-D, used as herbicides, are more biodegradable and less toxic than DDT. Therefore, they are less likely to accumulate in the food chain than DDT.

The safety levels of these pesticides for human health are set today on the basis of the effect of long-term exposure to these chemicals. Attempts have been made to use safety standards based on the concept of *ADI* (*acceptable daily intake*) and on that of *NOEL* (*no-observable-effect level*) of a contaminant. The NOEL or *health advisory level*, as called by Brady and Weil (1996), varies from 3 µg/L (3 parts per billion) for the herbicide atrazine to 200 µg/L for the insecticide malathion. The health advisory level is defined as the concentration level that will cause health problems in humans over a period of 70 years (Brady and Weil, 1996). This means that a daily intake of 3 µg/L, a minute amount, of atrazine will cause health problems when you reach 70 years of age. Though some of the information came from animal research studies, data to confirm all these on human health are unfortunately not available.

An additional environmental problem is the development of resistance by many organisms over time caused by the frequent use of pesticides. Generally, this forces even greater use of the chemicals leading to more contamination and pollution of the environment.

11.2.3 Crop Residue

Crop residues are generally very beneficial for soils and will cause no serious pollution when disposed of in a proper manner. They are the main source of food and energy for soil microorganisms. Used as mulch, they control loss of water by evaporation from soils, and protect the soil at the same time against erosion. They contain plant nutrients which will be released upon decomposition. The nutrient content is, however, very variable and depends on the type of plant material (Table 11.1). Generally the C/N ratio of plant residues varies from 80:1 in wheat straw to 20:1 in legume material. During the microbial decomposition process the C/N ratio is reduced to the narrowest ratio at which C and N can exist together in soils, i.e., around 10:1. However, because of continuous turnover of plant resi-due, this narrowest ratio is seldom reached in soils under vegetation (McCalla et al., 1977).

Differences in plant species may affect the quality of crop residues. Generally crop residues from legume plants are higher in N than those from nonlegume plants. Therefore, used as soil amendment or used as cattle feed, these crop residues can have different results. At one time, during 1970–1980, cattle fed with fescue grass acquired underdeveloped hind legs, a disease called *fescue foot disease*. The exact cause is still unknown, but the disease is believed to be caused by a fungus living in fescue, poisoning the cattle. The fungus produces alkaloids toxic to cattle. When cattle are raised on bermuda grass, they become excited. This disease is called *bermuda grass stagger*. The cows will recover if fed with grain-feed or rye grass (R. Wilkinson, USDA-ARS, personal communications).

Today, crop residues provide additional benefits as renewable resources for production of alternative fuel, earlier called *biofuel*. Researchers are pursuing ways to harvest fuel from crops and crop residues. As discussed previously, anaerobic decomposition of these

Table 11.1 Nutrient Composition of Leaves of Selected Crop

Crop	N	P	K	Ca	Mg	Fe	Cu	Zn	Mn	B
	-------------%------------					----------------mg/kg----------------				
Wheat	2.80	0.36	2.26	0.61	0.58	155	28	45	108	23
Corn	2.97	0.30	2.39	0.41	0.16	132	12	21	117	17
Peanuts	4.59	0.25	2.03	1.24	0.37	198	23	27	170	28
Soybeans	5.55	0.34	2.41	0.88	0.37	190	11	41	143	39
Potatoes	3.25	0.20	7.50	0.43	0.20	165	19	65	160	28
Sugar beets	3.76	0.38	4.01	0.78	0.68	126	26	40	086	53
Cotton	3.29	0.37	2.07	2.48	0.49	132	10	24	241	25
Alfalfa	4.63	0.48	2.76	2.38	0.66	140	21	46	065	56
Coastal bermuda	2.55	0.25	1.67	0.33	0.15	103	9	21	82	8
Tall fescue	2.78	0.33	3.01	0.43	0.32	098	8	20	135	8

Sources: Author's unpublished data; Anderson et al (1971); McCalla et al. (1977); Donahue et al. (1983); Jones et al. (1991).

residues may produce ethanol and methane. Carbohydrates from grasses, straw, and other plant residues are viable sources; in fact all kinds of farm refuse, such as chaff, corncobs, stalk, and even brewery grain residue, can be used to yield ethanol or methane. It will not be long now before fuel is harvested from US farmlands by cultivating rapidly growing bushes and trees for the needed cellulose, as is done now for the paper, pulp, and rayon industry.

11.2.4 Animal Waste

Animal wastes from agricultural operations include manure from cattle, swine, chickens, horses, sheep, and other types of animals. They vary considerably in chemical composition (Table 11.2), due to

Table 11.2   Wet-Mass Percentages of Nutrient Elements in
Animal Waste

Source	N	P	K	Ca	Mg	S	Fe
Dairy cattle	0.53	0.35	0.41	0.28	0.11	0.05	0.004
Beef cattle	0.65	0.15	0.30	0.12	0.10	0.09	0.004
Horse	0.70	0.10	0.58	0.79	0.14	0.07	0.010
Poultry	1.50	0.77	0.89	0.30	0.88	0.00	0.100
Sheep	1.28	0.19	0.93	0.59	0.19	0.09	0.020
Swine	0.58	0.15	0.42	0.57	0.08	0.14	0.020

Sources: Loehr (1974); Peterson et al. (1971); Tan et al. (1975),
    Brady (1990).

the type and amount of feed used and also the methods of collection, storage, and handling. For example, Donahue et al. (1983) reported figures for N content which were ten times higher than those listed in Table 11.2. This difference is caused perhaps by differences in analyses and calculations. The data presented by Donahue et al. (1983) were apparently percentages on a dry-mass basis.

As can be noticed from Table 11.2, poultry litter is the highest in N, P and Mg contents. When poultry litter is used as a fertilizer on fescue grass, this causes an unbalanced K/N ratio in poultry litter that was believed at one time to cause a disease in grazing cattle called *fescue foot disease*. As indicated previously, poisoning of cattle by the fungus living in fescue grass is considered a more likely reason for this disease. In general, the plant nutrient content in all types of manure is lower than that in commercial fertilizers. But manure is usually applied in large quantities, which then amounts to the addition of considerable quantities of nutrients. The data in Table 11.2 indicate that one metric ton of fresh (wet) dairy manure can supply approximately 5 kg N, 3.5 kg P, 4.1 kg K, 2.8 kg Ca, and 1.1 kg Mg.

Most of the nutrients in manure are not easily available to plants.

Their release depends on the rate of decomposition or mineralization. As such, they compare to the reaction of slow-release fertilizers. Crops are reported to seldom recover more than 2% of the soil nitrogen from humus (McCalla et al, 1977). This low availability is attributed to nitrogen's presence as organo-complexes, i.e., nitrogen is present as microbial protein and as complexes with humic acids or lignin-like substances.

Biologically, animal manures contain a large number of saprophytic, disease-carrying, and parasitic microorganisms, which need to be taken into consideration in the disposal of manure. Chicken manure is known to contain *Salmonella* sp., whereas swine manure may contain *Mycobacterium tuberculosis* (Azevedo and Stout, 1974). Feedlot waste was reported to contain 10^{10} organisms per gram, of which 10^9/g are various types of anaerobic organisms. It may also contain large amounts of *enterobacteria*, which is a health hazard if unsterilized waste is refed to animals (Hrubant et al, 1972). The unhealthy method of feeding animals with their waste resulted in *mad cow disease* in England.

11.2.5 Compost

Large piles of organic matter can be composted. Composting is a process resulting from controlled biological decomposition of organic residue into partially humified material. It seldom occurs naturally in soils, since large accumulation of vegetative residue in nature occurs where excessive amounts of water are present or where temperatures are low during most of the year. These conditions are not ideal for a composting process, which needs temperatures of 55°C (131°F). Moreover, anaerobic decomposition is not exactly a composting process, and may produce harmful by-products. The compost produced may contain toxic substances and won't be suitable for use as a soil amendment.

The materials for composting vary considerably, from materials high in N to materials low in N but high in C. For example, manure and sewage sludge are high in N, but dead leaves, grass clippings, straw, sawdust and wood chips or shavings are low in N but high in

C. Materials very rich in carbon (sawdust) are preferably combined with nitrogenous materials, such as stable manure, poultry litter, or sewage sludge in a 4:1 ratio. Practically any kind of organic material can be composted, but food wastes and animal tankage are very difficult to compost due to their gooey, viscous-like consistence. For use in composting, they require the addition of straw, leaves, and other fibrous-like material. During incubation the temperature of the mixture rises and levels off at 60-65°C (140–150°F) when the composting process is completed. These temperatures must be reached for a determined length of time to destroy pathogenic organisms. If temperatures decrease below 60°C, as often occurs at the edges of the piles, the time for the composting process can be increased to kill the pathogens. However, composting does not remove heavy metals from the materials, e.g., Pb, Cd, Hg, and As. These toxic metals may increase in concentration due to loss in weight of the organics. Hence caution should be executed in selecting material for composting. Sewage sludge from industrial sites is a poor source for producing compost because of its heavy metal content. Though not often, phytotoxins, teratogenic, and carcinogenic substances may be created by the decomposition process during composting. *Teratogenic* substances are substances that can cause deformities in plant growth.

Generally, compost is very beneficial for use as a soil amendment. It is the equivalent of humus. The composting process also reduces the amount of crop residues, hence is a good method for cleaning up the earth. For materials, known to be high in carbohydrates but deficient in some of the nutrient elements, the quality of the compost can be increased by adding the nutrients during the composting process. Water is also often added as required to keep the material moist.

11.2.6 Forest Waste

Forest waste is composed of leaves and twigs remaining from the harvest of forest trees. It also includes inferior and nonharvestable trees. The amount of waste from the harvest of forest trees, commonly known as *slash*, was estimated in the United States to be 23 million metric tons per year (Donahue et al., 1983). Due to problems in collecting and processing it for further use, little effort has been made to recycle this type of waste. In Africa, Southeast Asia, and South America, slash is removed by burning, as discussed in Chapter 8.

In general, forest waste has many properties in common with crop residues. As is the case with crop residue, forest waste also contains significant amounts of organic matter and plant nutrients. Therefore, it is a very valuable material for amending physical and chemical soil conditions. The nutrient content appears to not be affected by differences in tree species. Only small differences were noticed in C and N content between hardwood and softwood trees (Table 11.3). The nutrient content depends more on whether the forest waste is composed predominantly of leaves, twigs, branches, or bark. Lutz and Chandler (1947) reported that among hardwood trees, leaves had the highest nutrient content, followed by stems (Table 11.4). The bark of hardwood is generally lower in C, but higher in N content than bark of softwood trees.

Another type of waste related to the forest is the residue from the forest industry in the form of wood chips, bark, and sawdust. Totaling almost 63 million metric tons annually (Donahue et al., 1983), this

Table 11.3   Differences in C and N Contents Between Softwood and Hardwood Trees

Source	WOOD		BARK	
	C	N	C	N
	------%------		------%------	
Softwood[*]	48.5	0.10	50.4	0.14
Hardwood[**]	47.1	0.09	45.3	0.24

[*] Average of 19 tree species, including cedar, cypress, hemlock, fir, spruce, and pine.
[**] Average of 9 tree species, including oak, hickory, red gum, chestnut and black walnut. Sources: Allison (1965) and McCalla et al. (1977).

Table 11.4   Percentages of Major Nutrient Elements in the Ash
Content of Selected Trees

Source	P	K	Ca	Mg	Fe	Si
			-------%-------			
Ash, *Fraxinus excelsion*						
Leaves	9.9	15.5	28.1	4.9	0.8	1.2
Stem	3.0	10.9	44.3	3.5	1.3	1.0
Bark	1.7	06.9	57.3	1.4	0.8	0.7
Aspen, *Populus tremula*						
Leaves	3.8	15.3	35.4	2.4	1.4	3.5
Stem	1.9	09.8	50.8	2.3	0.7	1.3
Bark	1.4	06.4	52.0	4.3	2.1	1.0
Beech, *Fagus sylvatica*						
Leaves	3.4	18.1	31.6	4.4	1.6	4.9
Stem	1.2	12.0	43.0	2.7	1.6	4.7
Bark	0.9	04.2	59.5	2.2	0.5	1.7
English Oak, *Quercus robur*						
Leaves	5.4	18.6	33.6	1.7	2.2	2.4
Stem	1.6	08.1	54.4	2.2	1.1	1.1
Bark	1.3	02.4	62.8	1.2	0.7	0.4

Sources: Lutz and Chandler (1947) and McCalla et al. (1977).

type of waste finds extensive application in industry and nurseries
and as soil amendment. Sawdust is used commercially for the production of pressed wood utilized in the furniture and housing industry.
Bark has been applied extensively as mulching material and potting
mixtures in nurseries. Both bark and wood chips are valued for
landscaping purposes. However, the use of sawdust and bark in soils
to improve physical and chemical soil conditions has received mixed
reactions. Depending on the type of wood, both sawdust and bark have

been noted to have a harmful effect on the germination and growth of
pea seedlings (Allison, 1965), and caution was expressed on the use of
this type of waste in crop production (McCalla et al., 1977). The use of
wood chips in landscaping often creates breeding grounds for
millipedes resulting in considerable infestation and nuisance to
homeowners.

11.3 INDUSTRIAL WASTE

A variety of industrial waste products is currently causing concerns
regarding proper disposal. This waste can be in the gas, liquid and
solid form. The most important gases are CO_2, CO, NO, NO_2, and SO_2.
They are produced by the combustion of fossil fuels in industry and
automobiles, and they pose a hazard to the environment. Liquid
waste, called *effluent*, and solid waste are produced by food processing
plants and other types of agricultural factories, such as vegetable oil
factories, cane sugar factories, and beer breweries. Much of this solid
waste is organic in nature and has properties almost similar to crop
residue. For example, soybean, peanut, cottonseed, and sugar cane
pulp have great recycling potentials. Other forms of solid waste are
also produced by other industrial operations. They can be introduced
in the soil or in the atmosphere as industrial sludge or as dust or very
fine particulates, such as elemental sulfur, S. This constituent is
linked to the formation of acid rain, which will be discussed in a
following section.

11.3.1 Carbon Monoxide

Carbon monoxide, CO, is the result of incomplete oxidation of fossil
fuel. Most of it (approximately 80%) is produced by the combustion of
oil by automobiles. During rush hours in large cosmopolitan cities, the
CO content in the air can reach levels of 50 to 100 mg/L (Manahan,
1975). High levels of carbon monoxide are also frequently detected in
the craters of volcanos. In contrast to CO_2, carbon monoxide does not

dissolve in water. This type of gas is hazardous to animals and humans because it reacts with hemoglobin to form CO-hemoglobin. This *carboxyhemoglobin* is unable to carry O_2, causing respiratory failures. It is perhaps more deadly than nitrite poisoning due to formation of *methemoglobin*, which is also unable to carry O_2 as discussed earlier. Carbon monoxide is an odorless and tasteless gas and gives no warning of its presence. Therefore, the hazard of CO poisoning from faulty home heaters and exhaust from automobiles is very great. However, in contact with soil, CO gas is rapidly either adsorbed or oxidized into CO_2 by certain types of microorganisms. Manahan (1975) stated that OH radicals in the atmosphere are also capable of converting CO into CO_2. In the atmosphere, CO is usually oxidized and converted into CO_2 by reaction with O_2, O_3, N_2O, or NO_2 gas. It burns readily with oxygen forming carbon dioxide according to the following reaction:

$$2CO + O_2 \rightarrow 2CO_2 + energy \tag{11.1}$$

This reaction indicates that CO can be used as a fuel source.

11.3.2 Carbon Dioxide

Carbon dioxide, or CO_2, is produced naturally by the decomposition of organic matter and respiration of plant roots and microorganisms. The chemical properties and environmental aspects of this gas have been discussed in Chapter 5. In natural conditions, the CO_2 content of atmospheric air is very small and by volume it constitutes only 0.031% of our air. Because of this low level, some scientists are not concerned about pollution of atmospheric air by CO_2. They believe that the low amounts are insufficient to affect photochemical reactions. Absorption of infrared radiation by CO_2 is not energetic enough to induce chemical reactions (Moore and Moore, 1976; Manahan, 1975). Moreover, photosynthesis by terrestrial green plants may be able to absorb excess CO_2 from the air. Another sink for CO_2 is the seawater of the oceans. These bodies of salt water exhibit high pH values and are able to absorb large amounts of CO_2 as indicated in Chapter 5. The huge amounts of seaweed and kelp, which are green plants, also contribute tremendously in absorbing the CO_2 for use in photosynthesis.

Under normal condition, the presence of CO_2 has a buffering effect on the pH of rain water. Carbon dioxide dissolves in rain water to produce carbonic acid, which gives to the water some acidity, stabilizing at a pH of 5.6-6.0 (see reactions 5.12 and 5.13).

Many environmentalists reported substantial increases in the CO_2 content of the atmosphere due to the Industrial Revolution and the burning of fossil fuel in automobiles. Large-scale burning of tropical rain forests to clear the land for crop production has also introduced large amounts of CO_2 into our atmosphere. The increase in CO_2 content in the Earth's atmosphere is believed to create a *greenhouse effect*. Such an increase in temperature may increase the rate of weathering of soil minerals and decomposition of organic matter, whereas doubling the CO_2 concentration of the air may increase the yield of C_3 plants, such as soybeans, by 10 to 50% (Wild, 1993). C_4 plants, e.g., corn and sugar cane, are less affected by an increase in CO_2 content. C_3 plants fix CO_2 according to the *Calvin cycle*. These are mostly temperate region grasses and trees. On the other hand, C_4 plants fix CO_2 with a different process (*Hatch-Slack cycle)* that enables the plants to store larger amounts of CO_2 in the chloroplast, which is the reason why they do not respond as well to an increase in CO_2 as C_3 plants. The growth rate of C_4 plants is consequently greater than that of C_3 plants.

The increase in plant growth due to the rise of CO_2 content in the atmosphere has also attracted the attention of other scientists. In a series of articles disputing global warming as a *myth*, one of which appeared in the Wall Street Journal on Thursday, December 4, 1997, A. B. Robinson and Z. W. Robinson, chemists at the Oregon Institute of Science and Medicine, reported that the CO_2 content of the Earth's atmosphere has risen from a low of 0.029 % (290 mg/L) at the start of the 20th century to a high of 0.036 % (360 mg/L) at the end of the century. This is more correctly a 24% increase, rather than the 82% increase that the authors displayed in their graph. The CO_2 content is projected to increase to 0.06 % (600 mg/L) in the next millennium due to human activities. In the authors' opinion, the fluctuation and rise in CO_2 content follow a warming trend in the atmospheric temperature rather than being the cause of global warming. Burning coal and fossil fuel, some believe, is in fact a *recycling of hydrocarbon* from below the earth into the atmosphere, which is in part true. As discussed in Chapter 3, CO_2 is a major player in the *carbon cycle*. It will be returned to the earth through absorption by plants by a process called *photosynthesis*. The increase in CO_2 content has major environmental benefits due to increasing plant growth, which indeed could translate into more food and fiber for mankind and animals. However, the return of plant and animal residues into coal and fossil fuel to complete the so-called *hydrocarbon cycle* remains a big question. Formation of coal and fossil fuel in nature usually involves a *metamorphism process* that requires drastic environmental changes over a period of a geologic time scale.

11.3.3 Nitrogen Oxides

Nitrogen, N_2, nitrous oxides, N_2O, nitric oxide, NO, and NO_2 gas are common constituents of atmospheric air. The mixture of NO and NO_2 is usually referred to as NO_x (Manahan, 1975; Brady, 1990). Nitrogen gas is part of the *nitrogen cycle*, whereas the nitrogen oxide gases are the products of oxidation of N_2 gas. Nitrogen is the most abundant constituent of atmospheric air. Its concentration in normal atmospheric air is approximately 78%. In contrast, the nitrogen oxide gases are present only in very small amounts. For example, under normal conditions the concentration of N_2O in atmospheric air is 0.25 mg/L. However, due to the Industrial Revolution the concentration of nitrogen oxide gases in atmospheric air has increased 10 times or more. It is estimated that production of nitrogen oxide gas in 1975 amounts to 13 million tons per year (Moore and Moore, 1976; Manahan, 1975), and has increased since then. The resulting increase in nitrogen oxide content in the atmosphere has created much concern over harmful effects to the environment because of the formation of the so-called *acid rain, global warming,* and *destruction of the ozone shield*. These environmental issues will be the topics of separate sections below.

In atmospheric air, nitrogen oxide gases are mostly produced by oxidation by lightning. They can also be introduced by emissions from combustion of fossil fuel in industry and automobiles. Supersonic air transport is also suspected of releasing large amounts of nitrogen oxide into the atmosphere. The three types of nitrogen oxide gases (nitrous oxide, nitric oxide, and dinitrogen oxide) will be discussed in more detail below and their production, emission, and implications for environmental quality will be examined.

Nitrous Oxide. - Nitrous oxide, N_2O, is a colorless gas with a mild, pleasant odor. It tastes sweet and was used in the past as an anesthetic by dentists. It is sometimes referred to as *laughing gas*. Under high temperatures, N_2O decomposes to N_2 and O_2 gases according to the reaction

$$2N_2O \rightarrow 2N_2 + O_2 + \text{energy} \tag{11.2}$$

Because the decomposition reaction yields oxygen, N_2O gas is frequently used today as a fuel source in atomic absorption spectroscopy.

This gas is a *greenhouse gas*, and may also be detrimental to the ozone shield in our atmosphere. The gas is formed naturally by microbiological processes in soils. Denitrification of NO_3^- yields N_2O. Agricultural operations, biomass burning, and combustion of fuel in power stations and automobiles are considered important *anthropogenic* sources of N_2O gas. It is estimated that about 1/3 of the total emissions of N_2O is of anthropogenic origin (Granli and Bϕckman, 1993). Other sources are waste-water treatment plants and medical and industrial use of N_2O. Because of these, the concentration of N_2O gas has increased by about 0.25% annually, which is hardly of concern at all. However, several scientists believe that such an increase in N_2O emission may bring about a 5% increase/year in the greenhouse effect (Granli and Bϕckman, 1993).

Nitric Oxide. - Nitric oxide, NO, is a colorless gas and is insoluble in water. It is also a pollutant gas produced by burning fossil fuels in power plants, industry, and automobiles. Although it can also release its oxygen, it is not used as a fuel source as N_2O gas. NO acts more often as an oxidizing or reducing agent. The oxidation reaction may convert NO into NO_2 gas according to the reaction:

$$2NO + O_2 \rightarrow 2NO_2 \qquad (11.3)$$

Such a reaction can occur spontaneously in the air and the NO_2 formed is a polluting gas contributing to the development of acid rain.

Nitrogen Dioxide. - Nitrogen dioxide, NO_2, is a red gas, and can rapidly form *dimers*, called dinitrogen tetroxide, N_2O_4. Nitrogen dioxide usually occurs in equilibrium with its dimer

$$2NO_2 \rightleftarrows N_2O_4 \qquad (11.4)$$

Nitrogen dioxide is a condensable gas, and will dissolve in water to form nitric and nitrous acids. For this reason, it is considered a polluting gas contributing to acid rain, which will be discussed in Section 11.7.2.

11.3.4 Sulfur and Sulfur Oxide

Sulfur occurs in soils in organic form, e.g., humus, and in inorganic form, e.g., in the minerals gypsum, $CaSO_4$, and pyrite, Fe_2S. As discussed in Chapter 5, pure elemental S is produced in the craters of active volcanos, and a large amount of S is also emitted into the air by these volcanos in the form of small particulate matter. Sulfur oxide, also called sulfur dioxide, SO_2, is the product of oxidation of S by natural processes or by the combustion of coal and oil in power plants and other industries. It is a colorless gas and soluble in water. It condenses rapidly into a liquid. This liquid boils at -10°C. Liquid SO_2 must be really dry when stored in steel tanks for shipping. A trace of moisture will convert it into sulfuric and sulfurous acid, which is very corrosive to the steel tankers. Because of its easy conversion into these acids, it is a polluting gas contributing to acid rain. Large amounts of sulfur dioxides are also produced by ore smelting in factories.

This S and its oxides produced by industry are frequently referred to as *anthropogenic sulfur* in contrast to the natural sulfur emitted by active volcanos. Like N_2, sulfur is an essential plant nutrient, and is necessary for the production of proteins, vitamins, and hormones.

The S content of soils varies considerably from 0.002% in highly weathered soils of the humid regions to 5% in the alkaline soils of arid regions (Stevenson, 1986). The element and its oxide are part of the *global sulfur cycle* (Manahan, 1975; Stevenson, 1986). Currently, a lot of concern has been expressed over the S and its oxides released in the atmosphere by industry. Sulfur can also enter the air by biological processes in the form of H_2S. The amount of SO_2 released in the atmosphere by industry is estimated to amount to 65 million tons per year (Manahan, 1975), and approximately 200 kg S per ha are deposited in soils from the air in highly industrialized areas of Europe, the eastern part of North America, and parts of east Asia. The amount of sulfur gases introduced in the air by industry is now perhaps one-half as much as from natural sources (Stevenson, 1986).

These sulfur gases are particularly harmful to the environment, plant growth and human health. Exposure to high levels of SO_2 gas may cause leaf necrosis in plants. The harmful effect increases when the stomata in the leaves are fully open. Some plants require acid soils for optimum growth, such as azaleas, rhododendrons, tea, and pine trees. The soils are usually acidified by applying S as discussed in a previous section.

Oxidation of sulfur and sulfur dioxide in the air and their subsequent reaction with raindrops forms sulfuric acid and contributes toward the formation of acid rain. Sulfuric acid dissolves in rainwater and dissociates its proton, H^+, decreasing in this way the pH of rainwater. The reactions and effect of acid rain on environmental quality will be discussed in a separate section below.

Both the salts of this acid and the SO_2 gas have been associated with the formation of turbid haze, fog, and smog, which at one time was dangerously covering many industrial towns in the Midwest, California, and England. Not only is this kind of air pollution corrosive, but it also causes eye irritation and respiratory difficulties in humans. The seriousness of the effect depends on the length of exposure. Prolonged exposures to smog containing sulfuric acid may damage lung tissue.

A number of methods have been used to remove S and SO_2 from coal. The sulfur in coal is usually present as pyrite, Fe_2S, and organic sulfur. Upon burning the coal, the pyrite minerals are oxidized and sulfur dioxide is produced. To prevent the formation of SO_2, pyrite minerals can be removed physically, e.g., by magnetic separation. Another method is to reduce the amount of SO_2 released in air after burning the coal by using extremely tall smokestacks and a series of *scrubbers*. Some of the scrubbers are dry, whereas others are wetted with water. In the *wet-limestone scrubbing* process, powdered limestone is injected into the oven or boiler along with powdered coal and the S or SO_2 is removed according to the following reactions:

$$CaCO_3 \rightarrow CaO + CO_2\uparrow \tag{11.5}$$
$$\text{limestone} \quad \text{burnt lime}$$

$$CaO + SO_2 \rightarrow CaSO_3 \tag{11.6}$$
$$\text{calcium sulfite}$$

$$CaSO_3 + SO_2 + \tfrac{1}{2}O_2 \rightarrow CaSO_4 \tag{11.7}$$
$$\text{calcium sulfate}$$

Calcium sulfate can be removed easily and is a valuable liming material.

Magnesium hydroxide can also be used in the same way as $CaCO_3$, and the SO_2 can be removed by catalytic oxidation. By any method, removing the S and SO_2 to keep SO_2 emissions at relatively low levels appears to be a costly process.

11.4 WASTE FROM FOOD PROCESSING PLANTS

Waste from the food processing industry can be in the form of solid waste or liquid waste (effluent).

11.4.1 Solid Waste

Solid waste from food processing plants varies considerably depending on the types of food processed ranging from plant products to animal products. Nevertheless these types of wastes are invariably organic, and are related in chemical composition to agricultural waste products.

Most of the solid wastes can be recycled by using them as organic soil amendments or as compost. However, their nutrient content, especially their high N and P content, makes their disposal in soils uneconomical. Instead, most of them, such as soybean, peanut, cottonseed, sugar cane, and sugar beet pulp are processed today for animal feed. Solid waste from the chicken, meat, and fish industries are especially high in protein, hence in N content. They are excellent sources for recycling into fish feed and cat and dog food. Bones from the meat industry contain large amounts of P in the form of $Ca_3(PO_4)_2$. They are processed into bonemeal, an excellent source of P fertilizer.

11.4.2 Liquid Waste (Effluent)

The *effluent* produced from the food processing industry is generally rich in inorganic nutrients, such as N, P, and K, as well as in dissolved organic compounds. They create a pollution hazard when disposed of improperly in streams and lakes. The excessive enrichment of stream and lake water with the above plant nutrients causes *eutrophication*, as discussed previously. The dissolved organic compounds reduce the oxygen content of soil, stream, and lake waters, because of an increase in the so-called *oxygen demand*. The definition and concept of oxygen demand have been discussed in Chapter 5. *The biological oxygen demand*, or *BOD*, values in the effluent of food processing plants can, depending on dilution, run as high as 10,000 ppm. Runoff entering streams and lakes is a hazard to environmental quality if the BOD exceeds 20 ppm. High levels of organic pollutants cause microorganisms, responsible for the oxidation processes, to consume all the dissolved oxygen. The oxygen deficiency created in the effluent creates an anaerobic condition, which is highly undesirable.

However, liquid wastes applied as soil amendments are noted to be beneficial to plant growth. Such a process is a form of recycling and a prudent use will decrease the chances of degradation of the soil or the environment. In North Sumatra, liquid waste from the palm oil factories, with BOD values > 15,000 ppm, is reported to increase growth of oil palm seedlings.

11.5 MUNICIPAL WASTE

Municipal waste is distinguished by McCalla et al. (1977) into *municipal refuse* and *sewage sludge*.

11.5.1 Municipal Refuse

Municipal refuse, also called *municipal garbage*, is composed of discarded material by people in the home and in industry. It is composed of paper, plastic, food, and plants. An average composition of municipal refuse is given in Table 11.5. Paper makes up the largest amount of this refuse, whereas plastic comprises only very little and is considered a miscellaneous item. With the increasing use of plastics recently, they are an increasingly important component of refuse today and their percentage of total refuse is expected to become as high as paper. Toxic household and agricultural chemicals, fertilizers, solvents, and medicines are additional components of importance in today's refuse, and create serious disposal problems.

Municipal garbage can be recycled by *composting* or by burning, or it can be disposed of in *landfills*. Burning is the least desirable method because of air pollution and because a large amount of the garbage is noncombustible. Composting has attracted increasing attention recently. For the purpose of composting, the organic components are of considerable value and have to be separated carefully from glass, metals, and other inorganic components. Composting nonionic organic waste is currently receiving increasing attention. It relieves the municipalities of some of the heavy volume of garbage,

Table 11.5 Average Composition of Municipal Refuse

Component	%
Paper	58.8
Garden refuse	10.1
Food residue	9.2
Glass, ceramics, and ash	8.5
Metals (cans, etc.)	7.5
Miscellaneous (plastic, rags, etc.)	5.9

Source: McCall et al. (1977); Donahue et al. (1983).

and the compost is a very valuable product as a soil amendment in agricultural and horticultural operations. The process of composting has been discussed in a previous chapter. In addition, organic wastes are today considered of value for the production of alternative fuel. Municipal organic waste from metropolitan cities in the Midwest appears to be an important source of income by producing methane.

11.5.2 Environmental Problems in Disposal of Municipal Refuse

Most municipal garbage is still discarded in *sanitary landfills*. Because of the many environmental problems created by sanitary landfills strict rules have been set up by the Environmental Protection Agency (Donahue et al., 1983). One of the problems in landfills is the generation of gases, such as methane, CH_4, gas. As discussed before, this gas is produced by the anaerobic decomposition of organic matter (see Chapter 4). It is a flammable gas that can cause explosion, hence is a potential hazard to public safety. Improving aeration in landfills to create aerobic conditions may perhaps decrease formation of methane gas. Some municipalities have taken advantage of this problem by using the organic fraction of the garbage for the commercial production of methane, as indicated earlier.

11.5.3 Sewage Sludge

Sewage sludge is the product of wastewater treatment plants. The materials processed in the treatment plants are domestic and industrial wastes. They are usually a liquid mixture, composed of solid and dissolved organic and inorganic material. The water is separated from the solid part, and undergoes a number of treatments before it is declared environmentally safe for discharge in streams and lakes. The solid part is composed of organic matter mixed with some inorganic compounds. It is biologically unstable, and is stabilized by a series of aerobic and anaerobic digestion processes. This solid waste product is usually referred to as *sewage sludge* (McCalla et al., 1977; King, 1986). Depending on the sources, sewage sludge varies in macro- and micronutrient content. The average macronutrient composition of municipal sewage sludge is given in Table 11.6. The data indicates that the N content of municipal, textile, and fermentation sludge is generally high. On the other hand, sludge from wood processing plants is lower in N, but higher in K and Ca than the other three sludges. Sludge cake from England is also low in N, though its N content is slightly higher than that in wood processing sludge.

Heavy Metals in Sewage Sludge. - Of great public concern is the heavy metal content in sewage sludge. A metal is defined as a substance that (1) has large electrical and heat conductivity, (2) has a *metallic luster*, (3) is malleable, and (4) is ductile (can be drawn into wires). Eighty of the elements in the periodic system are considered metals. They can be distinguished into *alkali metals* (Li, Na, K, Rb, and Cs), *alkali-earth metals* (Be, Mg, Ca, Sr, Ba and Ra), and *transition metals* (Al, Fe, Mn, Cu, Co, and Cu). However, a definition of *heavy metal* is not available in pure chemistry, but Wild (1993) indicated that heavy metals are by definition metals with a density of >5—6 g/cm^3. In the literature, the term is generally used very loosely, since it includes other metals that are not officially considered heavy metals. Often the term refers to the elements Al, As, Cd, Co, Cr, Cu, Fe, Hg, Mn, Ni, Pb, and Zn. Some of them are required in trace amounts for plants, animals and humans, whereas others are hazard-

Table 11.6 Average Macronutrient Content of Sewage Sludges
from the Southern United States and England

Origin	N	P	K	Ca	Mg
	------------------- % -------------------				
Municipal sludge (USA)	3.0	1.8	0.2	1.5	0.2
Textile sludge (USA)	4.1	1.1	0.2	0.5	0.2
Fermentation sludge (USA)	4.1	0.4	0.1	4.5	0.1
Wood processing sludge (USA)	0.8	0.1	1.9	3.3	0.2
Sludge cake (England)	1.2	0.6	0.3	0.0	0.0

Sources: King et al (1986); Wild (1993).

ous to human health. Present in even minute concentrations, these
heavy metals tend to accumulate in biological systems. In passing
through the food chain associated with plant and animal life, their
concentrations may increase to harmful levels in the top members of
the food chain, e.g., fish, eagles, and humans.

The heavy metal content in sewage sludge varies considerably
depending on the contribution of industry (Table 11.7). In general,
municipal sludge is very high in Al, Fe, Zn, Cr, and Cu content. The
Pb, Ni and Cd content is on the average relatively small, whereas Hg
is hardly detected in sewage sludge. Cadmium, Pb, Hg, and Ni are
not considered plant nutrients, but these metals can potentially be
taken up by plants together with Fe, Cu, Mn and Zn. Aluminum is
officially not recognized as a plant nutrient, but plants develop
necrotic spots when grown in Al deficient media. Much concern has

Table 11.7 Average Metal Composition of Sewage Sludge in the Southern United States and United Kingdom

Origin	Al	Fe	Mn	Cu	Zn	Pb	Ni	Cd	Cr	Hg
	--------------------------- mg/kg ---------------------------									
Southern United States:										
Municipal sludge	7280	2370	150	565	2220	520	100	28	1040	5
Textile sludge				390	864	129	63	4	2490	0
Fermentation sludge				81	255	29	18	2	117	0
Wood processing sludge				53	122	42	119	2	81	0
England and Wales:										
Sludge				800	3000	700	80	0	250	0

Sources: Tan et al (1971); King and Giordano (1986); Stevenson (1986); Wild (1993).

been expressed about the harmful effects of Pb, Cd, and Hg on animals and humans, and these three elements will be discussed in more details in the following sections.

Lead. - Lead, Pb, is a soft metal with a dull gray color and low tensile strength. It is the heaviest of the metals, except for gold and mercury. The element is used for covering electric cables, and as insulation in x-ray instruments and was used in the past for the layering of water pipes. It is used as a paint pigment in the form of lead carbonate,

$Pb_3(OH)_2(CO_3)_2$, which is called *white lead*. Because of its high toxicity, its use in paint is now replaced by ilmenite. The organometal chelate of lead, called *lead tetraethyl*, $Pb(C_2H_5)_4$, has been used as an *antiknock* additive in gasoline (to prevent knock in automobile engines). Its use as an antiknock additive has now been replaced by other less toxic compounds. The oxides of lead, e.g., lead monoxide, also called *litharge*, and *red lead*, Pb_3O_4, are used in making lead crystals and lead glass. Red lead is also used in red paint for protecting iron against corrosion. These applications have been discontinued because of the hazard of Pb toxicity.

Lead, Pb, is physiologically not a nutrient and its accumulation in soil may cause contamination and pollution. This metal is implicated in causing liver damage. However, no adverse effect has been reported by the intake of Pb below the threshold values.

Cadmium. – Another element considered a liability to the environment (Lagerwerff, 1972) is cadmium, Cd. Cadmium, a silvery metal resembling Zn, is a rare metal and is present in soils and minerals in very low concentrations. The mineral source of Cd is *greenockite*, CdS, a mineral related to *sphalerite*, ZnS. The Cd concentration of phosphate fertilizers, another likely source of Cd in soils, is approximately 7 µg/g. The Cd content is slightly higher in sewage sludge, but still substantially lower compared to Al and Fe contents.

The element is used in electroplating iron and steel and for the production of a number of alloys. Cadmium-plated iron has a better appearance and is more resistant to corrosion than galvanized iron. Galvanized iron is coated (plated) with zinc. Cadmium is also used for making standard cells for measuring electrical potentials, whereas Cd rods find application in nuclear reactors to control chain reactions or absorb neutrons.

Cadmium is also not required as a nutrient, but can replace Zn in plant and animal nutrition and may cause renal and testicular damage in animals. Cigarette smoking is one reason for a high daily intake of Cd. The application of Cd in industry as a stabilizing agent in the production of polyvinyl plastics and its use in electroplating of metals and in the production of batteries and pigments have alarmed environmentalists worried about its harmful accumulation in soils, though very few human health problems have been noted because of Cd toxicity. The only case of Cd fatality was reported in 1950 in Japan, due to the consumption of rice containing high amounts of Cd. The paddy fields had been irrigated with water polluted by materials high in Cd from a neighboring zinc mine (Wild, 1993). Zinc smelters are notorious for the environmental damage they cause by emitting large amounts of ZnO and CdO fumes.

In view of the fact that the Cd content in sewage sludge is relatively low, concern for the pollution hazards of Cd and Hg, when sludge is applied in soils for crop production, is perhaps somewhat overstated. Cadmium and Hg are indeed very toxic to humans and animals, and their content in soils may build up with repeated application of sewage sludge over time. However, a build-up of Cd to dangerous levels may take a long time and only with very large amounts of sewage sludge application. It must be realized that huge amounts of materials are required to accumulate minute concentrations of the hazardous metals in soil. For example, Rainey (1967) points out that 5000 tons of material must be dissolved to reach a concentration of 1 µg/L (1 ppb) of a polluting metal in Lake Michigan (4871 km^3). With sewage sludge, Stevenson (1986) reported that an annual application of 20 metric tons/ha of digested sludge for 20 years would result in an increase of 8 mg/kg of Co, 180 mg/kg of Cu, 270 mg/kg of Pb, and 890 mg/kg of Zn. No data on Cd were presented, but since the Cd content was, for example, 35 times smaller than the Cu content, theoretically only 5 mg Cd/kg is expected to accumulate in soils with the above treatment of sludge in 20 years.

Therefore, of greater concern should be the Al content, which tops all the other metal concentrations in sewage sludge. Large amounts of Al are also toxic to plants, humans, and animals. Large amounts of Al intake have been implicated in damage to brain cells in humans. The presence of 50 mg Al/kg has been noted to create Al toxicity in plants. Much of the Al taken up is, however, retained in the roots, especially in the root epidermis in the form of organic chelates (Tan and Binger, 1986). It should also be realized that the metals in sludge are bound strongly to the organic matter by complex and chelation reaction. They will be released gradually as free ions as the sludge decomposes over a period of years. Even then, the presence of fulvic and humic acids in soils may regulate the metal concentrations in the soil solution to the extent that toxicity is suppressed (Tan and Binger, 1986). Its entry into the food chain can, therefore, be kept to a minimum.

Mercury. - Mercury, Hg, is a third element that has created a lot of concern among environmentalists. It is the only metal which is in a liquid state at room temperature. Because of its presence in the liquid state and its silvery luster, the metal is often called *quicksilver*. It exhibits a high boiling point and a uniform expansion and contraction. Therefore, mercury is an excellent substance for making thermometers, barometers, and the like. All metals, except Fe and Pt, will dissolve in Hg, including gold. Mercury vapor emits light, hence Hg is also used in mercury vapor and fluorescence lamps. Mercury is also useful in the production of explosives, detonators, and percussion caps. In the form of HgS, it is used as a paint pigment, called *vermillion*.

Mercury and all mercury compounds are toxic in large amounts. However, in small amounts they are used in medicines. Ointments for treating infection of open wounds contain Hg, such as *mercurychrome*. The critical amount of Hg in the form of $HgCl_2$ for Hg poisoning is approximately 0.3 g. The Hg^{2+} is reported to destroy the kidney. Mercury poisoning is often treated by giving milk and/or egg whites. These substances inactivate the Hg ions by forming organo-Hg chelates, which precipitate in the stomach. As indicated above, the amount of Hg detected in sewage sludge of the southern United States is so small that it would take tons of sewage sludge to accumulate significant amounts of Hg in soils. Theoretically, if it takes 20 years for 5 mg Cd to accumulate per kg of soils with an annual application of 20 metric tons of sewage, it will take 100 years (5 times longer) to accumulate an equivalent amount of Hg. As indicated in Table 11.7, the Hg concentration is 1/5 or less than that of Cd in sewage sludge of the southern United States. As indicated earlier, Al, Fe, and Zn top all the other elements in concentrations. An oversupply of soluble Al, Fe, and Zn ions is also harmful to plants, since they are needed only as micronutrients.

11.6 DETOXIFICATION OF WASTE

As indicated in Chapter 7, soil has the ability to resist changes in soil reaction, called *buffer capacity*, which is related to its cation exchange capacity. It appears that the soil's capacity to resist changes is not limited to buffering soil pH only. The soil also has the ability to adsorb, neutralize and detoxify a number of organic wastes, which, if released in streams and lakes, would pollute the environment. This unique detoxifying capacity is attributed to many soil characteristics. Soil water acts as a solvent, whereas the electrically charged surfaces of clay and humic matter provide for binding sites. Humic acid, especially, is known to be able to chelate large amounts of heavy metals in solution, reducing in this way the concentration and hence chemical activity of the metal ions. The presence of static electricity in clay minerals can produce an electrical discharge, which has been reported to be strong enough to destroy carbon compounds which are not too complex in molecular structure. Enzymes produced by microorganisms also contribute toward decomposing many organic wastes.

Biodegradation. – The term *biodegradable compounds* is used today to refer to materials that can be destroyed by microbial enzymatic decomposition, e.g., plastics and pesticides. The enzymes produced by microorganisms are used to attack natural organic matter in soils with the purpose of obtaining food and energy. Such an enzymatic decomposition depends on the complexity of the molecular structure of the organic material. When a foreign substance is introduced into the soil, the enzyme will be able to attack it if the chemical nature of the foreign substance does not differ too much from that of the natural compound normally broken down by the enzyme. In some cases, the soil appears to need some time to accomplish the detoxifying effect. This happens, especially, with compounds that have not previously been introduced into soils. The delay in detoxification reactions is attributed to the length of time the microorganisms need to increase their number sufficiently to produce enough enzymes, and is called the *lag time*. The process of cleaning, neutralizing, adsorbing, and destroying soil pollutants is called by different names, e.g., decontamination, detoxification, and soil remediation.

11.6.1 Soil Remediation

Soil remediation is by definition the process of correcting, counteracting, or cleaning the soil from hazardous contaminants or pollutants. It involves *physical*, *chemical*, and *biological* remediation.

Physical Remediation. – Physical remediation is the process of correcting the problem by physical means. An example is the process of *containment*, by which movement of pollutants from contaminated sites is prevented. This is done sometimes by covering the polluted site with a layer impermeable to rain water. Clay or cement was used in the past to provide an impermeable cap preventing water from reaching the pollutants and leaching them away. Landfills for waste disposal are in essence operating on this principle. Often, cement is pumped underneath the pollutants to keep them from reaching the ground water. This is not exactly getting rid of the pollutant. It is just being immobilized. Perhaps the process known as *in situ cleanup* is a better form of physical remediation. In situ cleanup involves in situ removal of pollutants by soil excavation, cleaning the soil, and returning the soil to the site. However, this is considered a very costly procedure. In situ cleanup applied to the spoiled beaches in the Prince William Sound of Alaska, due to the crude oil spill of the Exxon-Valdez, though successful, was very expensive. Bioremediation, which will be discussed below, was less expensive and yielded faster results.

Chemical Remediation. – Chemical remediation is correcting soil pollution by chemical means. A good example is the reclamation of soils contaminated by mine spoil. As discussed in Section 4.3.1, acid mine spoil, a very harmful contaminant of soils and streams, is controlled by liming the area and growing pioneer plants that can survive in an acid soil environment.

Biological Remediation. – Biological remediation, better known as *bioremediation,* is a process of cleaning soils and the environment with the assistance of microorganisms that are capable of breaking down pollutants. This capability of organisms to decompose organic residue has been discussed many times in preceding sections. It is a process that attracted special attention when PCBs, creosote and other petroleum products created problems in the 1980's in soils and lake waters making fish unsuitable for consumption. These products are in fact not strictly nonbiodegradable, having a composition different from that of food sources the microorganisms are accustomed to. A dramatic example of a successful bioremediation process is the cleanup of the coast of Alaska damaged heavily by the Exxon-Valdez disastrous oil spill on March 24, 1989. The crude oil destroying the beaches around Prince William Sound was sprayed with an N+P mixed fertilizer in the hope of stimulating the activity of indigenous bacteria that would consume crude oil. The results indicated that within 3 months large stretches of the beach on Green Island had recovered dramatically (Hodgson, 1990; Brady and Weil, 1996). Once the organisms are found or activated, it is not difficult to decompose even the pollutants most toxic to microorganisms. Persistent pesticides, such as DDT, 2,4,5-T , and dioxin, presumed to be very toxic to microorganisms, are reported to succumb to the attack of a recently discovered white rot fungi species, *Phanerochoete chrysosporium* (Bumpus, 1993).

Phytoremediation. – In addition to microorganisms, higher plants appear also to be able to neutralize harmful pollutants. Sunflower (*Helianthus anuus* L) is one of the plants that is reported to be capable of selective absorption of heavy metals (Miller and Gardiner, 1998). Attempts are underway to test removal of radioactive uranium, in particular from contaminated water of the Chernobyl disaster in 1986 in Russia by using higher plants. This process, called *phytoremediation*, is in essence decreasing the toxic concentration in soils or water, but is not exactly alleviating the problems of toxic metal pollution. The toxic metals are accumulated in the plant tissues, and thus the plants contain the hazardous materials. Only the soil and water are decontaminated by phytoremediation. Hence, this is only a form of the process earlier called *containment*, and perhaps the term *phytodecontamination* reflects the process more accurately. The process is not limited to removal of heavy metals only, but organic compounds can also be taken up and hence accumulated in the plant tissues.

Remediation by letting higher plants absorb and accumulate harmful compounds in their tissues is perhaps of advantage when large areas of soils and fast bodies of water are involved. However, for cleaning contaminated water from nuclear reactors, the necessity of using hydroponics for growing the plants is time-consuming and very expensive. A more efficient and cost-effective method is using humic acid or smectite. Of the two, humic acid is of special importance because it has a higher cation exchange capacity than smectite. Its chelation capacity, which permits it to hold the pollutants with stronger bonds than the usual electrostatic attraction by inorganic colloids, is an additional benefit. Results from experiments with water, containing 10 kBq/mL of either ^{137}Cs, ^{90}Sr, ^{169}Yb, or ^{51}Cr, indicated that humic acid has completely removed the radioisotopes from the solution, and radioactivity in the water was reduced for 99.0 to 99.8% (Gerse et al., 1994). Radioactive water from the Chernobyl nuclear reactors would have been cleaned rapidly had humic acids been used.

11.6.2 Natural Attenuation

This is a remediation process promoting the naturally occurring physical, chemical, and biological reactions in the environment. It is closely related to and often difficult to distinguish from soil remediation, as discussed in the foregoing section 11.6.1. Perhaps, it is correct to say that natural attenuation is a special case of soil remediation that takes advantage of the constituents or reactions naturally present in the soil. Hence, soil remediation is a broader method that may include natural attenuation, but can also involve many other methods to achieve its purpose. Though the term attenuation implies that only the severity of the degradation process is reduced, in several instances the soil properties can be modified so that the proper conditions are induced for eliminating completely the pollutant. Remediation of the environment by natural attenuation is usually utilized to restore the quality of soil and groundwater when no risk to public health and the environment is known to exist at a site, or when restoration by other means is not cost-effective, is impractical, or is not warranted due to site-specific conditions. Three types of natural attenuation can be distinguished on the basis of inducing or promoting changes in physical, chemical, and biological properties of soils: physical, chemical and biological attenuation.

Physical Attenuation. – An example of physical attenuation is the removal of toxic concentrations of salts from soils by enhancing the replacement of the soil solution by natural rain water or by water containing high concentrations of nontoxic solutes. This includes providing adequate amounts of pore spaces by promoting good structural development and the like for an uninterrupted flow of leachates. Since the soil solution and rain water are completely miscible with each other, the process of replacing the soil solution with a solution of a different composition is called *miscible displacement*. The rate of replacement decreases with time, since the displacing solution is of a lower concentration than the soil solution. However, by increasing the solute concentration of rain water, by for example liming the soil, the rate of replacement will increase with time. When plotted, these two displacement processes yield a so-called *breakthrough curve*, describing the solutes in the displacing water coming to an equilibrium with the soil (Taylor and Ashcroft, 1972). Miscible displacement is affected by a number of factors, e.g., hydrodynamic dispersion, and molecular diffusion. Hydrodynamic dispersion, or just dispersion, is influenced by charge densities of the solutes, double layers, and salt concentration. For example, the presence of a *flocculating concentration* of salt, as discussed earlier, may assist in granulation of the soil, beneficial in promoting development of pore spaces. On the other hand, molecular diffusion is the result of the intrinsic molecular movement due to concentration gradients. Diffusion would result in an even distribution of the molecules throughout the solution even though no flow has occurred.

Chemical Attenuation. – This is a process by which the natural chemical reactions are enhanced to achieve the objectives of decreasing and/or eliminating toxic pollutants. Abiotic decomposition by hydrolysis can be induced by decreasing soil pH, since protons, H^+, are essential in the reactions as indicated by reactions 2.6, 2.7, and 2.8. Anaerobic decomposition yielding many hazardous compounds can be controlled by changing the soil redox conditions. Soil aeration, water content and soil pH can be adjusted so that the redox potentials permit the aerobic decomposition of organics into H_2O and CO_2.

Biological Attenuation. – This is remediation of soils and the environment through promoting the growth and the activity of the organisms naturally present in the environment. It can also be called *bioattenuation*, and the process is closely related to bioremediation. The process of cleaning up the crude oil spill of the Exxon-Valdez, discussed above, is a good example of both bioattenuation or bioremediation, and a sharp distinction between the two is not present. However, in the strict sense of bioattenuation, the Exxon oil spill cleanup fits the definition of bioattenuation better. Finding the most effective native organisms and encouraging them to do the job in situ appear to be a crucial point in bioattenuation. As is the case in chemical attenuation, the soil properties can be changed to meet the requirements for a better growth of the organisms. To speed up the population growth, the indigenous organisms can be collected and cultured in the laboratory for seeding in the soil. A number of organic pollutants are positively charged and will be adsorbed by the negatively charged soil colloids, protecting them in this way against microbial decomposition. Changing the soil chemical reactions and/or adding materials to induce desorption may increase the destruction of the organic pollutants by the microbes. Changing soil pH may aid in promoting the growth of bacteria or higher fungal activity. The introduction of foreign microorganisms, as well as the introduction of new *superbugs*, produced by genetic engineering, that are most effective in degrading a wide range of chemical pollutants, as discussed earlier, can be very helpful, though such an introduction fuses the concept of bioattenuation with bioremediation.

Since the goal is to destroy organic pollutants into CO_2 and H_2O, an aerobic soil condition must be provided for such a microbial mineralization process. In anaerobic conditions, methanogenesis and fermentation produce a series of by-products, e.g., methane, ethanol, and other foul-smelling toxic substances (see Section 3.4.2). Therefore, promoting the growth of aerobic organisms is a crucial point for the success of bioattenuation. These are just a few examples and many more can be given. Not much is known on all these aspects, and the need for more research data is apparent to confirm the contentions above.

11.7 ENVIRONMENTAL IMPACT OF AGRICULTURAL AND INDUSTRIAL WASTE

11.7.1 Greenhouse Effect

The greenhouse effect is a phenomenon that causes the Earth's temperature to increase. Carbon dioxide and the other so-called *greenhouse gases*, CH_4, N_2O, and chlorofluorocarbons, emitted from the Earth's surface, move to the upper atmosphere, where they are heated by the sunlight. Energy from the sun's radiation is absorbed by these gases and emitted to the surface of the earth, resulting in an increase in temperature of the Earth's surface. The effect is similar to the warming effect attributed to the glass windows in *Florida rooms* or greenhouses. The warming of the interior of a car after standing in the sun is also a form of the greenhouse effect. In this case, the metal roof of the automobile is heated by the sun and becomes very hot. The heat absorbed increases the temperature of the interior of the car. Therefore, the greenhouse effect is not the result of blocking the sun's rays; rather the increase in temperature of the air on the Earth's surface is attributed to the invisible hot shield of CO_2 cloud. This increase in temperature not only may have a disastrous effect on the polar ice caps, but also many areas may be converted to desert lands.

The resulting global warming attributed to the greenhouse effect has been disputed recently. From the results of the most recent observations, some scientists indicate that the greenhouse effect does not uniformly affect the whole Earth, but only certain parts where the production of the pollutant gases is most intense, such as in regions where large industrial centers are located. It is contended that the increase in temperature is not long lasting, so that the long-term effect of global warming is negligibly small. A more serious challenge disputing the effect of CO_2 on global warming was discussed in preceding chapters. The contention was that the Earth's temperature today is still lower than that of the 3000-yr average, regardless of the fact that the amounts of greenhouses gases are on the rise.

Nevertheless, global warming and acid rain, which will be discussed in the next section, have become an environmental issue of global proportions. However, the fact that it has polarized the nations of the Northern and Southern Hemispheres is not exactly right. A misconception exists that large-scale burning of the rain forest occurred only in the "South," producing the CO_2. Most of the industrialized nations are indeed located in the Northern Hemisphere, but not all of the rain forest is in the Southern Hemisphere. A large part of the tropical rain forest is on the north site of the equator and is also cleared for crop production. It is an issue to be solved by both the nations that cleared the forest and those that are highly industrialized. The industry is perhaps also a bigger producer of CO_2. "Finger-pointing" will not solve the problem.

11.7.2 Acid Rain

The term *acid rain* was first used in the 19th century for the rain in the industrialized region of northwest England that contained acidic compounds (Wild, 1993). Acid rain is in fact a natural phenomenon occurring during any thunderstorm accompanied by heavy lightning or can result from volcanic eruptions. However, it is the *accelerated pollution* of the air with nitrogen- and sulfur-containing gases, emitted by industry, power stations, planes, and automobiles, that produces acid rain which has alarmed many people because of the increasing potential hazard for environmental degradation.

Acid rain is caused by the presence of sulfuric acid and nitric and nitrous acid in the rain drops. It is better called *acid precipitation*, because these acids can also be present in snow, ice, sleet, moisture in fog and smog, and other types of precipitation. The nitric and nitrous acids are formed by the reaction of NO_2 gas with moisture in the atmosphere. In the atmosphere, the NO_x gas is converted into NO_2 gas as illustrated by the following reactions:

$$2N_2O + O_2 \rightarrow 4NO \tag{11.8}$$

$$2NO + O_2 \rightarrow 2NO_2 \tag{11.9}$$

The NO_2 gas formed dissolves in raindrops yielding nitric and nitrous acids:

$$2NO_2 + H_2O \rightarrow HNO_2 + HNO_3 \tag{11.10}$$
$$\text{nitrous acid and nitric acid}$$

Nitric and nitrous acids dissolve in rainwater and dissociate their protons, H^+. Because of this, the pH of rainwater or water in fog decreases to 2.0 or lower.

Emission of S particulates in the air is another reason for the development of acid rain. The oxidation of S and formation of sulfuric acid are illustrated by the following reactions:

$$S + O_2 \rightarrow SO_2 \tag{11.11}$$

$$2SO_2 + O_2 \rightarrow 2SO_3 \tag{11.12}$$

$$2SO_2 + 2H_2O \rightarrow 2H_2SO_4 \tag{11.13}$$
$$\text{sulfuric acid}$$

Acid rain has become of increasing concern today in Europe and North America because of *die-back* of forest trees, a harmful process also known as *forest decline*. In Germany, Czechoslovakia, and Poland the die-back of the silver fir (*Abies alba*) forest is reported to be caused by acid rain. Observations by the author in the pine forests of northeastern France indicated that the trees affected by acid rain showed symptoms of severe Al toxicity. The question arises now whether acid rain was the direct reason for the forest decline, or whether the increased soil acidity has increased the Al concentrations to toxic levels killing the plants. A similar argument was presented for fish kill by acid rain as follows. Reports have indicated that the increased acidity of lake water in the Adirondack Mountains of the United States, because of acid rain, has contributed to *fish kill*. The disappearance of brown trout (*Salmo trutta*) in mountain lakes of Scandinavia is another example of the harmful effect of acidification of lake water by acid rain.

Two opposing theories are found in the literature explaining the damage to aquatic life by acidification of lake water. One group of scientists believes that fish kill is not caused by the increased acidity itself, but by increased Al concentration in lake water. The latter is produced by the accelerated weathering of rocks and minerals by acidic water. Survival and growth of the fry of brown trout have been noted to decrease in laboratory conditions at concentrations of 250 µg Al/L or lower (Wild, 1993). In contrast, the other theory indicates that fish are sensitive to water pH and will be harmfully affected by a water pH < 4.5 (Brady, 1990). The aquatic organisms perhaps, serving as food sources for the fish, are destroyed by the low pH. This may have contributed to the fish kill.

Acid rain is also corrosive to metals. The salts of these acids have been implicated in the formation of turbid haze, smog, and fog covering much of the industrial towns in the Midwest and California.

Today, the problem seems to be a global environmental issue. Acid rain and global warming have created a confrontation between the people in the Northern Hemisphere and those in the Southern Hemisphere, especially in the tropics. As discussed previously, the industries are located mostly in the North, whereas the tropics in the South are mostly inhabited by people from developing nations who cut the rain forests to make a living. As noted before, the industrial nations blame the South for global warming attributed to emission of polluting gases by large-scale burning of the rain forests. In turn, the South accuses the North of being the primary producers of pollutant gases responsible for acid rain.

The effect of acid rain is of a less serious nature in soils. Soils exhibit a cation exchange capacity (CEC) that provide them with a buffering capacity to adsorb the excess protons from acid rain. However, a prolonged impact by acid rain can saturate this buffer capacity in soils. From this point on, further addition of acid rain will increase soil acidity. Sandy soils are especially critical in this matter, since these soils are weakly buffered by nature, because of their low cation exchange capacity. Two methods have been suggested to control the effect of acid rain on soil acidity: (1) cleaning emissions in industry and motor vehicles and (2) liming soils and lakes. Combining the two methods is perhaps the best alternative.

11.7.3 Destruction of the Ozone Shield

Destruction of ozone in the stratosphere is another environmental issue resulting from accelerated emission of pollutant gases. It is due more to the effects of industrial- and especially transportation-related air pollution than to those of agricultural pollution of the air. Supersonic air transport is suspected of contributing to the destruction of ozone, because of CO_2 and H_2O emissions from combustion of hydrocarbon fuel.

Ozone is a form of oxygen with the formula O_3. It is a blue gas with a characteristic smell and its name is derived from the Greek term *ozein*, which means *to smell*. It is usually formed in air by a photolytic reaction, or the passage of electric sparks or arcs. It is a stronger oxidizing agent than ordinary oxygen, O_2, and it is capable of oxidizing organic substances in the atmosphere.

Ozone is a normal constituent of the stratosphere and is present as an invisible cloud at approximately 15 miles (24 km) above the Earth, called the *ozone layer*, shielding the Earth from harmful radiation from the sun. Under normal conditions, ozone is being destroyed and at the same time being reformed, maintaining in this way a photochemical equilibrium in the stratosphere. The destruction of ozone in nature takes place by the reaction of O_3 with hydroxyl, OH, groups derived from water vapor in the stratosphere. The process can be represented as follows (Moore and Moore, 1976):

$$O + H_2O \rightarrow 2OH \tag{11.14}$$

$$OH + O_3 \quad \rightarrow \quad HOO + O_2 \tag{11.15}$$

The biogeochemical importance of the ozone layer results from its capacity to absorb ultraviolet radiation at wavelengths between 360 and 240 nm. Such an absorption introduces a photochemical reaction, which decomposes the ozone molecule into ordinary oxygen as follows:

$$O_3 + hv \quad \rightleftarrows \quad O_2 + O \tag{11.16}$$

In natural conditions such a reaction is an equilibrium reaction, meaning that O_2 can react with O to form ozone again. In the reaction above, hv is the energy of UV radiation, which is used for the decomposition of ozone. In this way UV radiation from the sun is used, and blocked from reaching the earth surface. Ultraviolet radiation is destructive for many organic compounds, and if this ultraviolet light were permitted to reach the earth, life in its present form could not exist. Increased levels of UV radiation are linked to increased incidence of human skin cancer.

Since ozone reacts rapidly with organic compounds, the introduction of such compounds into the air damages the photochemical equilibrium of importance in formation and decomposition of ozone. Reaction (11.16) is then skewed toward the right, toward the decomposition of ozone. Combustion of fossil fuel and oxidation of organic compounds on earth yield H_2O and CO_2. Since H_2O can dissociate its OH ion (see reaction 11.14), the effect of H_2O is to accelerate the destruction of ozone. Other gases, responsible for decomposing ozone, are N_2O, NO, CH_3, CH_4, and gases from organic synthetics, such as $CFCl_3$ (freon) and CFCs (chlorofluorocarbons) used as refrigerants and aerosols. The decomposition of ozone by these gases can be illustrated by the following reactions:

$$Cl + O_3 \quad \rightarrow \quad ClO + O_2 \tag{11.17}$$
from freon

$$ClO + O \quad \rightarrow \quad Cl + O_2 \tag{11.18}$$

As can be noticed, the ClO produced in the decomposition of ozone again yields Cl. In this way, the destruction reaction can be perpetuated until all ozone is converted into O and O_2. It must be emphasized that the examples of the destruction of ozone above are simplified for an easy comprehension of the problem. In nature, the reaction of ozone with Cl, creating the *ozone hole* above Antarctica, is very complex. Today this hole was reported to have decreased in size due to conservation efforts in curtailing the use of freon and in decreasing the production of other pollutant gases.

On the Earth's surface, incomplete oxidation products of hydrocarbons from automobiles can react with NO_2 to form peroxyacetyl nitrate, the so-called *PAN*. When present in smog at concentrations larger than 0.3 ppm, PAN and ozone are severe irritants to humans.

The soil is a natural source of these pollutant gases. Respiration of roots and microorganisms produces CO_2. Aerobic decomposition of organic matter produces H_2O and CO_2, whereas anaerobic decomposition of organic matter yields CH_4, H_2S and other pollutant gases. However, the natural production of these gases is small compared to that from combustion of fossil fuel and the large-scale burning of the rain forests.

APPENDIX A

Atomic Weights of the Major Elements in Soils

Element	Symbol	Atomic number	Atomic weight
Aluminum	Al	13	027.0
Antimony	Sb	51	121.8
Argon	Ar	18	039.9
Arsenic	As	33	074.9
Barium	Ba	56	137.3
Beryllium	Be	04	009.0
Bismuth	Bi	83	209.0
Boron	B	05	010.8
Bromine	Br	35	079.9
Calcium	Ca	20	040.1
Carbon	C	06	012.0
Cesium	Cs	55	132.9
Chlorine	Cl	17	035.5
Chromium	Cr	24	052.0
Cobalt	Co	27	058.9
Copper	Cu	29	063.5
Fluorine	F	09	019.0
Gallium	Ga	31	069.7
Germanium	Ge	32	072.6

Element	Symbol	Atomic number	Atomic weight
Gold	Au	79	197.0
Helium	He	02	004.0
Hydrogen	H	01	001.0
Iodine	I	53	126.9
Iridium	Ir	77	192.2
Iron	Fe	26	055.9
Krypton	Kr	36	083.8
Lanthanum	La	57	138.9
Lead	Pb	82	207.2
Lithium	Li	03	006.9
Magnesium	Mg	12	024.2
Manganese	Mn	25	054.9
Mercury	Hg	80	200.6
Molybdenum	Mo	02	095.9
Nickel	Ni	28	058.7
Nitrogen	N	07	014.0
Oxygen	O	08	016.0
Phosphorus	P	15	031.0
Platinum	Pt	78	195.1
Potassium	K	19	039.1
Radon	Rn	86	222.0
Radium	Ra	88	226.1
Rhodium	Rh	45	102.9
Rubidium	Rb	37	085.5
Selenium	Se	34	079.0
Silicon	Si	14	028.1
Silver	Ag	47	107.9
Sodium	Na	11	023.0
Strontium	Sr	38	087.6
Sulfur	S	16	032.1
Tantalum	Ta	03	180.9
Tellurium	Te	52	127.6

Element	Symbol	Atomic number	Atomic weight
Thallium	Tl	81	204.4
Thorium	Th	90	232.1
Tin	Sn	50	118.7
Titanium	Ti	22	047.9
Tungsten	W	74	183.9
Uranium	U	92	238.0
Vanadium	V	23	050.9
Xenon	Xe	54	131.3
Yttrium	Y	39	088.9
Zinc	Zn	30	065.4
Zirconium	Zr	40	091.2

APPENDIX B

International System of Units (SI)

Basic unit	Symbol
Ampere (electrical current)	A
Candela (luminous intensity)	cd
Kelvin (thermodynamic temperature)	K
Kilogram (mass)	kg
Meter (length)	m
Mole (amount of substance)	mol
Second (time)	s

Factors for Converting US Units into SI Units

US unit	SI unit	To obtain SI unit multiply US unit by
Acre	Hectare, ha	0.405
Acre	Square meter, m^2	4.05×10^3

US unit	SI unit	To obtain SI unit multiply US unit by
Ångstrom	Nanometer, nm	10^{-1}
Atmosphere	Megapascal, MPa	0.101
Bar	Megapascal, MPa	10^{-1}
Calorie	Joule, J	4.19
Cubic foot	Liter, L	28.3
Cubic inch	Cubic meter, m^3	1.64×10^{-5}
Curie	Becquerel, Bq	3.7×10^{10}
Degrees, °C (+273, temperature)	Degrees, K	1
Degrees, °F (-32, temperature)	Degrees, °C	0.556
Dyne	Newton, N	10^{-5}
Erg	Joule, J	10^{-7}
Foot	Meter, m	0.305
Gallon	Liter, L	3.78
Gallon per acre	Liter per ha	9.35
Inch	Centimeter, cm	2.54
Micron	Micrometer, µm	1
Mile	Kilometer, km	1.61
Mile per hour	Meter per second	0.477
Millimho per cm	decisiemens per m, dS m^{-1}	1
Ounce (weight)	Gram, g	28.4
Ounce (fluid)	Liter, L	2.96×10^{-2}
Pint	Liter, L	0.473
Pound	Gram, g	454
Pound per acre	Kilogram per ha	1.12
Pound per cubic foot	Kilogram per m^3	16.02
Pound per square foot	Pascal, Pa	47.9
Pound per square inch	Pascal, Pa	6.9×10^3
Quart	Liter, L	0.946
Square foot	Square meter, m^2	9.29×10^{-2}

US unit	SI unit	To obtain SI unit multiply US unit by
Square inch	Square cm, cm^2	6.45
Square mile	Square kilometer, km^2	2.59
Ton (2000 lbs)	Kilogram, kg	907
Ton per acre	Megagram per ha, Mg/ha	2.24

APPENDIX C

US Weights and Measures

LAND MEASURE

1 foot	12 inches
1 yard	3 feet
1 mile	5,280 feet
1 mile	1,760 yards
1 square foot	144 square inches
1 square yard	9 square foot
1 square mile	640 acres
1 acre	4,840 square yards
1 acre	43,560 square feet

AVOIRDUPOIS WEIGHT
(Used in weighing all articles except drugs, gold, silver and precious stones)

1 pound, lb	16 ounces
1 hundredweight (cwt)	100 lbs
1 ton	20 cwt
1 ton	2000 lbs
1 long ton	2240 lbs

TROY WEIGHT (Used in weighing gold, silver and precious stones)

1 pound, lb	12 ounces

DRY MEASURE

1 quart	2 pints
1 bushel	32 quarts

LIQUID MEASURE

1 quart	2 pints
1 gallon	4 quarts
1 barrel	31.5 gallons

FLUID MEASURE

1 fluid dram	60 minims
1 fluid ounce	8 fluid drams
1 pint	16 fluid ounce
1 gallon	8 pints

GRAIN WEIGHTS PER BUSHEL

Barley	48 lbs
Beans	60 lbs
Bran	20 lbs
Buckwheat	42-52 lbs
Clover seed	60 lbs
Corn (in the ear husked)	70 lbs
Corn (shelled)	56 lbs
Corn meal	48 lbs
Flax seed	56 lbs
Malt	30-38 lbs
Millet seed	50 lbs
Oats	32 lbs
Peas	60 lbs
Rye	56 lbs
Wheat	60 lbs

REFERENCES AND ADDITIONAL READINGS

Agassi, M., I. Shainberg, and J. Morin. 1981. Effect of electrolyte concentration and sodicity on infiltration rate and crust formation. Soil Sci. Soc. Am. J. 45: 848-851.

Ahmad, F., and K. H. Tan. 1986. Effect of lime and organic matter on soybean seedlings grown in aluminum toxic soil. Soil Sci. Soc. Am. J. 50: 656-661.

Aiken, G. R., D. M. McKnight, R. L. Wershaw, and P. MacCarthy (eds). 1985. *Humic Substances in Soil, Sediments, and Water.* Wiley Interscience., New York, NY.

Alexander, M. 1965. Nitrification. pp. 307-343. In: *Soil Nitrogen*, W. V. Bartholomew and F. E. Clark (eds). Agronomy series No.10, Am. Soc. Agronomy, Inc., Publ., Madison, WI.

Aldrich, S. R., W. O. Scott, and E. R. Long. 1975. *Modern Corn Production.* A & L Publications, Champaign, IL.

Allison, F. E. 1965. Decomposition of wood and bark sawdusts in soil, nitrogen requirements, and effects on plants. USDA-ARS Techn. Bull. 1332.

Allison, F. E., and J. H. Doetsch. 1951. Nitrogen gas production by the reaction of nitrites with amino acids in slightly acidic media. Soil Sci. Soc. Am. Proc. 15: 163-166.

Amano, Y. 1981. Phosphorus status of some Andosols in Japan. Japan Agric. Res. Quaterly. (JARQ), 15: 14-21.

Anderson, O. E., H. F. Perkins, R. L. Carter, and J. B. Jones. 1971. Plant Nutrient Survey of Selected Plants and Soils of Georgia. University of Georgia, College Agric. Expt. Stns. Research Report

102, Athens, GA.

Anonymous. 1990. Farewell phosphogypsum. Agric. Age 34: 17-18.

Arnold, R. W., and C. A. Jones. 1987. Soils and climate effects upon crop productivity and nutrient use. pp. 9-17. In: *Soil Fertility and Organic Matter as Critical Components of Production Systems*, R.F. Follett (chair). Soil Sci. Soc. Am., Inc., and Am. Soc. Agronomy, Inc., Publ., Madison, WI.

Ashworth, W. 1991. *The Encyclopedia of Environmental Studies*. Facts on File, New York, NY.

Azevedo, J., and P. R. Stout. 1974. Farm Animal Manures: An Overview of Their Role in the Agricultural Environment. University of California, Calif. Agric. Expt. Stn., Extension Service Manual 44.

Bajaj, Y. P. S. (ed). 1989. *Biotechnology in Agriculture and Forestry. Trees II*. Vol. 5. Springer-Verlag, Berlin, Heidelberg, New York, NY.

Baver, L. D. 1956. *Soil Physics*. 3rd edition. Wiley, New York, NY.

Baver, L. D. 1963. The effect of organic matter on soil structure. Pontificiae Academiae Scientiravm Scripta Varia, 32: 383-413.

Beaty, E. R., K. H. Tan, R. A. McCreery, J. H. Edwards, Jr., and R. L. Stanley. 1976. Returned clippings and N fertilization on bahia grass herbage production and nitrogen and organic matter contents of soil. Agron. J. 68: 384-387.

Bell, F. W., and E. R. Canterbery. 1976. *Aquaculture for the Developing Countries*. Ballinger Publishing Co., Cambridge, MA.

Bernhard-Reversat, F., C. Huttel, and G. Lemee. 1979. Structure et fonctionnement des écosystèmes de la forêt pluvieuse sempervirente de la Côte d'Ivoire. UNESCO-Paris, Rech. Res. Nat. 14: 605-625.

Birkeland, P. W. 1974. *Pedology, Weathering and Geomorphological Research*. Oxford University Press, London.

Birrell, K. S., and W. Fieldes. 1952. Allophane in volcanic ash soils. J. Soil Sci. 3: 156-166.

Bonga, J. M., and D. J. Durzan. 1982. *Tissue Culture in Forestry*. Martinus Nijhoff, Boston, MA.

Brady, N. C. 1984. *The Nature and Properties of Soils*. 9th edition. Macmillan Publ. Co., New York, NY.

Brady, N. C. 1990. *The Nature and Properties of Soils*. 10th edition. Macmillan Publ. Co., New York, NY.

Brady, N. C., and R. R. Weil. 1996. *The Nature and Properties of Soils*. 11th edition. Prentice Hall, Upper Saddle River, NJ.

Bremner, J. M. 1965. Organic nitrogen in soils. pp. 93-149. In: *Soil Nitrogen*, W. V. Bartholomew and F. E. Clark (eds.). Agronomy series 10. Am. Soc. Agronomy, Inc., Publ., Madison, WI.

Brewer, R., and J. R. Sleeman. 1988. *Soil Structure and Fabric*. C.S.I.R.O., Div. Soils, Adelaide, Australia.

Broadbent, F. E., and F. E. Clark. 1965. Denitrification. pp. 344-359. In: *Soil Nitrogen*, W. V. Bartholomew and F. E. Clark (eds). Agronomy series 10. Am. Soc. Agronomy, Inc., Publ., Madison, WI.

Bumpus, J. A. 1993. White rot fungi and their potential use in bioremediation processes. In: *Soil Biochemistry*, Jean-Marc Bollog and G. Stotzky (eds) 8: 65-100.

Buol, S. W., F. D. Hole, and R. J. McCracken. 1973. *Soil Genesis and Classification*. The Iowa State University Press, Ames, IA.

Burns, R. G. 1986. Interaction of enzymes with soil minerals. pp. 429-451. In: *Interaction of Soil Minerals with Natural Organics and Microbes*, P. M. Huang and M. Schnitzer (eds). Soil Sci. Soc. Am. Special Publ. No.17. Soil Sci. Soc. Am., Inc., Madison, WI.

Campbell, W. A., and F. F. Hendrix, Jr. 1974. Diseases of feeder roots. pp. 219-243. In: *The Plant Root and Its Environment*, E. W. Carson (ed). Univ. Press of Virginia, Charlottesville, VA.

Cannon, W. A., and E. E. Free. 1925. Physiological features of roots, with especial references to the relation of roots to aeration of the soil. Carnegie Inst. Publ. 368. Washington, DC.

CAST. 1980. Organic Farming and Technology. Council for Agricultural Science and Technology. Ames, IA.

Chandler, Jr., R. F. 1968. Dwarf rice - A giant in tropical Asia. pp. 252-255. In: *Science for Better Living*, USDA, The Yearbook of Agriculture. USDA, U.S. Govern. Printing Office, Washington, DC.

Chang, H. T., and W. E. Loomis. 1945. Effect of carbon dioxide on absorption of water and nutrients by roots. Plant Phys., 20: 221-232.

Chenu, C. 1995. Extracellular polysaccharides. An interface between microorganisms and soil constituents. pp. 217-229. In: *Environmental Impact of Soil Component Interactions. Natural and Anthropogenic Organics*, Vol. I, P. M. Huang, J. Berthelin, J.-M. Bollag, W. B. McGill, and A. L. Page (eds). CRC, Lewis Publ., Boca Raton, FL.

Clark, C. J., and M. B. McBride. 1984. Cation and anion retention by natural and synthetic allophane and imogolite. Clays Clay Mineral: 32: 291-299.

Clark, F. E., and K. H. Tan. 1969. Identification of polysaccharide ester linkage in humic acid. Soil Biol. Biochem. 1: 75-81.

Clarke, J. W. 1924. The data of geochemistry. U.S. Geol. Survey Bull. 770.

Cole, R. J. 1996. Toxin to get competition from a nontoxic cousin. Agric. Research 44: 23.

Conn, E. E., and P. K. Stumpf. 1967. *Outlines of Biochemistry*. 2nd edition. Wiley & Sons, Inc., New York, NY.

Cooper, A. 1976. *Nutrient Film Technique for Growing Crops*. Grower Books, London.

De Coninck, F. 1980. Major mechanisms in formation of spodic horizons. Geoderma 24: 101-128.

De Coninck, F., and J. A. McKeague. 1985. Micromorphology of spodosols. pp. 121-144. In: *Micromorphology and Soil Classification*, J. A. Douglas, and M. L. Thompson (eds). SSSA Special Publ. No. 15. Soil Sci. Soc. Am., Madison, WI.

Deevey, Jr., E. S. 1960. The human population. Scientific American, 203: 195-204.

Dell'Agnola, G., and S. Mardi. 1986. News about biological effect of humic substances. pp. 78-88. In: *Humic Substances. Effects on Soil and Plants,* Reda (ed). Ramo Editorial Degli Agricoltora, Milan.

Del Tredici, P. 1980. Legumes aren't the only nitrogen fixers. Horticulture, pp. 30-33.

Deshpande, T. L., D. J. Greenland, and J. P. Quirk. 1964. Role of iron oxides in the bonding of soil particles. Nature 201: 107-108.

Dobereiner, J., and J. M. Day. 1975. Nitrogen fixation in the rhizosphere of tropical grass. pp. 39-56. In: *Nitrogen Fixation by Free-*

living Micro-Organisms, W. D. P. Stewart (ed). Intern. Biol. Programme 6. Cambridge University Press, New York, NY.

Donahue, R. L., R. W. Miller, J. C. Shickluna. 1983. *Soils. An Introduction to Soils and Plant Growth*. Prentice-Hall, Inc., Englewood Cliffs, NJ.

Douglas, J. S. 1976. *Advanced Guide to Hydroponics*. Drake Publ., Inc., New York, NY.

Douchaufour, P. 1976. *Atlas Écologique des Sols du Monde*. Masson, Paris, New York, Barcelona, Milan.

Dudal, R. 1976. Inventory of the major soils of the world with special reference to mineral stress hazards. pp. 3-14. In: *Plant Adaptation to Mineral Stress in Problem Soils*, M. J. Wright (ed). Proc. Workshop Cornell University, Ithaca, NY. FAO-Food and Agriculture Organization. 1982. FAO Production Yearbook, Vol.36, U.N.-F.A.O., Rome.

Emerson, W. W., R. C. Foster, and J. M. Oades. 1986. Organo-mineral complexes in relation to soil aggregation and structure. pp. 521-548. In: *Interactions of Soil Minerals with Natural Organics and Microbes*, P. M. Huang, and M. Schnitzer (eds). SSSA Special Publ. No.17. Soil Sci. Soc. Am., Madison, WI.

Food and Agriculture Organization. 1987. Agriculture: Toward 2000. U.N.-F.A.O., Rome.

Fearnside, P. M. 1987. Causes of deforestation in the Brazilian Amazon. pp. 37-61. In: *The Geophysiology of Amazonia. Vegetation and Climate Interactions*, R. E. Dickenson (ed). Wiley & Sons, New York, NY.

Flaig, W. 1971. Organic compounds in soils. Soil Sci., 111: 19-33.

Flaig, W. 1975. An introductory review on humic substances: Aspects of research on their genesis, their physical and chemical properties, and their effect on organisms. pp. 19-42. In: *Humic Substances. Their Structure and Function in the Biosphere*, D. Povoledo and H. L. Golterman (eds). Centre Agri. Publ. and Documentation, Wageningen, The Netherlands.

Follett, R. F., S. C. Gupta, and P. G. Hunt. 1987. Conservation practices. Relation to the management of plant nutrients for crop production. pp.19-51. In: *Soil Fertility and Organic Matter as Critical Components of Production Systems*, R. F. Follett (chair).

Soil Sci. Soc. Am., Inc., and Am. Soc. Agronomy, Inc., Publ., Madison, WI.

Foth, H. D., 1990. *Fundamentals of Soil Science*. 8th edition. Wiley & Sons, New York, NY.

Frear, G. L., and J. Johnston. 1929. The solubility of calcium carbonate (calcite) in certain acqueous solutions at 25°. Amer. Chem. Soc. J., 51: 2082-2093.

Fuchsman, C. H. 1980. *Peat: Industrial Chemistry and Technology*. Academic Press, New York, NY.

Garrels, R. M., and C. L. Christ. 1965. *Solutions, Minerals and Equilibria*. Harper and Row, New York, NY.

Gerse, J., R. Kremo, J. Csicsor, and L. Pinter. 1994. Application of humic acids and their derivatives in environmental pollution control. pp. 1297-1302. In: *Humic Substances in the Global Environment and Implications on Human Health*, N. Senesi and T. M. Miano (eds). Proc. 6th Int. Meeting Int. Humic Acid Soc., Monopoli (Bari), Sept. 22-25, 1992. Elsevier, Amsterdam.

Genevini, P. L., G. A. Sacchi, and D. Borio. 1994. Herbicide effect of atrazine, diuron, linuron, and prometon after interaction with humic acids from coal. pp. 1291-1296. In: *Humic Substances in the Global Environment and Implications on Human Health*, N. Senesi, and T. M. Miano (eds). Proc. 6th Int. Meeting Int. Humic Acid Soc., Monopoli (Bari), Sept. 22-25, 1992. Elsevier, Amsterdam

Ghuman, B. S., and R. Lal. 1987. Effects of deforestation on soil properties and microclimate of a high rainforest in South Nigeria. pp. 225-244. In: *The Geophysiology of Amazonia. Vegetation and Climate Interactions*, R. E. Dickinson (ed). Wiley & Sons, New York, NY.

Gill, W. R., and R. D. Miller. 1956. A method for study of the influence of mechanical impedance and aeration on the growth of seedling roots. Soil Sci. Soc. Amer. Proc. 20: 154-157.

Gingrich, J. R., and M. B. Russell. 1956. Effect of soil moisture tension and oxygen concentration on the growth of corn roots. Agronomy J. 48: 517-520.

Goenadi, D. H., and K. H. Tan. 1991. The weathering of paracrystalline clays into kaolinite in Andosols and Ultisols in

Indonesia. Indonesian J. Trop. Agric., 2: 56-65.

Golley, F. B., J. T. McGinnis, R. G. Clements, G. I. Child, and M. J. Duever. 1975. *Mineral Cycling in a Tropical Moist Forest Ecosystem*. University of Georgia Press, Athens, GA.

Gorniok, A., T. Latour, and A. Nowacka. 1972. Proc. Fourth Int. Peat Congress, Otenieui, Finland, pp. 61-66.

Gortner, R. A. 1949. *Outlines of Biochemistry*. (Third edition, edited by R. A. Gortner, Jr., and W. A. Gortner). Wiley & Sons, Inc., New York, NY.

Granli, T., and O. C. Bøckman. 1993. Nitrous oxide (N_2O) from agriculture. Bull. Norsk Hydro Research Centre, Porsgrunn, Norway.

Greenland, D. J., and C. J. B. Mott. 1978. Surfaces of soil particles. pp. 321-353. In: *The Chemistry of Soil Constituents*, D. J. Greenland, and M. H. B. Hayes (eds). Wiley and Sons, New York, NY.

Grimshaw, R. W. 1971. *The Chemistry and Physics of Clays and Allied Ceramic Materials*. Wiley-Interscience, New York, NY.

Harley, J. L. 1969. *The Biology of Mycorrhiza*. 2nd edition. Leonard Hill, London.

Harpstead, D. D. 1975. Man-molded cereal-hybrid corn's story. pp. 213-224. In: *That We May Eat*. The Yearbook of Agriculture. USDA., U.S. Govern. Printing Office, Washington, DC.

Havelka, U. D., M. G. Boyle, and R. W. F. Hardy. 1982. Biological nitrogen fixation. pp. 365-422. In: *Nitrogen in Agricultural Soils*, F. J. Stevenson (ed). Am. Soc. Agronomy, Madison, WI.

Hays, S. M. 1996. Golden nematodes are anything but. Agric. Research 44: 16-17.

Henderson-Sellers, A. 1987. Effects of change in land use on climate in the humid tropics. pp. 463-496. In: *The Geophysiology of Amazonia. Vegetation and Climate Interactions*. R. E. Dickinson (ed). Wiley & Sons, New York, NY.

Hensley, D. L., and P. L. Carpenter. 1979. The effects of temperature on N_2-fixation (C_2H_2 reduction) by nodules of legume and actino-mycete-nodulated woody species. Bot. Gazette 140 (supplement): 558-564.

Herrera, R. A. 1979. Nutrient distribution and cycling in an Amazon

Caatinga forest on Spodosols in southern Venezuela. Thesis, University of Reading, England.

Hewitt, E. J. 1966. Sand and water culture methods used in the study of plant nutrition. Techn. Comm. No. 22 (revised). Commonwealth Agric. Bureaux, Maidstone, Kent, England.

Hillel, D. 1980. *Fundamentals of Soil Physics*. Academic Press, New York, NY.

Hillel, D. 1998. *Environmental Soil Physics*. Academic Press, San Diego, CA.

Hoagland, D. R., and D. I. Arnon. 1950. The water-culture method for growing plants without soil. University of California, Agric. Expt. Stn. Cir. 347. Berkeley, CA.

Hodgson, B. 1990. Alaska's big spill. Can the wilderness heal? Nat. Geogr. 177: 5-42.

Hrubrant, G. R., R. V. Daugherty, and R. A. Rhodes. 1972. Enterobacteria in feedlot waste and runoff. Appl. Microbiol., 24: 378-383.

Huang, P. M. 1995. The role of short-range order mineral colloids in abiotic transformation of organic components in the environments. pp. 135-167. In: *Environmental Impact of Soil Component Interactions. Natural and Anthropogenic Organics,* Vol. I, P. M. Huang, J. Berthelin, J.-M. Bollag, W. B. McGill, and A. L. Page (eds). CRC, Lewis Publ., Boca Raton, FL.

Huang, P. M., and A. Violante. 1986. Influence of organic acids on crystallization and surface properties of precipitation products of aluminum. pp. 159-221. In: *Interaction of Soil Minerals with Natural Organics and Microbes*, P. M. Huang, and M. Schnitzer (eds). SSSA Special Publ. No.17. Soil Sci. Soc. Am., Madison, WI.

Hubbell, D. S., and G. Staten. 1951. Studies on soil structure. New Mexico Agri. Expt. Sta. Tech. Bull. 363. New Mexico State University, Las Cruces, NM.

Hu Han and Shao Qiquan. 1981. Advances in plant cell and tissue culture in China. Advances in Agronomy, 34: 1-13.

Hunt, C. B. 1972. *Geology of Soils. Their Evolution, Classification, and Uses*. Freeman and Co., San Francisco, CA.

Hunter, C., and E. M. Rich. 1925. The effect of artificial aeration of the soil on *Impatiens balsamina* L. New Phytol., 24: 257-271.

Hurlbut, Jr., C. S., and C. Klein. 1977. *Manual of Mineralogy* (after

James D. Dana). Wiley & Sons, New York, NY.

Hutchinson, G.E. 1944. Nitrogen in the biochemistry of the atmosphere. Am. Scientist, 32: 178-195.

Inoue, K., and P. M. Huang. 1984. Influence of citric acid on the natural formation of imogolite. Nature (London), 308: 58-60.

Irianto, B. 1991. Effect of humic acid on tissue culture of slash pine (*Pinus elliottii* Engelm.). M.S. Thesis, University of Georgia, Athens, GA.

Irianto, B., and K. H. Tan. 1993. Effect of humic acid on callus culture of slash pine (*Pinus elliottii* Engelm). J. Plant Nutrition, 16: 1109-1118.

Jackson, T. A. 1975. Humic matter in natural waters and sediments. Soil Sci., 119: 56-64.

Jenny, H. 1941. *Factors of Soil Formation*. McGraw-Hill, New York, NY.

Jensen, H. L. 1965. Nonsymbiotic nitrogen fixation. pp. 436-480. In: *Soil Nitrogen*. W. V. Bartholomew and F. E. Clark (eds). Agronomy series 10. Am. Soc. Agronomy, Inc., Publ., Madison, WI.

Jones, J. B., Jr. 1983. *A Guide for the Hydroponic and Soilless Culture Grower*. Timber Press, Portland, Oregon.

Jones, J. B., Jr. 1997. *Hydroponics. A Practical Guide for the Soilless Grower*. St Lucie Press, Boca Raton, FL.

Jones, J. B., Jr., B. Wolf, and H.A. Mills. 1991. *Plant Analysis Handbook*. Micro-Macro Publ., Inc., Athens, GA.

Jones, R. A. 1983. Potential of acid resistant and fungicide resistant *Rhizobium japonicum* strains for improvement of nitrogen fixation in soybeans. Ph.D. dissertation, University of Georgia, Athens, GA.

Jordan, C. F. 1985. *Nutrient Cycling in Tropical Forest Ecosystems*. Wiley & Sons, New York, NY.

Jurcsik, I. 1994. Possibilities of applying humic acids in medicine (wound healing and cancer therapy). pp. 1331-1340. In: *Humic Substances in the Global Environment and Implications in Human Health*, N. Senesi, and T. M. Miano (eds). Proc. 6th Intern. Meetings Intern. Humic Subst. Soc., Monopoli (Bari), Italy, Sept. 20-25, 1992. Elsevier, Amsterdam.

Kafuku, T., and H. Ikenoue (eds). 1983. *Modern Methods of Aquaculture in Japan*. Kodansha Ltd., Tokyo, and Elsevier Sci. Publ., Co., Amsterdam.

Kardos, L. T., W. E. Sopper, and E. A. Myers. 1968. A living filter for sewage. pp. 197-201. In: *Science for Better Living*. The Yearbook of Agriculture, USDA, U.S. Govern. Printing Office, Washington, DC.

Keen, B. A. 1931. *The Physical Properties of the Soils*. Longmans, Green, New York, NY.

Keller, W. D. 1954. Bonding energies of some silicate minerals. Am. Mineral., 39: 783-793.

Kellogg, C. E. 1941. Climate and soil. pp. 265-291. In: *Climate and Man*. The Yearbook of Agriculture, USDA, U.S. Govern. Printing Office, Washington, DC.

Kennedy, I. R. 1992. *Acid Soil and Acid Rain*. Second edition. Res. Studies Press, Taunton, Somerset, England.

King, L. D. (ed). 1986. Agriculture use of municipal and industrial sludges in the Southern United States. South. Coop. Series Bull. 314. North Carolina State University, Raleigh, NC.

King, L. D., and P. M. Giordano. 1986. Effect of sludges on heavy metals in soils and crops. pp. 21-29. In: *Agricultural Use of Municipal and Industrial Sludges in the Southern United States*, L. D. King (ed). South. Coop. Series Bull. 314, North Carolina State University, Raleigh, NC.

King, L. D., R. W. Taylor, and J. W. Shuford. 1986. Macro-nutrients in municipal and industrial sludges and crop response to sludge applications. pp. 11-19. In: *Agricultural Use of Municipal and Industrial Sludges in the Southern United States*, L. D. King (ed). South. Coop. Series Bull. 314, North Carolina State University, Raleigh, NC.

Kloecking, H. P. 1994. Influence of humic acids and humic acid-like polymers in fibrinolytic and coagulation system. pp. 1337-1340. In: *Humic Substances in the Global Environment and Implications on Human Health*, N. Senesi and T. M. Miano (eds). Proc. 6th Intern. Meeting Intern. Humic Substances Soc., Monopoli, Bari, Italy, Sep.20-25, 1992. Elsevier, Amsterdam.

Kononova, M. M. 1975. Humus of virgin and cultivated soils. pp. 475-

526. In: *Soil Components*, Vol. 1, *Organic Components*, J. E. Gieseking (ed). Springer-Verlag, Berlin.

Krauskopf, K. B. 1956. Dissolution and precipitation of silica at low temperatures. Geochim. Cosmochim. Acta, 10: 1-26.

Kubiena, W.L. 1938. *Micropedology*. Collegiate Press, Inc., Ames, IA.

Kyte, L. 1983. *Plants from Test Tube. An Introduction to Micropropation*. Timber Press, Portland, OR.

Ladd, J. N., and R. B. Jackson. 1982. Biochemistry of ammonification. pp. 173-228. In: *Nitrogen in Agricultural Soils*, F. J. Stevenson (ed). Am. Soc. Agronomy, Madison, WI.

Lagerwerff, J. V. 1972. Lead, mercury, and cadmium as environmental contaminants. pp. 593-636. In: *Micronutrients in Agriculture*, J. J. Mortvedt, P. M. Giordano, and W. L. Lindsay (eds). Soil Sci. Soc. Am., Madison, WI.

LaRue, T. A., and T. G. Patterson. 1981. How much nitrogen do legumes fix? Advances in Agronomy, 34: 15-38.

Lavelle, P. 1984. The soil system in the humid tropics. Biol. Int., 9: 2-17.

Lavelle, P. 1987. Biological processes and productivity of soils in the humid tropics. pp. 175-214. In: *The Geophysiology of Amazonia. Vegetation and Climate Interactions*, R. E. Dickinson (ed). Wiley & Sons, New York, NY.

Lawton, K. 1945. The influence of soil aeration on the growth and absorption of nutrients by corn plants. Soil Sci. Soc. Am. Proc., 10: 263-268.

LeBaron, H. M. (ed). 1987. *Biotechnology in Agricultural Chemistry*. ACS Symposium Series 334. American Chemical Society, Washington, DC.

Levy, G. J. 1996. Soil stabilizers. pp. 267-299. In: *Soil Erosion, Conservation and Rehabilitation*, M. Agassi (ed). Marcel Dekker, New York, NY.

Lindsay, W. L., M. Peech, and J. S. Clark. 1959. Solubility criteria for the existence of variscite in soils. Soil Sci. Soc. Am. Proc., 23: 357-360.

Lobartini, J. C., G. A. Orioli, and K. H. Tan. 1997. Characteristics of soil humic fractions separated by ultrafiltration. Comm. Soil Sci. Plant Anal. 28: 787-796.

Lobartini, J. C., K. H. Tan, J. A. Rema, A. R. Gingle, C. Pape, and D. S. Himmelsbach. 1992. The geochemical nature and agricultural importance of commercial humic matter. The Science of the Total Environment, 113: 1-15.

Loehr, R. C. 1974. *Agricultural Waste Management: Problems, Processes and Approaches*. Academic Press, New York, NY.

Lutz, H. J., and R. F. Chandler. 1947. The organic matter of forest soils. pp.140-197. In: *Forest Soils*, R. F. Chandler (ed). Wiley & Sons, New York, NY.

Lyons-Johnson, D. 1996. Nematode takes on Japanese beetle grubs. Agric. Research 44: 12-13.

MacCarthy, P., and J. A. Rice. 1994. Industrial application of humus: An overview. pp. 1209-1223. In: *Humic Substances in the Global Environment and Implications on Human Health*, N. Senesi, and T. M. Miano (eds). Proc. 6th Int. Meeting Int. Humic Acid Soc., Monopoli (Bari), Sept. 22-25, 1992. Elsevier, Amsterdam.

Mackenzie, R. C. 1975. The classification of soil silicates and oxides. pp. 1-25. In: *Soil Components*, Vol. 2, *Inorganic Components*, J. E. Gieseking (ed). Springer-Verlag, New York, NY.

Manahan, S. E. 1975. *Environmental Chemistry*. Willard Grant Press, Boston, MA.

Martin, P. A. 1996. Banking on Bt. Agri. Research 44: 20-21.

Marx, D. H. 1969. The influence of ectotrophic mycorrhizal fungi on the resistance of pine roots to root pathogenic fungi and soil bacteria. Phytopathology, 59: 153-163.

Marx, D. H., and S. V. Krupa. 1978. Mycorrhizae. A. Ectomycorrhizae. pp. 373-400. In: *Interactions between Nonpathogenic Soil Microorganisms and Plants*, V. R. Dommerques and S. V. Krupa (eds). Elsevier Sci. Publ. Co., Amsterdam.

Mason, B. 1958. *Principles of Geochemistry*. Wiley & Sons, New York, NY.

McBride, J. 1979. Pushing plants to full potential. Agri. Research 28: 14-15.

McCalla, T. M., J. R. Peterson, and C. Lue-Hing. 1977. Properties of agricultural and municipal wastes. pp. 11-43. In: *Soils for Management of Organic Wastes and Waste Waters*, L. F. Elliot and F. J. Stevenson (eds). Soil Sci. Soc. Am., Am. Soc. Agronomy and

Crop Sci. Soc. Am., Madison, WI.

Meek, B. D., T. J. Donavan, and L. E. Graham. 1980. Summertime flooding effects on alfalfa mortality, soil oxygen concentration, and matric potential in a silty clay loam soil. Soil Sci. Soc. Am. J., 44: 433-435.

Miller, R. M., and D. T. Gardiner. 1998. *Soils in Our Environment*. 8th edition. Prentice Hall, Upper Saddle River, NJ.

Millot, G. 1964. *Geologie des Argiles: Alterations, Sedimentologie, Geochimie*. Masson, Paris.

Ming, D. W., and D. L. Henninger (eds). 1989. *Lunar Base Agriculture: Soils for Plant Growth*. Am. Soc. Agronomy, Inc., Crop Sci. Soc. Am., Inc., and Soil Sci. Soc. Am., Inc., Madison, WI.

Mohr, E. C. J., and F. A. Van Baren. 1960. *Tropical Soils*. Les Editions A. Manteau, S.A., Bruxelles.

Mohr, E. C. J., F. A. Van Baren, and J. Van Schuylenborgh. 1972. *Tropical Soils*. Mouton-Ichtiar Baru-Van Hoeve, The Hague.

Mori, S. A., and G. T. Prance. 1987. Species diversity, phenology, plant-animal interactions, and their correlation with climate, as illustrated by the Brazil nut family (Lecythidaceae). pp. 69-101. In: *The Geophysiology of Amazonia. Vegetation and Climate Interactions*, R. E. Dickenson (ed). Wiley & Sons, New York, NY.

Moore, J. W., and E. A. Moore. 1976. *Environmental Chemistry*. Academic Press. New York, NY.

Murashige, T., and F. Skoog. 1962. A revised medium for rapid growth and bioassays with tobacco tissue cultures. Physiol. Plant., 15: 473-497.

Nutman, P. S. 1965. Symbiotic nitrogen fixation. pp. 360-383. In: *Soil Nitrogen*. W. V. Bartholomew and F. E. Clark (eds). Agronomy Series 10. Am. Soc. Agronomy, Inc., Publ., Madison, WI.

Odum, H. T., and R. F. Pigeon. 1970. *A Tropical Study of Irradiation and Ecology at El Verde*. U.S. Atomic Energy Commission, Washington, DC.

Ollier, C. D. 1975. *Weathering*. Longman Group Ltd., London.

Orlov, D. S. 1985. *Humus Acids in Soils*. English translation, K. H. Tan (ed). USDA and NSF Publ., Amerind Publ. Co., New Delhi.

Osmond, D. A. 1958. Micropedology. Soils and Fertilizers 21: 1-6.

Page, H. J. 1930. Studies on the carbon and nitrogen cycles in the soil.

J. Agric. Sci. 20: 455-459.

Pais, I., and J. B. Jones, Jr. 1997. *The Handbook of Trace Elements.* St. Lucie Press. Boca Raton, Fl.

Papendick, R. I., and C. S. Campbell. 1981. Theory and measurement of water potential. pp. 1-22. In: *Water Potential Relations in Soil Microbiology*, J. F. Parr, W. R. Gardner, and L. F. Elliot (eds). SSSA Publ. No. 9. Soil Sci. Soc. Am., Madison, WI.

Paul, E. A., and F. E. Clark. 1989. *Soil Microbiology and Biochemistry.* Academic Press, Inc., San Diego, CA.

Persley, G. J. 1990. *Agricultural Biotechnology. Opportunities for International Development.* C.A.B. International, Wallingford, Oxon, UK.

Peterson, J. R., T. M. McCalla, and G. E. Smith. 1971. Human and animal wastes as fertilizers. pp. 557-596. In: *Fertilizer Technology and Use*, R. A. Olson (ed-in-chief). Soil Sci. Soc. Am., Inc., Madison, WI.

Ponnamperuma, F. N., E. Martinez, and T. Loy. 1966. Influence of redox potential and partial pressure of carbon dioxide on pH value and the suspension effect of flooded soils. Soil Sci. 101: 421-431.

President's Science Advisory Committee Panel on World Food Supply. 1967. *The world food problem.* Volumes I-III. The White House, Washington, DC.

Putnam, A. R. 1994. Phytotoxicity of plant residue. pp. 284-314. In: *Managing Agricultural Residues*, P. W. Unger (ed). Lewis Publ., Ann Arbor, MI.

Rainey, R. H. 1967. Natural displacement of pollution from the Great Lakes. Science 155: 1242-1243.

Raman, K. V., and M. L. Jackson. 1964. Vermiculite surface morphology. Clays Clay Miner., 12: 423-429.

Reitz, L. P. 1968. Short wheats stand tall. pp. 236-239. In: *Science for Better Living.* The Yearbook of Agriculture. USDA, U.S. Govern. Printing Office, Washington, DC.

Ries, S. K. 1985. Regulation of plant growth with triontanol. Crit. Rev. Plant Sci. 2: 239-285.

Rovira, A. D., and C. B. Davey. 1974. Biology of the rhizosphere. In: *The Plant Root and Its Environment*, E.W. Carson (ed). University Press of Virginia, Charlottesville, VA.

Rovira, A. D., and B. M. McDougall. 1967. Microbiological and biochemical aspects of the rhizosphere. pp. 417-463. In: *Soil Biochemistry*, A. D. McLaren and G. H. Peterson (eds). Marcel Dekker, New York, NY.

Russell, E. J., and E. W. Russell. 1950. *Soil Conditions and Plant Growth*. 8th ed., recast and rewritten by E. W. Russell. Longmans, Green and Co., London.

Ryther, J. H., and J. E. Bardach. 1968. The Status and Potential of Aquaculture. Clearinghouse for Federal Scientific and Technical Information, P.B. 177. Springfield, VA.

Sander, D. 1969. *Orchids and Their Cultivation*. Blandford Press, London.

Savoini, G. 1986. General conclusions on the application of humic substances in agriculture. pp. 40-53. In: *Humic Substances. Effect on Soil and Plants*, Enichem (ed). Congress Proc. Enichem Agricoltora, Milano, 1986. Ramo Editoriale Degli Agricoltora, Milano, Italy.

Schnitzer, M. and D. A. Hindle. 1980. Effect of peracetic acid oxidation on N-containing components of humic materials. Can. J. Soil Sci. 60: 541-548.

Schnitzer, M., and S. U. Khan. 1972. *Humic Substances in the Environment*. Marcel Dekker, Inc., New York, NY.

Schnitzer, M., and H. Kodama. 1976. The dissolution of mica by fulvic acids. Geoderma 15: 381-391.

Shell, E. W. 1975. A fish story pans out and world is better fed. pp. 149-156. In: *That We May Eat*. The Yearbook of Agriculture. USDA, House document No. 94-4. U.S. Govern. Printing Office, Washington, D.C.

Shoji, S., M. Nanzyo, and R. A. Dahlgren. 1993. *Volcanic Ash Soils. Genesis, Properties, and Utilization*. Elsevier, Amsterdam.

Singer, A. 1979. Clay mineral formation. pp. 76-82. In: *The Encyclopedia of Soil Science*, Part 1, R. W. Fairbridge and C. W. Finkl, Jr. (eds). Dowden, Hutchinson and Ross, Stroudsburg, PA.

Singer, M.. J., and D. N. Munns. 1996. *Soils. An Introduction*. 3rd edition. Prentice Hall, Upper Saddle River, NJ.

Smith, J., and C. Lewis. 1991. *Biotechnology in the Food and Agro-*

Industries. Special Report No. 234. The Economist Intelligence Unit, London.

Soil Survey Staff. 1951. *Soil Survey Manual*. USDA Agric. Handbook No. 18. U.S. Govern. Printing Office, Washington, DC.

Soil Survey Staff. 1962. *Soil Survey Manual*. USDA Agric. Handbook No. 18. U.S. Govern. Printing Office, Washington, DC.

Soil Survey Staff. 1966. *Keys to Soil Taxonomy*. USDA Natural Resources Conservation Service. U.S. Govern. Printing Office, Washington, DC.

Soil Survey Staff. 1975. *Soil Taxonomy. A Basic System of Soil Classification for Making and Interpreting Soil Surveys*. USDA Agric. Handbook No. 436. U.S. Govern. Printing Office, Washington, DC, Soil Survey Staff. 1990. *Keys to Soil Taxonomy*. AID, USDA, Soil Management Support Services, Monograph No. 19. Virginia Polytech. and State University, Blacksburg, VA.

Soil Survey Staff. 1992. *Keys to Soil Taxonomy*. AID, USDA, SCS, SMSS Techn. Monograph No. 19, fifth edition. Pocahontas Press, Inc., Blacksburg, VA.

Spangler, M. G., and R. L. Handy. 1982. *Soil Engineering*. 4th edition. Harper and Row, New York, NY.

Sposito, G. 1989. *The Chemistry of Soils*. Oxford University Press, New York, NY.

Stevenson, F. J. 1967. Organic acids in soils. pp. 119-146. In: *Soil Biochemistry*, Vol. 1, A. D. McLaren and G. W. Peterson (eds). Marcel Dekker, Inc., New York, NY.

Stevenson, F. J. 1982. *Humus. Chemistry, Genesis, Composition, Reactions*. Wiley Interscience, New York, NY.

Stevenson, F. J. 1986. *Cycles of Soil Carbon, Nitrogen, Phosphorus, Sulfur, Micronutrients*. Wiley Intersci., New York, NY.

Stevenson, F. J. 1994. *Humus Chemistry, Genesis, Composition, Reactions*. 2nd edition. Wiley & Sons, New York, NY.

Stewart, J. W. B., R. F. Follett, and C. V. Cole. 1987. Integration of organic matter and soil fertility concepts into management decisions. pp. 1-8. In: *Soil Fertility and Organic Matter as Critical Components of Production Systems*, R. Follett (chair). Soil Sci. Soc. Am., Inc., and Am. Soc. Agronomy, Inc., Publ., Madison, WI.

Sticher, H., and R. Bach. 1966. Fundamentals in the chemical weathering of silicates. Soils Fertilizers (Harpenden), 29: 321-325.

Stolzy, L. H. 1974. Soil atmosphere. pp. 335-361. In: *The Plant Root and Its Environment*, E. W. Carson (ed). Univ. Press Virginia, Charlottesville, VA.

Stolzy, L. H., and J. Letey. 1964. Measurement of oxygen diffusion rates with the platinum electrode. III. Correlation of plant response to oxygen diffusion rates. Hilgardia 35: 567-576.

Stoops, G., and H. Eswaran (eds). 1986. *Soil Micromorphology*. Van Nostrand Reinhold Co., New York, NY.

Swain, F. M. 1963. Geochemistry of humus. pp. 81-147. In: *Organic Geochemistry*, J. A. Breger (ed). Pergamon Press, New York, NY.

Tan, K. H. 1965. The Andosols in Indonesia. Soil Sci., 99: 375-378.

Tan, K. H. 1968. The genesis and characteristics of paddy soils in Indonesia. Soil Sci. Plant Nutr. 14: 117-121.

Tan, K. H. 1976. Complex formation between humic acid and clays as revealed by gel filtration and infrared spectroscopy. Soil Biol. Biochem. 8: 235-239.

Tan, K. H. 1978. Formation of metal-humic acid complexes by titration and their characterization by differential thermal analysis and infrared spectroscopy. Soil Biol. Biochem. 10: 123-129.

Tan, K. H. 1980. The release of silicon, aluminum, and potassium during decomposition of soil minerals by humic acids. Soil Sci. 129: 5-11.

Tan, K. H. 1981. *Basic Soils Laboratory*. Burgess Publ. Co. Minneapolis, MN.

Tan, K. H. 1982. *Principles of Soil Chemistry*. Marcel Dekker, Inc., New York, NY.

Tan, K. H. 1984. *Andosols*. Van Nostrand Reinhold Co., New York, NY.

Tan, K. H. 1986. Degradation of soil minerals by organic acids. pp. 1-27. In: *Interactions of Soil Minerals with Natural Organics and Microbes*, P. M. Huang and M. Schnitzer (eds). Soil Sci. Soc. Am. Special Publ. No.17. Soil Sci. Soc. Am., Inc., Madison, WI.

Tan, K. H. 1993. *Principles of Soil Chemistry*. 2nd edition. Marcel Dekker, Inc., New York, NY.

Tan, K. H. 1998a. *Andosol. Capita Selecta*. Program Studi Ilmu

Tanah. Program Pascasarjana, University of North Sumatera, Medan, Indonesia.

Tan, K. H. 1998. *Principles of Soil Chemistry*. 3rd edition. Marcel Dekker, New York, NY.

Tan, K. H., and A. Binger. 1986. Effect of humic acid on aluminum toxicity in corn plants. Soil Sci., 14: 20-25.

Tan, K. H., and F. E. Clark. 1968. Polysaccharide constituents in fulvic and humic acids extracted in soils. Geoderma 2: 245-255.

Tan, K. H., L. D. King, and H. D. Morris. 1971. Complex reactions of zinc with organic matter extracted from sewage sludge. Soil Sci. Soc. Am. Proc., 35: 748-751.

Tan, K. H., and R. A. McCreery. 1975. Humic acid complex formation and intermicellar adsorption by bentonite. pp. 629-641. In: *Proc. Intern. Clay Conf.*, S. W. Bailey (ed). Mexico City, Mexico. Applied Publ. Ltd., Wilmette, IL.

Tan, K. H., V. G. Mudgal, and R. A. Leonard. 1975. Adsorption of poultry litter extracts by soil and clay. Env. Sci. & Techn., 9: 132-135.

Tan, K. H., and J. Van Schuylenborgh. 1961. On the classification and genesis of soils developed over acid volcanic materials under humid tropical condition. Neth. J. Agri. Sci. 9: 41-54.

Tan, K. H., and J. Van Schuylenborgh. 1959. On the classification and genesis of soils derived from andesitic volcanic material under a monsoon climate. Neth. J. Agri. Sci. 9: 174-180.

Tan, K. H., and P. S. Troth. 1981. Increasing sensitivity of organic matter and nitrogen analysis using soil separates. Soil Sci. Soc. Am. J. 45: 574-577.

Taylor, S. A., and G. L. Ashcroft. 1972. *Physical Edaphology*. Freeman and Co., San Francisco, CA.

Theng, B. K. G. (ed). 1980. *Soils with Variable Charge*. New Zealand Soil Sci. Soc., Soil Bureau, D.S.I.R., Lower Hutt, New Zealand.

Thiel, K. D., R. Kloecking, H. Schweizer, and M. Sproessig. 1977. Zentralblatt fuer Bakterilogie, Parasitenkunde, Infektion-Skrankheiten und Hygiene, Abteilung I; Originale Reih A: Mediziche Mikrobiologie und Parasitologie 234: 304-321.

Thorp, J., and G. D. Smith. 1949. Higher categories of soil classification. Soil Sci., 67: 117-126.

Tiddens, A. 1990. *Aquaculture in America. The Role of Science, Government, and the Entrepreneur.* Westview Press, Boulder, CO.

Tisdale, S. L., and W. L. Nelson. 1975. *Soil Fertility and Fertilizers.* Macmillan Publ. Co., Inc., New York, NY.

Tisdale, S. L., W. L. Nelson, J. D. Beaton, J. L. Havlin. 1993. *Soil Fertility and Fertilizers.* Macmillan Publ. Co., New York, NY.

Troeh, F. R., J. A. Hobbs, and R. L. Donahue. 1980. *Soil and Water Conservation for Productivity and Environmental Protection.* Prentice-Hall, Inc., Englewood Cliffs, NJ.

Uribe, I. 1975. George Harrar sets off the green revolution. pp. 312-322. In: *That We May Eat.* The Yearbook of Agriculture. USDA, U.S. Govern. Printing Office, Washington, DC.

U.S. Environmental Protection Agency. 1972. Glossary of solid waste management. Publ. GP-1972-3. National Center for Resource Recovery, Inc., Washington, DC.

Van Wijk, W. R., and A. J. W. Borghorst. 1963. Turbulent transfer in air. pp. 236-276. In: *Physics of Plant Environment*, W. R. Van Wijk (ed). North Holland Publ. Co., Amsterdam.

Van Wijk, W. R., and D. A. DeVries. 1963. The atmosphere and the soil. pp. 17-61. In: *Physics of Plant Environment*, W. R. Van Wijk (ed). North Holland Publ. Co., Amsterdam.

Van Schuylenborgh, J., and P. L. Arens. 1950. The electro-kinetic behaviour of freshly prepared γ- and α-FeOOH Rec. Trav. Chim. Pays-Bas, 69: 1557-1565.

Van Schuylenborgh, J., and A. M. H. Sanger. 1949. The electro-kinetic behaviour of iron- and aluminium-hydroxides and oxides. Rec. Trav. Chim. Pays-Bas, 68: 999-1010.

Veihmeyer, D. S., and A. H. Hendrickson. 1948. Soil density and root penetration. Soil Sci. 65: 487-493.

Wada, K. 1977. Allophane and imogolite. pp. 603-639. In: *Minerals in Soil Environment*, J. B. Dixon, S. B. Weed, J. H. Kittrick, M. H. Milford, and J. L. White (eds). Soil Sci. Soc. Am., Madison, WI.

Wada, K., and D. J. Greenland. 1970. Selective dissolution and differential infrared spectroscopy for characterization of amorphous constituents in soil clays. Clay Mineral. 8: 241-245.

Waksman, S. A. 1936. *Humus.* Williams and Wilkins, Baltimore, MD.

Waksman, S. A. 1938. *Humus - Origin, Chemical Composition and*

Importance in Nature. 2nd edition. Williams and Wilkins, Baltimore, MD.

Wallace, J. M., and L. E. Elliott. 1979. Phytotoxins from anaerobically decomposing wheat straw. Soil Biol. Biochem. 11: 325-330.

Wiegand, C. L., and E. R. Lemon. 1958. A field study of some plant-soil relations in aeration. Soil Sci. Soc. Am. Proc., 22: 216-221.

Wild, A. 1993. *Soils and the Environment.* Cambridge University Press, New York, NY.

Wilding, L. P., and K. W. Flach. 1985. Micropedology and soil taxonomy. pp. 1-16. In: *Micromorphology and Soil Classification*, L.A. Douglas, and ML. Thompson (eds). SSSA Special Publ. No.15. Soil Sci. Soc. Am., Madison, WI.

Wisaksono, M. W., and K. H. Tan. 1964. *Ilmu Tanah.* Part II. Pradnja-paramita II Publ., Jakarta.

World Resources. 1987. A report of the International Institute for Environment and Development, and World Resources Institute, Washington, D.C.

Yoshinaga, N., and S. Aomine. 1962. Imogolite in Some Ando soils. Soil Sci. Plant Nutr., 8: 6-13.

Young, J. A. 1988. Dyers woad wages war in the West. Agric. Research 36: 4.

INDEX

Abalone, *Haliotis (Nordotis)*, 332
Abiotic chemical weathering,
  carbonation, 49
  hydration and dehydration, 47-48
  hydrolysis in, 46-47
  oxidation and reduction, 48-49
Abiotic physical weathering,
  differential stress in, 45
  effect of temperature in, 45-46
  lateral stress in, 46
Accelerated pollution, 386
Achromobacter, 121
Acid loving plants, 166
Acid mine spoil, 155
  reclamation of soils containing, 155
Acid precipitation, see Acid rain
Acid rain, 117, 122, 368, 386
  concept, definition of, 386-387
  effect on environment of, 387-389
  effect on soils, 389
  effect on weathering of rocks and minerals, 387, 388
  gases contributing to, 387
  global effect of, 388
  significance of, 387-389
Actinomycetes, 103
  allelophatic importance of, 104
  effect of pH on, 104
  N_2-fixing, 104
  scab disease by, 104
  pharmaceutical importance of, 104
Activation energy barrier, 109
Active movement, 162
Adhesion, 169

421